Analog Circuits: Concepts, Devices and Systems

Analog Circuits: Concepts, Devices and Systems

Edited by **Gus Winters**

New York

Published by NY Research Press,
23 West, 55th Street, Suite 816,
New York, NY 10019, USA
www.nyresearchpress.com

Analog Circuits: Concepts, Devices and Systems
Edited by Gus Winters

International Standard Book Number: 978-1-63238-506-2 (Hardback)

Printed in the United States of America.

Contents

Permissions

List of Contributors

Preface

This book has been an outcome of determined endeavour from a group of educationists in the field. The primary objective was to involve a broad spectrum of professionals from diverse cultural background involved in the field for developing new researches. The book not only targets students but also scholars pursuing higher research for further enhancement of the theoretical and practical applications of the subject.

Analog circuits are very sophisticated designs that can be either complex or simple like combination of two resistors to make a voltage divider. These are characterized by the use of continuous time voltages and currents. The components of analog circuits involve capacitors, inductors, wires, resistors, diodes and transistors. This book will prove to be immensely beneficial to students and researchers involved in this area. The topics covered herein offer the readers new insights in the field of analog circuits. This book brings forth some of the most innovative concepts and elucidates the unexplored aspects of analog circuits. Coherent flow of topics, reader-friendly language and extensive use of examples make this book an invaluable source of knowledge.

It was an honour to edit such a profound book and also a challenging task to compile and examine all the relevant data for accuracy and originality. I wish to acknowledge the efforts of the contributors for submitting such brilliant and diverse chapters in the field and for endlessly working for the completion of the book. Last, but not the least; I thank my family for being a constant source of support in all my research endeavours.

Editor

Design and Implementation of Android Based Wearable Smart Locator Band for People with Autism, Dementia, and Alzheimer

Isha Goel[1] and Dilip Kumar[2]

[1]Department of Academic and Consultancy Services Division, C-DAC, Mohali 160071, India
[2]SLIET, Longowal, Sangrur 148106, India

Correspondence should be addressed to Isha Goel; er.ishagoel@yahoo.com

Academic Editor: Gianluca Traversi

A wearable smart locator band is an electronic device which can be worn on the wrist of the children to monitor and keep an eye on them. As the number of mishaps with children is increasing, it is a must to keep them safe. This also helps reducing crime rates. The research study proposed the development of a wearable smart locator band that helps keeping track of kids. The developed device includes an AVR microcontroller (ATmega8515), global positioning system (GPS), global system for mobile (GSM), and switching unit and the monitoring unit includes Android mobile device in parent's hand with web based Android application as well as location indicated on a Google Map. This development is very useful for senior people and individuals suffering from memory diseases. This device, hence, behaves as a communication interface between wearer and caregiver.

1. Introduction

Safety concern is a major issue these days. Incidents of kidnapping, child abuse, lost persons, and misbehaviour with children, adults, and aged people are increasing day by day. The wearable smart locator helps to maintain an eye on our beloved ones [1]. It has been found that 30 million individuals in the world are suffering from autism (nervous breakdown), dementia (short term and long term memory loss), and Alzheimer (loss of brain function). Then, wearing this device helps locating these diseased individuals in emergency conditions by pressing emergency buttons. It will set off an automatic location beacon of the wearer through message and will ring the emergency contacts [2, 3].

Even this device includes a virtual radius which can be set around a destination by which a notification will be sent to a caregiver's number whenever their wearer enters into or leaves it and an LED glows as an indication in the device [4, 5]. This device is a continuous monitoring embedded application of wearer's location, whether still or moving, and reports the status of that location to caregiver's mobile. The caregiver can even talk to the wearer with a two-way calling facility, including audio channel in GSM with microphone and speaker which help in calling.

It is worn by wearer all the time and does not require the person to operate this device in any way except in case of emergency. This focuses on better security and safety of persons by improving the feasibility and reliability of available locating devices which face obstacles in communication and adapting or cognitive performance as well as emphasizing the relative merits and limitations of existing technologies. To boot, it seeks to gain an enhanced understanding of the complex realities related to the carrying out of locating technology.

2. Literature Survey

In the past years, various tracking/monitoring systems had been designed and these systems are generally implemented in the form of children or vehicle tracking systems.

Mammone in 2005 caves in a method for nearby people in a crisis situation with rapid soliciting. This invention allows a parent to alert other people who are present in a fixed radius

[3]. In 2007, Kennedy designed an alert notification which is text messaging based on the Amber Alert system which helped in child kidnapping [4].

In the year 2005 King and Yancey had given an attack warning for vehicle and location scheme. Emergency vehicles will get to a destination by travelling quickly and safely and are indicated along the map with an icon with different visual characteristics with respect to the position of the vehicle [5].

Then Curran et al. proposed a method in 2012 for defining the devices entering into a 2D geographic zone area with a user alert [6]. Pankaj and Bhatia in the year 2013 also have given their thought to implement GPS/GSM based vehicle tracking system and track the vehicle on Google Map and also provide the shortest route to reach vehicle easily in minimal time [7].

As these tracking systems are utilized for tracking children or vehicles, this sort of technique is likewise employed for people who are suffering from diseases like autism, dementia, and Alzheimer and elderly individuals.

3. Device Architecture

3.1. Description of Transmission Unit (Figure 1)

3.1.1. AVR (ATmega8515). The ATmega8515 belongs to AVR (enhanced RISC architecture) family which is a low power (7.5 mW), high performance device as it works at crystal frequency of 4 MHz and executes powerful instructions in a single clock cycle (it achieves throughputs approaching 1MIPS per MHz) and In-System Self-Programmable Flash 8-bit microcontroller. It has 512-byte SRAM and 512-byte EEPROM internal memories. It is as well recognized as the centre of this system. It mostly works as an interface between a GPS receiver and GSM module. It has a feature of three power saving modes: idle, power-down, and standby. This microcontroller initiates and sends the wearer's information, message, and voice calling details to mobile phone through the GSM chip.

3.1.2. GPS Technology. The GPS is based on a global navigation satellite system to determine speed, position, direction, and time. It utilizes a constellation of 24/32 active satellites in Earth orbit that transmit an accurate microwave signal and enable GPS receiver. A GPS receiver needs at least three or four satellites to calculate the distance as shown in Figure 2 and figure out its two dimensions, that is, latitude and longitude, or three dimensions, that is, latitude, longitude, and altitude positions [8, 9].

3.1.3. GSM Technology. The GSM modem which acts as a mobile phone accepts any GSM network operator SIM card with its own unique phone number. This SIM900A GSM modem can communicate and develop embedded application of SMS based remote control, for example, to send/receive SMS and make/receive voice calls [10].

It can also be used for data logging application which connects to internet with GPRS mode. It is dual band 850/1900 MHz which makes it a flexible plug and makes it suitable for long distance data transmission. Its international roaming capability is an advantage, with improved battery life and data up to 9600 bps baud rate.

3.2. Monitoring Unit. The monitoring unit illustrated in Figure 3 includes an Android GSM mobile with an internet plan and a web based Android application supporting it. The GSM mobile will receive an SMS which includes the automatic location beacon of the wearer (longitude and latitude) and another SMS which includes the virtual radius entering and leaving information [11, 12].

By opening that SMS it will directly connect to the Android application within a second and open the Google Map with a pointer pointing towards the coordinates which is the exact current location of the wearer.

4. Hardware Description

This hardware design is used to locate the wearer and help him in case of need with the help of pushbuttons S1, S2, S3, and S4 that even has two-way calling facility by using GPS and GSM which are serially interfaced with ATmega8515 controller for continuous monitoring and message sending from the device.

4.1. Switching Unit. As shown in Figure 4, it consists of four switches which are used by device, that is, smart locator band for calling and sending SMS. Four different operations are performed by each switch as follows:

  S1: for calling the two emergency numbers one after the other,

  S2: for disconnecting the call,

  S3: for answering the incoming call,

  S4: for sending SMS of location to both emergency numbers.

4.2. GPS Module. Figure 5 illustrates GPS module which works on 3.3 V supply. It continuously senses the current position of the wearer and sends it automatically to the microcontroller. GPS parameter and specifications are shown in Table 1.

4.3. GSM Module. In Figure 6 GSM module is used for communication, that is, two-way calling that includes dialling, receiving call with the help of microphone and speaker, and sending SMS which contains the current location of the wearer and virtual radius entering and leaving information. GSM parameter and specifications are shown in Table 2.

4.4. Battery. When using a combination of GPS with GSM technology, approximately 100 mAh is required. Taking these factors into consideration the rechargeable batteries with the capacity of 7.2 V and 2200 mAh or 4.2 V and 1900 mAh (three in the series) among which one can be used. On an average it can provide a backup of 22 hours a day and it can last up to 2-3 days depending on usage.

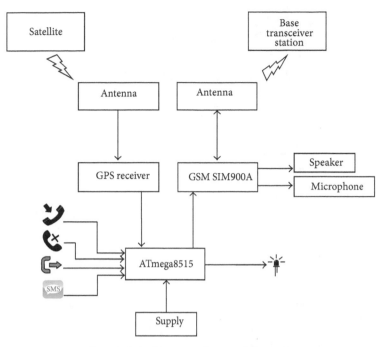

FIGURE 1: Architecture of transmitting unit.

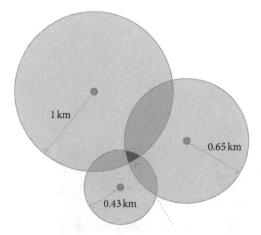

FIGURE 2: GPS signal working.

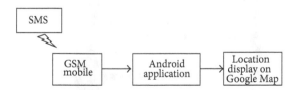

FIGURE 3: Monitoring unit architecture.

FIGURE 4: Prototype model of device.

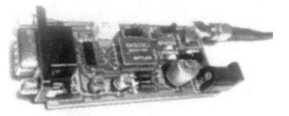

FIGURE 5: GPS module.

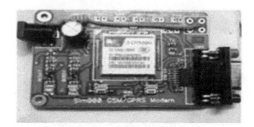

FIGURE 6: GSM module.

5. Software Description

5.1. MicroC Pro for AVR. This software is primarily utilized to activate AVR (ATmega8515) microcontroller according to the input received by it. "Embedded C" code is written using this programming tool. In this project, coding is written for GPS, GSM, and switching unit which is interfaced with

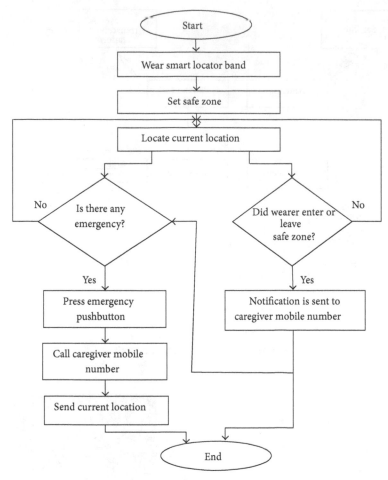

FIGURE 7: Flowchart of the device.

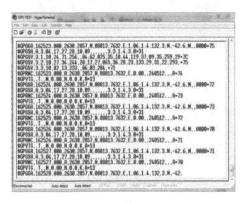

FIGURE 8: GPS output.

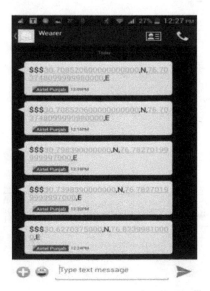

FIGURE 9: SMS received on smartphone.

AVR microcontroller at the transmitter end. As per the code embedded in the controller, the interfaced modules generate appropriate output at the receiving terminal.

5.2. ECLIPSE (LUNA 4.2). Application in Android mobile device is created using ECLIPSE software. It is elastic and provides compatibility to create novel applications in Android mobile devices. The Java language is preferred as

the basic platform for application creation. In this project, an application named "SMS Receiver" is created using ECLIPSE which enables the caregiver at receiving end to visually see

30°42'30.7"N 76°42'13.5"E

FIGURE 10: Pointer in GMAP.

TABLE 1: GPS parameter and specifications.

GPS module	Chipset	SKYLAB SKG13C
	Receiver type	22 tracking/66 acquisition channels
	Sensitivity	Ultrahigh, −165 dBm
	Protocol type	NMEA-0813
	Operating temperature range	−40 to +85°C
	Power consumption	Lower, 45 mA @ 3.3 V
	Operating voltage	Typical 3.0 V to 4.2 V
	Command statements	$GPGSA, $GPGLL, $GPGGA, $GPGSV, $GPZDA, $GPVTG, $GPRMC
	Dimensions	15 × 13 × 2.7 mm

TABLE 2: GSM parameter and specification.

GSM module	Frequency band	850/1900 MHz dual band
	Baud rate	9600 bps
	Transmission power	2 W @ 850 MHz 1 W @ 1900 MHz
	Power supply	3.2 to 4.8 V.
	Operating temperature	Restricted operation: −40 to −30°C and +80 to +85°C
		Normal operation: −30 to +80°C
		Storage temperature: −45 to +90°C
	Dimensions	24 × 24 × 3 mm

the place on Google Map corresponding to the position of the wearer at the transmitting end [12, 13].

5.3. Design Methodology. Figure 7 describes the flowchart/design methodology steps as follows.

(i) The wearer will wear the smart locator band.

(ii) Then, a safe zone is defined or set by the caregiver for safety of wearer.

(iii) The device will continuously monitor the latest location of the wearer, generate an SMS alert, and give birth to the possibility of two conditions:

    (a) whether the wearer enters or leaves the secure zone;

    (b) whether there is an emergency situation or not.

(iv) When the wearer crosses the safe zone, then SMS notification will be sent; otherwise it will not be sent.

(v) When an emergency situation is noticed then wearer presses emergency button S1 for calling and S4 for sending location via SMS.

6. Result and Discussion

6.1. GPS Output. As shown in Figure 8 the controller filters and selects the incoming one packet, that is, $GPRMC from continuous six packets of GPS data, and extracts the current location of the wearer by forwarding only the latitude and longitude values to GSM.

It has been found that the shadowing and multipath effect due to tall buildings in urban canyons or even indoors like public malls cause difficulty in detecting the position of the wearer. To overcome this problem an alternative approach is evaluated which uses the internal memory, that is, SRAM of the Atmega8515 microcontroller, to keep the code data and EEPROM is used to save last valid position data (up to 20 values) in the device itself. Therefore, when the GPS gives the invalid location of the wearer in the process of continuous position sensation, then the last saved location in EEPROM is automatically sent by the device on the caregiver side so that approximate location of the wearer can be identified.

6.2. SMS Receiver Application Installed on Android Phone. The Android application which is web based named as "SMS Receiver" represents the complete output of the device created in caregiver's Android mobile, which is opened directly when received SMS is opened with the working Internet pack. This application then points towards the exact location on Google Map retrieved by the latest Lat., Long. values sent by GSM modem [14, 15].

Characteristics of SMS receiver application are as follows.

(i) It is activated on special message received by the device with $$$ unique starting keyword.

(ii) It will extract the location from the message.

(iii) It displays current location of the wearer.

(iv) SMS receiver uses Internet connection and automatically searches the location on Google Map and points to it with a cursor.

(v) It is easy to understand and is user friendly.

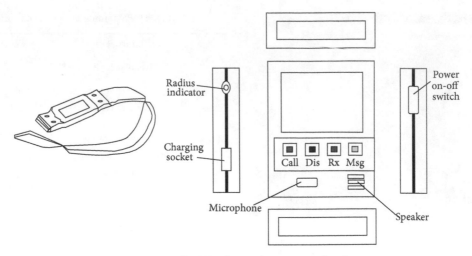

FIGURE 11: 3D view of smart locator wrist band.

TABLE 3: Testing of developed prototype on different subjects at different locations.

Subjects with diseases	Location coordinates	Coordinates on Google	Name of the location	Error (in meter)
1	30.7021052, 76.850094	30.7021049, 76.80091	Industrial area phase-1 Chandigarh (India)	2
2	30.71447700, 76.7148930	30.71447800, 76.7148927	Phase-5 Mohali (India)	0
3	30.719425700, 76.781259600	30.719325700, 76.781259600	Sector-20 Chandigarh (India)	7
4	30.739833900, 76.783207999	30.739833900, 76.782707999	Sector-17 Chandigarh (India)	3
5	30.708171000, 76.718559000	30.708191000, 76.718559000	Phase-3B-2 Mohali (India)	2
6	30.681173000, 76.746737800	30.681073000, 76.746637800	Phase-11 Mohali (India)	13
7	30.707679600, 76.747769780	30.707669600, 76.747669780	Sector-44 Chandigarh (India)	5
8	30.707679600, 76.752279400	30.707669600, 76.752179400	Sector-45 Chandigarh (India)	8
9	30.70544444, 76.709112222	30.70541350, 76.709119999	Sector-71 Chandigarh (India)	3
10	30.712486500, 76.73656500	30.712487300, 76.73659800	Sector-52 Chandigarh (India)	3

6.3. Received SMS Containing Latitude-Longitude on Smartphone. When switch S4 is pressed by the wearer, GPS module of the system senses the position of the wearer with longitude-latitude and this value is sent by using the GSM module to the caregiver's mobile. Figure 9 shows that the message is received on caregiver's smartphone.

6.4. Location Indication with Pointer on GMAP. Figure 10 shows that when SMS is received, the SMS receiver application directly opens and the longitude-latitude values received in SMS are located by the pointer which are showing the current location (position) of the wearer whenever there is an emergency message on caregiver's Android smartphone, that is, at receiving end, and the output of GMAP is obtained with the help of Android application [16].

6.5. Testing of Device on Different Subjects. The complete prototype developed was tested on different subjects at different locations. When subject (suffering from autism or dementia or Alzheimer) enters or leaves the zone defined or call is done or received by the GSM modem it sends the coordinates received by the GPS module to user in SMS (short

message services) form. So, the accuracy of device is nearly perfect. Table 3 describes the testing of the prototype.

6.6. Ergonomic Study of Wrist Shape. For designing the wrist band for wearer the following ergonomic design steps need to be followed.

(1) The size of the wrist strap must be equal to the circumference of the wrist of the wearer and is calculated by taking measurements of different subjects (at least 10), and an average value is calculated which can be adjusted according to wearers need.

(2) The device placed on the wrist is in rectangular shape and its dimensions need not be greater than wrist of the wearer.

(3) Four different colour switches are placed on the top of the device for call, disconnect, receive, and message.

(4) A microphone and a speaker are also available on the front panel for the wearer communication.

(5) A radius indicator, on-off switch, and the charging socket are provided on the sides of the device.

TABLE 4: Comparison of work with existing devices.

Parameters/devices	Amber Alert GPS	Pocket finder	Spark nano	Live view GPS	E zoom	Zoom bak	Itrack 2.0	5 Star	Mei track	Itrack GPS tracker	Smart locator band
Volume (inches³)	2.8 × 1.5 × 0.8	2.25 × 1.62 × 0.62	2.66 × 1.57 × 0.83	3.5 × 2.09 × 0.75	2.76 × 1.73 × 0.85	2.87 × 1.69 × 0.82	2.2 × 1.4 × 0.4	2.9 × 1.7 × 0.7	2.29 × 1.5 × 0.7	3 × 2 × 0.5	3 × 3 × 1
Weight (grams)	61.5	40	100	113	85.0	70.8	297.7	51.0	65.2	48.2	70
Battery life (hours)	40	50	336	96	72	72	1440	96	24	144	72
Battery (rechargeable)	Y	Y	Y	Y	Y	Y	Y	Y	Y	Y	Y
Water (resistant/proof) protection	Res	Proof	Res	Res	Proof	Res	Res	Res	Proof	Res	Proof
Panic button	Y	N	Y	N	Y	N	Y	Y	Y	Y	Y
Two-way calling	Y	N	N	N	N	N	Y	Y	Y	N	Y
Zone alert	Y	Y	Y	Y	Y	Y	Y	N	Y	Y	Y
Text notification	Y	Y	Y	Y	Y	Y	Y	N	Y	Y	Y
Emergency contacts	Y	Y	Y	Y	Y	Y	Y	Y	Y	Y	Y
Tracking history	Y	Y	Y	Y	Y	Y	Y	N	Y	Y	Y
Web based user interface	Y	Y	Y	Y	Y	Y	Y	Y	Y	Y	Y
Android application	Y	Y	N	Y	Y	Y	N	Y	Y	N	Y
Status indication	Y	Y	Y	Y	Y	N	N	N	N	N	Y
Device price (Rupees)	7620	7920	9110	12130	6100	4880	12190	3050	5490	7870	2900
Activation charges (Rupees)	1220	Free	Free	1220	1830	920	—	2140	—	—	Free
Monthly price (Rupees)	920	790	1830	2440	1220	1220	—	920	—	—	Free
Contract length (yearly/monthly)	Y	M	M	M	Y	Y	—	M	—	—	Life time

Figure 11 represents the proposed ergonomic design of a smart locator band including the front, back, and side views of the device. All the above-mentioned ergonomic design steps are considered while designing this band. This designed band is adjustable in size and it can be locked so the wearer who may not agree to wear it, he himself cannot remove it easily.

6.7. Comparison with the Existing Devices. A comparison work is also made between the existing devices and the developed prototype system. This is performed on the basis of different required parameters as described in Table 4.

The indigenous developed prototype can overcome the several limitations of imported commercial devices; that is, some device manufacturer charges extra monthly rental for using their online application services to access the GPS location, there is some activation charges to activate the device, with a less license validity period for using their device (for one month or a few months not for a lifetime). Even the existing devices do not include a panic button (that can be used in case of emergency), some of these devices are water resistant but not waterproof and they do not include two-way calling facility in the device for the wearer to stay in contact with the caregiver in case of need.

The novelty of this device is that it overcomes all the limitations of the existing commercial devices with an additional low cost product developed. After all, the developed device main advantage is that for accessing a location in existing devices the caregivers will have to open the application and access the wearer location themselves, but in this device an Android based application is also developed, which automatically opens the location of the wearer on Google Map within the fraction of seconds when it receives the message containing latitude and longitude and the caregivers have no need to open it manually like in other existing devices.

7. Conclusion and Future Scope

Operational and testing results of prototype system indicate that the system worked efficiently. If this device is fabricated into a wrist band, no one would be able to evaluate whether it is a safety locator band or a wrist band/watch.

This work is of low cost, very effective, and productive. But there is always room for improvement. This merchandise has been designed as a prototype and requires further developments for using it in assorted applications. This system can be further expended in developing a Windows application which can support windows phone and the wearer device must be small and unobtrusive in the form of compact watch and it should not label people.

Conflict of Interests

The authors declare that there is no conflict of interests regarding the publication of this paper.

References

[1] C. Yamagata, J. F. Coppola, M. Kowtko, and S. Joyce, "Mobile app development and usability research to help dementia and Alzheimer patients," in *Proceedings of the 9th Annual Conference on Long Island Systems, Applications and Technology (LISAT '13)*, pp. 1–6, Farmingdale, NY, USA, May 2013.

[2] M. Randolph-Gips, "Autism: a systems biology disease," in *Proceedings of the 1st IEEE International Conference on Healthcare Informatics, Imaging and Systems Biology (HISB '11)*, pp. 359–366, San Jose, Calif, USA, July 2011.

[3] R. Mammone, "Child locator apparatus and method," U.S. Patent Application 10/689,216, 2003.

[4] P. J. Kennedy, "Mobile phone Amber alert notification system and method," U.S. Patent No. 7,228,121, 2007.

[5] B. King and D. A. Yancey, "GPS-based vehicle warning and location system and method," U.S. Patent No. 6,895,332, May 2005.

[6] D. Curran, J. Demmel, and R. A. Fanshier, "Geo-fence with minimal false alarms," U.S. Patent no. 8,125,332, February 2012.

[7] V. Pankaj and J. S. Bhatia, "Design and development of GPS-GSM based tracking system with Google map based monitoring," *International Journal of Computer Science, Engineering and Applications*, vol. 3, no. 3, 2013.

[8] P. Wang, Z. Zhao, C. Xu, Z. Wu, and Y. Luo, "Design and implementation of the Low-Power Tracking System based on GPS-GPRS Module," in *Proceedings of the 5th IEEE Conference on Industrial Electronics and Applications (ICIEA '10)*, pp. 207–210, June 2010.

[9] M. Kunal, S. Mandeep, and J. Neelu, "Real time vehicle tracking system using GSM and GPS technology—an anti-theft tracking System," *International Journal of Electronics and Computer Science Engineering*, vol. 1, no. 3, 2012.

[10] M. A. Al-Khedher, "Hybrid GPS-GSM localization of automobile tracking system," *International Journal of Computer Science and Information Technology*, vol. 3, no. 6, pp. 75–85, 2011.

[11] P. B. Fleischer, A. Y. Nelson, R. A. Sowah, and A. Bremang, "Design and development of GPS/GSM based vehicle tracking and alert system for commercial inter-city buses," in *Proceedings of the IEEE 4th International Conference on Adaptive Science and Technology (ICAST '12)*, pp. 1–6, IEEE, October 2012.

[12] J.-H. Liu, J. Chen, Y.-L. Wu, and P.-L. Wang, "AASMP-Android Application Server for Mobile Platforms," in *Proceedings of the IEEE 16th International Conference on Computational Science and Engineering (CSE '13)*, pp. 643–650, 2013.

[13] J. Saranya and J. Selvakumar, "Implementation of children tracking system on android mobile terminals," in *Proceedings of the International Conference on Communications and Signal Processing (ICCSP '13)*, pp. 961–965, April 2013.

[14] N. Chadil, A. Russameesawang, and P. Keeratiwintakorn, "Real-time tracking management system using GPS, GPRS and Google Earth," in *Proceedings of the 5th International Conference on Electrical Engineering/Electronics, Computer, Telecommunications and Information Technology (ECTI-CON '08)*, vol. 1, pp. 393–396, Krabi, Thailand, May 2008.

[15] Y. A. N. G. Mei, "Application design and implementation of GPS-GPRS location system vehicle terminals," *Telecommunication Engineering*, vol. 3, article 024, 2004.

[16] L. Rui, Z. Minjian, L. Wenhu, and H. Tao, "Designed and implementation of the positioning and tracking system based on GSM SMS module," *Journal of Electronic Measurement and Instrument*, no. z1, pp. 283–286, 2008.

InAs/GaSb Type-II Superlattice Detectors

Elena A. Plis[1,2]

[1] *Center for High Technology Materials, Department of Electrical and Computer Engineering, University of New Mexico, Albuquerque, NM, USA*
[2] *Skinfrared, LLC, Lobo Venture Lab 801, University Boulevard, Suite 10, Albuquerque, NM 87106, USA*

Correspondence should be addressed to Elena A. Plis; elena.plis@gmail.com

Academic Editor: Meiyong Liao

InAs/(In,Ga)Sb type-II strained layer superlattices (T2SLs) have made significant progress since they were first proposed as an infrared (IR) sensing material more than three decades ago. Numerous theoretically predicted advantages that T2SL offers over present-day detection technologies, heterojunction engineering capabilities, and technological preferences make T2SL technology promising candidate for the realization of high performance IR imagers. Despite concentrated efforts of many research groups, the T2SLs have not revealed full potential yet. This paper attempts to provide a comprehensive review of the current status of T2SL detectors and discusses origins of T2SL device performance degradation, in particular, surface and bulk dark-current components. Various approaches of dark current reduction with their pros and cons are presented.

1. Introduction

Since proposed in 1980s [1–3], the InAs/(In,Ga)Sb T2SL has gained a lot of interest for the infrared (IR) detection applications. Focal plane arrays (FPAs) based on T2SL and operating in mid-wave IR (MWIR, 3–5 μm) and long-wave IR (LWIR, 8–12 μm) are of great importance for a variety of civil and military applications. Currently market dominating technologies are based on bulk mercury cadmium telluride (MCT) and InSb [4–6], and GaAs/AlGaAs quantum well IR photodetectors (QWIPs).

While MCT detectors have very large quantum efficiency (>90%) and detectivity, they are still plagued by nonuniform growth defects and a very expensive CdZnTe substrate that is only available in limited quantities by a foreign manufacturer. There has been significant progress on development of MCT on silicon substrates, but good performance has been limited to the MWIR band only. Moreover, MCT is characterized by low electron effective mass resulting in excessive leakage current [7]. The InSb detectors do not cover the LWIR spectral range. QWIPs are based on III-V semiconductors and their mature manufacturing process enables them to be scaled to large format FPAs with a high degree of spatial uniformity [8–10]. However, due to polarization selection rules for electron-photon interactions in GaAs/AlGaAs QW, this material system is insensitive to surface-normal incident IR radiation resulting in poor conversion quantum efficiency. In addition, their large dark currents lower the operating temperature and increase the operating cost of the imager. The development of FPAs based on mature III-V growth and fabrication technology and operating at higher temperatures will result in highly sensitive, more reliable, lighter, and less costly IR sensors than currently available ones.

The InAs/(In,Ga)Sb T2SL material system is characterized by a broken-gap type-II alignment schematically illustrated in Figure 1 with electron and hole wavefunctions having maxima in InAs and GaSb layers, respectively. The overlap of electron (hole) wave functions between adjacent InAs (GaSb) layers result in the formation of an electron (hole) minibands in the conduction (valence) band. Optical transition between the highest hole (heavy-hole) and the lowest conduction minibands is employed for the detection of incoming IR radiation. The operating wavelength of the T2SLs can be tailored from 3 μm to 32 μm by varying thickness of one or two T2SL constituent layers [11–13]. Some parameters of T2SL constituent materials, InAs and GaSb, are shown in Table 1.

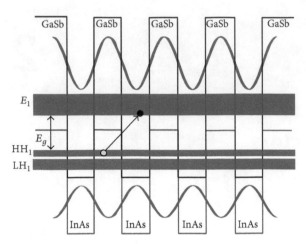

FIGURE 1: Type-II band alignment of InAs/GaSb T2SL system.

TABLE 1: Some band structure parameters for InAs and GaSb (0 K).

Parameter	InAs	Reference	GaSb	Reference
E_g^{Γ} (eV)	0.417	[14]	0.812	[15]
E_g^{X} (eV)	1.433	[16]	1.141	[17]
E_g^{L} (eV)	1.133	[18]	0.875	[19]
m_e^* (Γ)	0.026	[20]	0.039	[16]
m_l^* (L)	0.640	[21, 22]	1.30	[23, 24]
m_t^* (L)	0.050	[21, 22]	0.100	[23, 24]
m_l^* (X)	1.130	[21, 22]	1.510	[23, 24]
m_t^* (X)	0.160	[21, 22]	0.220	[23, 24]

InGaSb layers of InAs/InGaSb T2SL are subjected to biaxial compression strain causing splitting of light hole and heavy-hole minibands in the T2SL band structure and, therefore, suppression of Auger recombination rates relative to bulk MCT detectors [25, 26]. However, the majority of the research in the past ten years has focused on the binary InAs/GaSb system. This is attributed to the critical thickness limitations imposed on strained material grown with the large mole fraction of In. The scope of this paper is also limited to the InAs/GaSb T2SL devices.

1.1. Characterization of T2SL Material System. The physics behind the T2SL material system is not yet very well understood. Different theoretical methods have been applied to understand the band structure, electronic, and optical properties of superlattices. For example, Flatté et al. have undertaken extensive theoretical modeling of the band structure of superlattices [27–29], including investigation of electronic structure of dopants [30]. Features of T2SL photoabsorption spectra and optical properties of T2SL detectors were studied by Livneh et al. [31] and Qiao et al. [32], respectively, using k · p tight-binding model [33]. Empirical pseudopotential method, in its canonical shape [34–36] and four-component variation that includes interface layers [37], was successfully utilized for the heterojunction design of T2SL devices. Bandara et al. [38] have modeled the effect of doping on the Shockley-Read-Hall (SRH) lifetime and the

dark current; Pellegrino and DeWames [39] have performed extensive modeling to extract the SRH lifetime from dark-current measurements.

Background carrier concentration is one of the fundamental properties of the absorber layer of T2SL detector since it determines the minority carrier lifetime and diffusion lengths. Transport measurements in T2SL are difficult because of the lack of semi-insulating GaSb substrates. Several techniques have been reported to measure and analyze the electrical properties of T2SL by different groups. Magneto-transport analysis [40] was performed on T2SL structures grown on top of electrically insulating AlGaAsSb buffer in order to suppress parasitic conduction. Hall [41], capacitance-voltage, and current-voltage measurements [42] of T2SL structures grown on semi-insulating GaAs substrate directly or with the interfacial misfit (IMF) dislocation arrays technique [43] were also reported. Variable magnetic field geometric magnetoresistance measurements and a mobility spectrum analysis, (MSA) technique for data analysis, have been employed by Umana-Membreno et al. [44] to study vertical minority carrier electron transport parameters in T2SL structures. Works of Christol et al. [45, 46], Haugan et al. [47], and Szmulowicz et al. [48, 49]are concerned with the influence of T2SL composition and growth conditions on background carrier concentration and mobility.

Since performance of T2SL device is strongly dependent on T2SL structural perfection, the information on interfacial roughness, compositional profile (i.e., interfacial intermixing), and interfacial bonding across the noncommon anion layers of InAs/GaSb T2SL is very important. Growth conditions of T2SLs have been optimized by various research groups to improve the interface quality [50–54]. Steinshnider and colleagues [55–58] utilized the cross-sectional scanning tunneling microscopy (XSTM) to identify the interfacial bonding and to facilitate direct measurements of the compositional grading at the GaSb/InAs heterojunction. *In situ* study of origins of interfacial disorder and cross-contamination in T2SL structures [59, 60] revealed importance of Arsenic (As) background pressure control during the GaSb layers growth. Luna et al. [61] proposed the method of systematic characterization of InAs-on-GaSb and GaSb-on-InAs interfaces in T2SL with resolution less than 0.5 nm.

1.2. InAs/GaSb T2SL Detectors. T2SL diodes are predicted to have a number of advantages over bulk MCTs, including lower tunneling current, since the band edge effective masses in T2SL are not directly dependent on the band gap energy and are larger than HgCdTe at the same band gap [3]. The band-engineered suppression of Auger recombination rates [25, 26] leads to improved temperature limits of spectral detectivities. In contrast with QWIPs, normal incidence absorption is permitted in T2SLs, contributing to high conversion quantum efficiencies. Moreover, the commercial availability of substrates with good electrooptical homogeneity, and without large cluster defects, also offers advantages for T2SL technology. Thorough comparisons between MCT, InSb, QWIP, and T2SL technologies can be found in the literature [62–65].

TABLE 2: Properties of MWIR and LWIR T2SL detectors at 77 K [88, 89].

Parameter	MWIR T2SL $\lambda_{\text{Cut-off}}$ = 5 μm	LWIR T2SL $\lambda_{\text{Cut-off}}$ = 10 μm
Quantum efficiency (%)	~70	~70
$R_0 A$ ($\Omega \cdot$cm^2)	10^6	10^3
Detectivity (Jones) FOV = 0	1×10^{14}	6×10^{11}

TABLE 3: Properties of MWIR and LWIR T2SL FPAs at 77 K [90–92].

Parameter	MWIR T2SL $\lambda_{\text{Cut-off}}$ = 5 μm	LWIR T2SL $\lambda_{\text{Cut-off}}$ = 10 μm
Format	320 × 256	1024 × 1024
Quantum efficiency (%)	~50	~50
NEDT (mK)	>15	~30

High performance InAs/GaSb T2SL detectors have been reported for MWIR [66–68], LWIR [12, 69–72], and very-long wave IR (VLWIR) [73, 74] spectral regions. Moreover, mega-pixel FPAs, that is, FPAs of sizes up to 1024 × 1024, have been demonstrated [75, 76]. Multiband T2SL structures were realized, including short-wave IR (SW)/MWIR [77], MW/MWIR [78], MW/LWIR [79, 80], LW/LWIR [81], and SW/MW/LWIR [82] devices. Low-dark-current architectures with unipolar barriers such as M-structure [83], complementary-barrier infrared detector (CBIRD) [70], W-structure [69, 84], N-structure [85], nBn [86, 87], and pBiBn [12] have been designed and fabricated into single-pixel detectors and FPAs at university laboratories (Northwestern University, Arizona State University, University of Oklahoma, University of Illinois, Georgia Tech University, Bilkent University (Turkey), University of New Mexico), federal laboratories (JPL, NRL, ARL, NVESD, and SNL), and industrial laboratories (Raytheon, Teledyne Imaging Systems, Hughes Research Laboratories, QmagiQ LLC, etc.). Tables 2 and 3 summarize properties of MWIR and LWIR T2SL detectors and FPAs at 77 K.

2. Limitations of T2SL Technology

Despite the numerous technological and theoretically predicted advantages T2SLs offer over present-day detection technologies, the promise of superior performance of T2SL detectors has not been yet realized. The T2SL detectors are approaching the empirical benchmark of MCT's performance level, Rule 07 [93]; however, the dark-current density demonstrated by the T2SL detectors is still significantly higher than that of bulk MCT detectors, especially in the MWIR range, as illustrated in Figure 2.

To understand the reasons of high dark-current levels demonstrated by the T2SL detectors the origins of dark current have to be analyzed. Generally, dark current in detectors based on narrow band gap semiconductors may be differentiated into "bulk" and "surface" currents. The most important "bulk" dark currents are (i) generation-recombination (G-R) current associated with the SRH process in the depletion region of the detector and (ii) thermally generated diffusion current associated with Rogalski [94] or radiative process in both the n- and p-extrinsic regions of the detector [95].

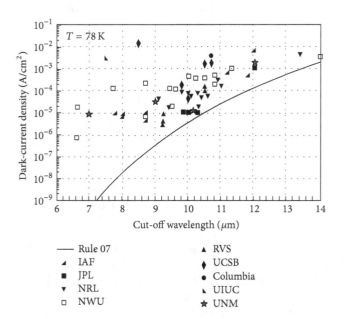

FIGURE 2: Dark-current density of T2SL detectors compared with Rule 07 [93]. Abbreviations for the different institution working on T2SL detectors: Fraunhofer-Institut (IAF), Jet Propulsion Laboratory (JPL), Naval Research Laboratory (NRL), Northwestern University (NWU), Raytheon Vision Systems (RVS), University of California, Santa Barbara (UCSB), Columbia University (Columbia), University of Illinois, Urbana-Champaign (UIUC), and University of New Mexico (UNM).

The SRH G-R process occurs through the trap levels within the energy gap thus limiting lifetime of the minority carriers. The origins of SRH centers are not well understood. According to the statistical theory of the SRH process, the SRH rate approaches a maximum as the energy level of the trap center approaches midgap. Thus, the most effective SRH centers are those located near the middle of the band gap [96]. Analysis of the defect formation energy of native defects dependent on the location of the Fermi level stabilization energy has been performed by Walukiewicz [97], who reported that, in bulk GaAs and GaSb, the stabilized Fermi level is located near either the valence band or the midgap, whereas in bulk InAs the stabilized Fermi level is located above the conduction band edge. From this observation, the midgap trap levels in GaAs and GaSb are available for SRH recombination, whereas in InAs they are inactive for the SRH process, suggesting a longer carrier lifetime in bulk InAs than in bulk GaSb and GaAs materials. Experimentally measured values of carrier lifetimes yielded ~325 ns for bulk InAs and ~100 ns for bulk GaSb, thereby confirming the initial observation [98]. It may then be hypothesized that native defects associated with GaSb are responsible for the SRH-limited minority-carrier lifetimes observed in InAs/GaSb T2SL.

Several methods have been employed to measure lifetime of photogenerated carriers of T2SL, including optical modulation response [38, 47], time-resolved PL [99–102], and photoconductive response variation measurements [103], to name just a few. Some of them provide more direct measures of lifetime, while others rely on assumptions or further measurements to perform extraction of lifetime [39, 104, 105]. Overall, the lifetimes reported for MWIR and LWIR InAs/GaSb T2SL range from 0.13 ns [106] to about 100 ns [107]. These values are significantly lower compared to the MCT devices operating in the same wavelength range [107].

A "surface" dark current component is associated with the surface states in the junction. During the individual pixel isolation process, the periodic crystal structure terminates abruptly resulting in formation of unsatisfied (dangling) chemical bonds at the semiconductor-air interface responsible for generation of surface states within the band gap and pinning of the Fermi level. Moreover, etch by-products, surface contaminant associated with the fabrication procedure, and differential etching also create additional interfacial states contributing to the dark current. Scaling of the lateral dimensions of a T2SL detector (e.g., typical mesa dimensions of individual FPA pixels are $20 \, \mu m \times 20 \, \mu m$) makes FPA performance strongly dependent on surface effects due to a large pixel surface/volume ratio.

This paper aims to review various ways of improving performance of T2SL detectors in order for T2SLs to be the technology of choice for high-performance IR imaging systems. Proposed solutions for the reduction of "bulk" and "surface" dark-current components as well as improvement of detector signal-to-noise ratio and operating temperature limits will be discussed in detail.

3. Proposed Solutions for the Improvement of T2SL Detector Performance

3.1. Reduction of "Bulk" Dark Currents. To overcome the carrier lifetime limitations imposed by the GaSb layer in an InAs/GaSb T2SL, the type-II Ga-free SL, that is, InAs/InAsSb SL, may be utilized for IR detection. A significantly longer minority carrier lifetime has been obtained in an InAs/InAsSb SL system as compared to an InAs/GaSb T2SL operating in the same wavelength range (at 77 K, ~412 ns, and ~100 ns, resp.) [100, 108]. Such increases in minority carrier lifetimes, along with demonstrated band gap adjustability [109] and suppressed Auger recombination rates [110], suggest lower dark currents for InAs/InAsSb SL detectors in comparison with their InAs/GaSb T2SL

counterparts. However, performance, in particular, signal-to-noise ratio, of InAs/InAsSb SL-based detectors with pin [111] and nBn [112] architectures was not superior to T2SL-based devices operating in the same wavelength range. This may be attributed to the increased tunneling probability in InAs/InAsSb SL system due to the smaller band offsets [111] and significant concentration of SRH centers in this material [113].

Thermally generated diffusion currents may be significantly suppressed by the incorporation of barriers into conduction and valence bands to impede the flow of carriers associated with dark current (noise) without blocking photocurrent (signal). The improved performance of these T2SL devices is credited to better confinement of the electron wavefunctions, reduced tunneling probability, increased electron effective mass in modified T2SL structures, and reduction in dark-current through the use of current blocking layers that reduce one or more dark-current component. nBn [86, 87], pBiBn [12], M-structure [83], W-structure [84], CBIRD [70], and N-structure [85] are examples of T2SL detectors with barrier architecture.

The band-offset tunability is critical parameter for the realization of barrier devices. Barrier layers are selected such that the hole-blocking layer offers an unimpeded electrons flow while blocking holes and electron-blocking layer fulfill the opposite function. Hence, one requires hole- (electron) blocking layers to have zero valence (conduction) band offsets with the absorber layer. Moreover, for the efficient barrier structure design, complete macroscopic simulations are required to get a good assessment of actual dark current and photocurrent. This simulation may as well help with design optimization of barrier structures, in particular, selecting an optimal barriers thickness, composition, and doping concentration.

The extension of concept of heterostructure barrier engineering in T2SL resulted in realization of interband cascade infrared photodetectors (ICIPs) [114–118]. In ICIP detectors each cascade stage is comprised of an absorber region, relaxation region, and interband tunneling region. While photocurrent is limited to the value produced in an individual absorber, adding of extra- stages benefits the signal-to-noise ratio, since the noise current in such devices scales inversely with the total number of stages. Ability to change number of stages with different absorber thicknesses is important for the design of T2SL detectors with maximized signal-to-noise ratio. The drawbacks of ICIPs are associated with the complicated structure of these devices. In particular, due to the number of layers and interfaces in the structure, some of the fundamental device physics is still unclear and MBE growth procedure is challenging.

Ability to heteroengineer the band structure of the T2SL devices stipulates realization of one more type of low-noise T2SL detectors, avalanche photodiodes (APDs) [119–121]. Control of individual layer thickness and composition offers great flexibility in engineering of the electron band structure to initiate single-carrier ionization. Moreover, APDs with either electron or hole dominated avalanching may be fabricated by engineering the higher lying T2SL energy levels. It should be noted that an APD device with hole dominating avalanching is expected to have lower noise due to reduced tunneling of heavier holes. Existence of hole dominated avalanching structure also opens up possibility of combining separate electron and hole multiplication regions in a single device achieving very high gain with low excess noise factor.

3.2. Reduction of "Surface" Dark Currents.

Despite numerous efforts of various research groups devoted to the development of effective passivation schemes for T2SL detectors, there is still no well-established and generally acknowledged procedure for passivation of such devices. Part of the problem is the complexity of T2SL system, composed by the hundreds of relatively thick (several monolayers, MLs) InAs, GaSb, and, sometimes, AlSb layers, and thin (typically, less than 1 ML) interfacial GaAs and InSb layers [55, 57, 58, 122]. Passivation should satisfy dangling bonds of all these T2SL constituent materials, originated at exposed device sidewalls after mesa definition process, and prevent formation of interface states in the T2SL band gap.

The great advantage of T2SL system, band gap tunability, allowing realization of detectors spanning wide IR range, serves as a disservice for the passivation development. Interface states cause the pinning of Fermi level with the bands bend towards lower energy near the surface. This band bending induces accumulation or type inversion of charge resulting in surface tunneling currents along sidewalls. As was shown by Delaunay et al. [123], the narrow band gap devices (LWIR and VLWIR, with band gap of 120 meV or lower) are more susceptible to the formation of charge conduction channels along the sidewalls. Consequently, the same passivation may be suitable for the T2SL MWIR and inefficient for the T2SL detectors with longer operating wavelength.

Moreover, passivation should exhibit thermal and long term stability. In other words, passivation layer must not undergo any change in its constitutional, physical, and interfacial properties at variable temperatures (30–300 K) during the lifetime of the T2SL detector (typically, 10,000 hrs). Finally, passivation has to be easily integrated into the FPA fabrication process.

In addition, since passivation applied on rough surfaces, or surfaces contaminated by native oxides, and foreign particles will result in little or no improvement of device performance, we spent some time discussing the surface preparation issues. To achieve minimal surface leakage, the device sidewalls must be smooth, with no patterns of preferential etch presented, and clean, with removed native oxides and etch by-products. Moreover, vertical etch profile is essential for the realization of high-fill factor, small pixel pitch, and large format T2SL FPAs. The thorough comparisons of various surface preparation and passivation techniques of T2SL detectors are out of the scope of this review article and can be found in literature [124, 125]. Next two sections aim to familiarize the reader with various mesa definition and passivation methods developed for T2SL devices.

3.2.1. Surface Preparation.

Definition of nearly vertical mesa sidewalls that are free of native oxide and defects is the crucial step in InAs/GaSb T2SL detector fabrication process [126, 127]. Presence of elemental antimony on the etched T2SL device sidewalls [128] may result in the conduction channel parallel to the interface, which leads to increasing of surface component of dark current. Unwanted native oxides are usually removed prior to or during the pixel isolation process with immersion in ammonium sulfide [127], phosphoric or hydrochloric acid based solutions [129]. Introduction of BCl_3 gas into the plasma chemistry is also effective in removal of native oxides and redeposited by-products [130].

Nowadays, high-density plasma etch processes are commonly utilized for InAs/GaSb T2SL material in spite of inevitable degradation of sidewall surface electronic properties due to ion bombardment or unwanted deposition of etch by-products [131, 132]. Plasma chemistry usually consists of chlorine-based precursors (BCl_3, Cl_2, or $SiCl_2$) due to high volatilities of gallium, indium, antimony, and arsenide chlorides providing fast etch rates and smooth morphologies [133]. The resulting etch profiles are vertical due to the plasma sheath and the ionized gas directionality. Damage produced during the dry etch may be partially restored by subsequent chemical treatment [134]. Due to the ability of wet etches to cause virtually no surface electronic damage, a chemical etch attracts attention of researchers for single-pixel T2SL device fabrication [135–140]. However, the isotropic nature of wet etch process resulting in concave sidewall profile and an unavoidable tendency to undercut etch masks making precise dimensional control more difficult stipulates limited application of wet etches for T2SL FPA fabrication.

3.2.2. Passivation.

Conventional passivation methods of T2SL devices include encapsulation of device sidewalls, by thick layer of dielectric or organic material, and sulfidization. Dielectric passivation of T2SL detectors is compatible with current T2SL FPA fabrication procedures and, consequently, very appealing to the T2SL scientists and engineers. Numerous reports on passivation of MWIR and LWIR T2SL detectors by silicon oxide or silicon nitride have been published over the last fifteen years [74, 81, 130, 141–144]. Dielectric passivation, though shown to be effective, presents the challenges of developing high-quality, low fixed, and interfacial charges density dielectrics at process temperatures substantially lower that the InAs/GaSb T2SL growth temperature to prevent the T2SL period intermixing. Moreover, native fixed charges presented in dielectric passivation layer can either improve or deteriorate the device performance [143]; consequently, the dielectric passivation may not passivate the low band gap materials as effectively as high band gap materials.

T2SL passivation with organic materials, which are polyimide or various photoresists (PRs), is emerging alternative to the dielectric passivation approach [134, 135, 145–150]. PRs are commonly deposited at room temperature and thus the T2SL thermal budget is not exerted. Moreover, PRs equally effectively passivate T2SL detectors with different operating wavelengths.

Chalcogenide passivation, or saturation of unsatisfied bonds on semiconductor surface by sulfur atoms, has been employed from early 1990s for the passivation of bulk III-V materials [151–162]. The enhanced photoluminescence (PL) and reduced diode leakage current were credited to the formation of III-S bond responsible for the reduction of surface states within band gap.

The simplest sulfidization scheme of T2SL detectors is device immersion in aqueous solution of ammonium sulfide. No native oxide removal step is required prior to passivation because the native oxides are etched by $(NH_4)OH$ formed in water solution of ammonium sulfide. Short-term benefits for the MWIR and LWIR T2SL device performance have been reported [127, 129, 163] and the necessity for a suitable capping layer to preserve good passivation quality in the long term was reaffirmed. Thioacetamide (TAM, C_2H_5NS) [124, 164] and octadecanethiol (ODT, $CH_3[CH_2]_{17}SH$) [85] treatments offer formation of more stable bonds between sulfur and T2SL constituent elements (Ga, In, As, and Sb) compared to weaker III (V)-oxygen-S bonds formed after ammonium sulfide treatment.

One of relatively new sulfidization methods is electro-chemical passivation (ECP) [129, 165] that is saturation of dangling bonds with sulfur through electrolysis in S-containing solution. Though effective, sulfur layer deposited through ECP may oxidize easily and additional encapsulation is required. Electron-beam evaporated ZnS satisfies the dangling bonds with S-atoms simultaneously acting as an encapsulant [119–121, 166].

Recently, several research groups reported the "combined" approach for the passivation of T2SL detectors. For example, Zhang et al. [167] noticed that the anodic sulfide passivation combined with the SiO_2 significantly improved the performance of MWIR T2SL detectors. DeCuir Jr. et al. [147] found that the sulfide chemical treatment followed by the SU-8 treatment inhibits the formation of native surface oxides, satisfies dangling bonds, and prevents the sulfide layer degradation over time.

4. Other Methods of T2SL Detector Performance Improvement

The bulk components of the dark current (SRH and thermally generated diffusion current) in T2SL detector may be significantly diminished by scaling thickness of the device. The abridged quantum efficiency (QE) of such device may be restored through plasmon assisted coupling of incident electromagnetic radiation while maintaining low dark-current level. Transmission enhancement and QE increase through subwavelength metal hole array [168] and corrugated metal surface structure [169], respectively, have been reported for MWIR T2SL detectors.

Surface currents may be suppressed by reduced exposure of narrow gap materials to the environment, for example, as a result of encapsulation of etched sidewalls with wide band gap material [133, 170] or buried architecture [84] that isolates the neighboring devices but terminates within a wider band gap layer. The former passivation approach requires very careful surface cleaning prior the overgrowth procedure, whereas latter is subjected to the possible crosstalk issues in FPAs due to the uncertainty of the lateral diffusion length of minority

carriers. If the values of lateral diffusion length are larger than the distance between neighboring pixels in the FPA, crosstalk between the FPA elements can be encountered that leads to the degradation of image resolution.

Another approach for the realization of high performance T2SL sensors is growth of T2SL structures on high-index plane GaSb [171]. The thickness of the T2SL detector grown on the GaSb (111) substrate is reduced due to the natural difference of lattice parameters in the (111) and (100) directions, whereas heavy hole confinement is increased by a factor of three [172]. This translates into thinner detector structures for a given detection wavelength and absorption coefficient realized on (111) GaSb substrate, resulting in shorter growth times. This also means decreased costs and material usage, both of which are highly desirable. Moreover, the decreased detector volume results in an improved signal-to-noise ratio, since the number of thermally generated carriers is correspondingly reduced.

5. Summary

This work provides a review of the current status and limitations of IR detectors based on an InAs/GaSb T2SLs. It should be noted that applications of T2SL system are not limited to the IR detection only. Low thermal conductivity of T2SL identifies it as a prospective material for low-temperature Peltier coolers [173]. Spatially separated confinement of electrons and holes, signature of type-II band alignment, initiated InAs/GaSb core-shell nanowires realization [174, 175]. Field-effect transistors (FETs) [176, 177] and thermo photovoltaic (TPV) [178] T2SL devices are another examples of unconventional applications of T2SL material system.

Despite the numerous theoretically predicted advantages that T2SLs offer over MCT, InSb, and QWIP-based detectors, intensive heterostructure engineering efforts and development of epitaxial growth and fabrication techniques, the promise of superior performance of T2SL detectors has not been yet realized. The dark-current density demonstrated by the T2SL detectors is still significantly higher than that of bulk MCT detectors, especially in the MWIR range.

The complexity of T2SL system, along with the intricate detector architectures, results in no universal solution for the suppression of dark currents. Different approaches that address suppression of either bulk or surface dark current components in order for T2SL to be the technology of choice for high-performance imaging systems have been presented.

The SRH and thermally generated diffusion currents may be significantly reduced by exclusion of GaSb layer from InAs/GaSb T2SL stack, that is, Ga-free T2SL, and by the incorporation of barriers device structure to impede the flow of carriers associated with dark current (noise) without blocking photocurrent (signal), respectively. Passivation treatment of the exposed device sidewalls decreases the surface currents. However, development of effective passivation technique is hindered by the ease of native oxide formation and requirements to the etched surface. In addition, the same passivation may be suitable for the T2SL MWIR and inefficient for the T2SL detectors with longer operating wavelength. Finally, one of the most effective passivation approaches, saturation of unsatisfied chemical bonds with sulfur atoms, results in formation of passivation layer with poor long-term stability, and additional encapsulation is required.

Integration of T2SL detectors with surface plasmon couplers and utilization of high-index plane GaSb substrates are recent alternatives for the improvement of T2SL detector performance. Despite the promising preliminary results, both of these directions require additional investigation.

In conclusion, unique combination of band structure engineering flexibility and material properties of InAs/GaSb T2SL provide a prospective benefit in the realization of next generation IR imagers. Performance of MWIR and LWIR T2SL detectors has not achieved its theoretically predicted limit. To fully realize the T2SL potential methods of suppression of various dark current components have to be developed. Up-to-date techniques of dark current reduction include not only traditional passivation, but advanced heterostructure engineering and integration of T2SL with nanostructures as well.

Conflict of Interests

The author declares that there is no conflict of interests regarding the publication of this paper.

Acknowledgments

This work was supported by the AFOSR FA9550-10-1-0113 and AFRL FA9453-12-1-0336 Grants.

References

[1] G. A. Sai-Halasz, R. Tsu, and L. Esaki, "A new semiconductor superlattice," *Applied Physics Letters*, vol. 30, no. 12, pp. 651–653, 1977.

[2] L. Esaki, "InAs-GaSb superlattices-synthesized semiconductors and semimetals," *Journal of Crystal Growth*, vol. 52, no. 1, pp. 227–240, 1981.

[3] D. L. Smith and C. Mailhiot, "Proposal for strained type II superlattice infrared detectors," *Journal of Applied Physics*, vol. 62, no. 6, pp. 2545–2548, 1987.

[4] J. Rothman, E. D. Borniol, P. Ballet et al., "HgCdTe APD-Focal plane array performance at DEFIR," in *Infrared Technology and Applications XXXV*, vol. 7298 of *Proceedings of SPIE*, pp. 729835–729845, April 2009.

[5] J. M. Peterson, D. D. Lofgreen, J. A. Franklin et al., "MBE growth of HgCdTe on large-area Si and CdZnTe wafers for SWIR, MWIR and LWIR detection," *Journal of Electronic Materials*, vol. 37, no. 9, pp. 1274–1282, 2008.

[6] O. Nesher, I. Pivnik, E. Ilan et al., "High resolution 1280×1024, 15 μm pitch compact InSb IR detector with on-chip ADC," in *Infrared Technology and Applications XXXV*, vol. 7298 of *Proceedings of SPIE*, April 2009.

[7] A. Rogalski, "HgCdTe infrared detector material: history, status and outlook," *Reports on Progress in Physics*, vol. 68, no. 10, pp. 2267–2336, 2005.

[8] H. Schneider and H. C. Liu, *Quantum Well Infrared Photodetectors*, Springer Series in Optical Sciences, Springer, 2007.

[9] A. Soibel, S. V. Bandara, D. Z. Ting et al., "A super-pixel QWIP focal plane array for imaging multiple waveband temperature sensor," *Infrared Physics and Technology*, vol. 52, no. 6, pp. 403–407, 2009.

[10] A. Nedelcu, V. Guériaux, A. Bazin et al., "Enhanced quantum well infrared photodetector focal plane arrays for space applications," *Infrared Physics and Technology*, vol. 52, no. 6, pp. 412–418, 2009.

[11] C. Cervera, I. Ribet-Mohamed, R. Taalat, J. P. Perez, P. Christol, and J. B. Rodriguez, "Dark current and noise measurements of an InAs/GaSb superlattice photodiode operating in the midwave infrared domain," *Journal of Electronic Materials*, vol. 41, pp. 2714–2718, 2012.

[12] N. Gautam, H. S. Kim, M. N. Kutty, E. Plis, L. R. Dawson, and S. Krishna, "Performance improvement of longwave infrared photodetector based on type-II InAs/GaSb superlattices using unipolar current blocking layers," *Applied Physics Letters*, vol. 96, no. 23, Article ID 231107, 2010.

[13] Y. Wei, A. Gin, M. Razeghi, and G. J. Brown, "Advanced InAs/GaSb superlattice photovoltaic detectors for very long wavelength infrared applications," *Applied Physics Letters*, vol. 80, no. 18, pp. 3262–3264, 2002.

[14] Y. Lacroix, C. A. Tran, S. P. Watkins, and M. L. W. Thewalt, "Low-temperature photoluminescence of epitaxial InAs," *Journal of Applied Physics*, vol. 80, no. 11, pp. 6416–6424, 1996.

[15] P. S. Dutta, H. L. Bhat, and V. Kumar, "The physics and technology of gallium antimonide: an emerging optoelectronic material," *Journal of Applied Physics*, vol. 81, no. 9, pp. 5821–5870, 1997.

[16] M. Levinshtein, S. Rumyantsev, and M. Shur, Eds., *Handbook Series on Semiconductor Parameters*, vol. 1 and 2, World Scientific, 1996.

[17] A. Joullié, A. Z. Eddin, and B. Girault, "Temperature dependence of the $L_6^c - \Gamma_6^c$ energy gap in gallium antimonide," *Physical Review B*, vol. 23, no. 2, pp. 928–930, 1981.

[18] E. Adachi, "Energy band parameters of InAs at various temperatures," *Journal of the Physical Society of Japan*, vol. 24, no. 5, p. 1178, 1968.

[19] C. Alibert, A. Joullie, and A. M. Joullie, "Modulation-spectroscopy study of the GaAlSb band structure," *Physical Review B*, vol. 27, p. 4946, 1983.

[20] M. B. Thomas and J. C. Woolley, "Plasma edge reflectance measurements in GaInAs and InAsSb alloys," *Canadian Journal of Physics*, vol. 49, p. 2052, 1971.

[21] D. C. Tsui, "Landau-level spectra of conduction electrons at an InAs surface," *Physical Review B*, vol. 12, no. 12, pp. 5739–5748, 1975.

[22] R. A. Stradling and R. A. Wood, "The temperature dependence of the band-edge effective masses of InSb, InAs and GaAs as deduced from magnetophonon magnetoresistance measurements," *Journal of Physics C: Solid State Physics*, vol. 3, no. 5, article 005, pp. L94–L99, 1970.

[23] C. Ghezzi, R. Magnanini, A. Parisini et al., "Optical absorption near the fundamental absorption edge in GaSb," *Physical Review B*, vol. 52, no. 3, pp. 1463–1466, 1995.

[24] H. Arimoto, N. Miura, R. J. Nicholas, N. J. Mason, and P. J. Walker, "High-field cyclotron resonance in the conduction bands of GaSb and effective-mass parameters at the L points," *Physical Review B*, vol. 58, no. 8, pp. 4560–4565, 1998.

[25] E. R. Youngdale, J. R. Meyer, C. A. Hoffman et al., "Auger lifetime enhancement in InAs-Ga$_{1-x}$In$_x$Sb superlattices," *Applied Physics Letters*, vol. 64, no. 23, pp. 3160–3162, 1994.

[26] G. M. Williams, "Comment on 'Temperature limits on infrared detectivities of InAs/In$_x$Ga$_{1-x}$Sb superlattices and bulk Hg$_x$Cd$_{1-x}$Te' [J. Appl. Phys. 74, 4774 (1993)]," *Journal of Applied Physics*, vol. 77, no. 8, pp. 4153–4155, 1995.

[27] M. E. Flatté, C. H. Grein, H. Ehrenreich, R. H. Miles, and H. Cruz, "Theoretical performance limits of 2.1-4.1 μm InAs/InGaSb, HgCdTe, and InGaAsSb lasers," *Journal of Applied Physics*, vol. 78, no. 7, pp. 4552–4561, 1995.

[28] K. Abu El-Rub, C. H. Grein, M. E. Flatte, and H. Ehrenreich, "Band structure engineering of superlattice-based short-, mid-, and long-wavelength infrared avalanche photodiodes for improved impact ionization rates," *Journal of Applied Physics*, vol. 92, no. 7, pp. 3771–3778, 2002.

[29] M. E. Flatté and C. H. Grein, "Theory and modeling of type-II strained-layer superlattice detectors," in *Quantum Sensing and Nanophotonic Devices VI*, vol. 7222 of *Proceedings of SPIE*, January 2009.

[30] M. E. Flatté and C. E. Pryor, "Defect states in type-II strained-layer superlattices," in *Quantum Sensing and Nanophotonic Devices VII*, vol. 7608 of *Proceedings of SPIE*, January 2010.

[31] Y. Livneh, P. Klipstein, O. Klin et al., "kp model for the energy dispersions and absorption spectra of InAs/GaSb type-II superlattices," *Physical Review B*, vol. 86, Article ID 235311, 2012.

[32] P.-F. Qiao, S. Mou, and S. L. Chuang, "Electronic band structures and optical properties of type-II superlattice photodetectors with interfacial effect," *Optics Express*, vol. 20, no. 3, pp. 2319–2334, 2013.

[33] P. C. Klipstein, "Operator ordering and interface-band mixing in the Kane-like Hamiltonian of lattice-matched semiconductor superlattices with abrupt interfaces," *Physical Review B*, vol. 81, no. 23, Article ID 235314, 2010.

[34] G. C. Dente and M. L. Tilton, "Pseudopotential methods for superlattices: applications to mid-infrared semiconductor lasers," *Journal of Applied Physics*, vol. 86, no. 3, pp. 1420–1429, 1999.

[35] R. Kaspi, C. Moeller, A. Ongstad et al., "Absorbance spectroscopy and identification of valence subband transitions in type-II InAs/GaSb superlattices," *Applied Physics Letters*, vol. 76, no. 4, pp. 409–411, 2000.

[36] G. C. Dente and M. L. Tilton, "Comparing pseudopotential predictions for InAs/GaSb superlattices," *Physical Review B*, vol. 66, no. 16, Article ID 165307, 2002.

[37] J. M. Masur, R. Rehm, J. Schmitz, L. Kirste, and M. Walther, "Four-component superlattice empirical pseudopotential method for InAs/GaSb superlattices," *Infrared Physics & Technology*, vol. 61, pp. 129–133, 2013.

[38] S. Bandara, P. G. Maloney, N. Baril, J. G. Pellegrino, and M. Z. Tidrow, "Doping dependence of minority carrier lifetime in long-wave Sb-based type II superlattice infrared detector materials," *Optical Engineering*, vol. 50, no. 6, Article ID 061015, 2011.

[39] J. Pellegrino and R. DeWames, "Minority carrier lifetime characteristics in type II InAs/GaSb LWIR superlattice n+πp+ photodiodes," in *Infrared Technology and Applications XXXV*, vol. 7298 of *Proceedings of SPIE*, April 2009.

[40] L. Bürkle, F. Fuchs, J. Schmitz, and W. Pletschen, "Control of the residual doping of InAs/(GaIn)Sb infrared superlattices," *Applied Physics Letters*, vol. 77, no. 11, pp. 1659–1661, 2000.

[41] X. B. Zhang, J. H. Ryou, R. D. Dupuis et al., "Improved surface and structural properties of InAs/GaSb superlattices on (001) GaSb substrate by introducing an InAsSb layer at interfaces," *Applied Physics Letters*, vol. 90, no. 13, Article ID 131110, 2007.

[42] X. B. Zhang, J. H. Ryou, R. D. Dupuis et al., "Metalorganic chemical vapor deposition growth of high-quality InAsGaSb type II superlattices on (001) GaAs substrates," *Applied Physics Letters*, vol. 88, no. 7, Article ID 072104, 2006.

[43] A. Jallipalli, G. Balakrishnan, S. H. Huang et al., "Structural analysis of highly relaxed GaSb grown on GaAs substrates with periodic interfacial array of 90° misfit dislocations," *Nanoscale Research Letters*, vol. 4, no. 12, pp. 1458–1462, 2009.

[44] G. Umana-Membreno, B. Klein, H. Kala et al., "Vertical minority carrier electron transport in p-type InAs/GaSb type-II superlattices," *Applied Physics Letters*, vol. 101, no. 25, Article ID 253515, 2012.

[45] P. Christol, L. Konczewicz, Y. Cuminal, H. Aït-Kaci, J. B. Rodriguez, and A. Joullié, "Electrical properties of short period InAs/GaSb superlattice," *Physica Status Solidi C*, vol. 4, no. 4, pp. 1494–1498, 2007.

[46] C. Cervera, J. B. Rodriguez, J. P. Perez et al., "Unambiguous determination of carrier concentration and mobility for InAs/GaSb superlattice photodiode optimization," *Journal of Applied Physics*, vol. 106, no. 3, Article ID 033709, 2009.

[47] H. J. Haugan, S. Elhamri, F. Szmulowicz, B. Ullrich, G. J. Brown, and W. C. Mitchel, "Study of residual background carriers in midinfrared InAsGaSb superlattices for unfcooled detector operation," *Applied Physics Letters*, vol. 92, no. 7, Article ID 071102, 2008.

[48] F. Szmulowicz, S. Elhamri, H. J. Haugan, G. J. Brown, and W. C. Mitchel, "Demonstration of interface-scattering-limited electron mobilities in InAs/GaSb superlattices," *Journal of Applied Physics*, vol. 101, no. 4, Article ID 043706, 2007.

[49] F. Szmulowicz, S. Elhamri, H. J. Haugan, G. J. Brown, and W. C. Mitchel, "Carrier mobility as a function of carrier density in type-II InAs/GaSb superlattices," *Applied Physics Letters*, vol. 105, Article ID 074303, 2009.

[50] K. Mahalingam, H. J. Haugan, G. J. Brown, and A. J. Aronow, "Strain analysis of compositionally tailored interfaces in InAs/GaSb superlattices," *Applied Physics Letters*, vol. 103, no. 21, Article ID 211605, 2013.

[51] H. Kim, Y. Meng, J. L. Rouviere, D. Isheim, D. N. Seidman, and J. M. Zuo, "Atomic resolution mapping of interfacial intermixing and segregation in InAs/GaSb superlattices: a correlative study," *Journal of Applied Physics*, vol. 113, no. 10, Article ID 103511, 2013.

[52] Z. Xu, J. Chen, F. Wang, Y. Zhou, C. Jin, and L. He, "Interface layer control and optimization of InAs/GaSb type-II superlattices grown by molecular beam epitaxy," *Journal of Crystal Growth*, vol. 386, pp. 220–225, 2014.

[53] B. Arikan, G. Korkmaz, Y. E. Suyolcu, B. Aslan, and U. Serincan, "On the structural characterization of InAs/GaSb type-II superlattices: the effect of interfaces for fixed layer thicknesses," *Thin Solid Films*, vol. 548, pp. 288–291, 2013.

[54] Y. Ashuach, Y. Kauffmann, C. Saguy et al., "Quantification of atomic intermixing in short-period InAs/GaSb superlattices for infrared photodetectors," *Journal of Applied Physics*, vol. 113, no. 18, Article ID 184305, 2013.

[55] J. Steinshnider, M. Weimer, R. Kaspi, and G. W. Turner, "Visualizing interfacial structure at non-common-atom heterojunctions with cross-sectional scanning tunneling microscopy," *Physical Review Letters*, vol. 85, no. 14, pp. 2953–2956, 2000.

[56] J. Steinshnider, J. Harper, M. Weimer, C.-H. Lin, S. S. Pei, and D. H. Chow, "Origin of antimony segregation in GaInSb/InAs strained-layer superlattices," *Physical Review Letters*, vol. 85, no. 21, pp. 4562–4565, 2000.

[57] R. Kaspi, J. Steinshnider, M. Weimer, C. Moeller, and A. Ongstad, "As-soak control of the InAs-on-GaSb interface," *Journal of Crystal Growth*, vol. 225, no. 2–4, pp. 544–549, 2001.

[58] R. Kaspi, "Compositional abruptness at the InAs-on-GaSb interface: optimizing growth by using the Sb desorption signature," *Journal of Crystal Growth*, vol. 201-202, pp. 864–867, 1999.

[59] P. M. Thibado, B. R. Bennett, M. E. Twigg, B. V. Shanabrook, and L. J. Whitman, "Origins of interfacial disorder in GaSb/InAs superlattices," *Applied Physics Letters*, vol. 67, pp. 3578–3580, 1995.

[60] E. M. Jackson, G. I. Boishin, E. H. Aifer, B. R. Bennett, and L. J. Whitman, "Arsenic cross-contamination in GaSb/InAs superlattices," *Journal of Crystal Growth*, vol. 270, no. 3-4, pp. 301–308, 2004.

[61] E. Luna, B. Satpati, J. B. Rodriguez, A. N. Baranov, E. Tourní, and A. Trampert, "Interfacial intermixing in InAs/GaSb short-period-superlattices grown by molecular beam epitaxy," *Applied Physics Letters*, vol. 96, no. 2, Article ID 021904, 2010.

[62] M. A. Kinch, "Fundamental physics of infrared detector materials," *Journal of Electronic Materials*, vol. 29, no. 6, pp. 809–817, 2000.

[63] A. Rogalski, "Third-generation infrared photon detectors," *Optical Engineering*, vol. 42, no. 12, pp. 3498–3516, 2003.

[64] A. Rogalski, "Infrared detectors: status and trends," *Progress in Quantum Electronics*, vol. 27, no. 2-3, pp. 59–210, 2003.

[65] C. Downs and T. E. Vanderveld, "Progress in infrared photodetectors since 2000," *Sensors*, vol. 13, pp. 5054–5098, 2013.

[66] M. Walther, J. Schmitz, R. Rehm et al., "Growth of InAs/GaSb short-period superlattices for high-resolution mid-wavelength infrared focal plane array detectors," *Journal of Crystal Growth*, vol. 278, no. 1-4, pp. 156–161, 2005.

[67] Y. Wei, A. Hood, H. Yau et al., "Uncooled operation of type-II InAs/GaSb superlattice photodiodes in the midwavelength infrared range," *Applied Physics Letters*, vol. 86, no. 23, Article ID 233106, pp. 1–3, 2005.

[68] E. Plis, J. B. Rodriguez, H. S. Kim et al., "Type II InAsGaSb strain layer superlattice detectors with p-on-n polarity," *Applied Physics Letters*, vol. 91, no. 13, Article ID 133512, 2007.

[69] I. Vurgaftman, E. H. Aifer, C. L. Canedy et al., "Graded band gap for dark-current suppression in long-wave infrared W-structured type-II superlattice photodiodes," *Applied Physics Letters*, vol. 89, no. 12, Article ID 121114, 2006.

[70] D. Z.-Y. Ting, C. J. Hill, A. Soibel et al., "A high-performance long wavelength superlattice complementary barrier infrared detector," *Applied Physics Letters*, vol. 95, no. 2, Article ID 023508, 2009.

[71] P. Y. Delaunay and M. Razeghi, "Spatial noise and correctability of type-II InAs/GaSb focal plane arrays," *IEEE Journal of Quantum Electronics*, vol. 46, no. 4, pp. 584–588, 2010.

[72] N. Gautam, S. Myers, A. V. Barve et al., "Barrier engineered infrared photodetectors based on type-II InAs/GaSb strained layer superlattices," *IEEE Journal of Quantum Electronics*, vol. 49, no. 2, pp. 211–217, 2013.

[73] Y. Wei, A. Gin, M. Razeghi, and G. J. Brown, "Type II InAs/GaSb superlattice photovoltaic detectors with cutoff wavelength approaching 32 μm," *Applied Physics Letters*, vol. 81, no. 19, pp. 3675–3677, 2002.

[74] A. Hood, M. Razeghi, E. H. Aifer, and G. J. Brown, "On the performance and surface passivation of type II InAs/GaSb superlattice photodiodes for the very-long-wavelength infrared," *Applied Physics Letters*, vol. 87, no. 15, Article ID 151113, pp. 1–3, 2005.

[75] S. D. Gunapala, D. Z. Ting, C. J. Hill et al., "Demonstration of a 1024 × 1024 Pixel InAsGaSb superlattice focal plane array," *IEEE Photonics Technology Letters*, vol. 22, no. 24, pp. 1856–1858, 2010.

[76] A. Haddadi, S. Ramezani-Darvish, G. Chen, A. M. Hoang, B.-M. Nguyen, and M. Razeghi, "High operability 1024 × 1024 long wavelength type-II superlattice focal plane array," *IEEE Journal of Quantum Electronics*, vol. 48, no. 2, pp. 221–228, 2012.

[77] A. M. Hoang, G. Chen, A. Haddadi, and M. Razeghi, "Demonstration of high-performance biaselectable dual-band short-mid-wavelength infrared photodetectors based on type-II InAs/GaSb/AlSb superlattices," *Applied Physics Letters*, vol. 102, Article ID 011108, 2013.

[78] R. Rehm, M. Walther, F. Rutz et al., "Dual-color InAs/GaSb superlattice focal-plane array technology," *Journal of Electronic Materials*, vol. 40, no. 8, pp. 1738–1743, 2011.

[79] A. Khoshakhlagh, J. B. Rodriguez, E. Plis et al., "Bias dependent dual band response from InAsGa (In) Sb type II strain layer superlattice detectors," *Applied Physics Letters*, vol. 91, no. 26, Article ID 263504, 2007.

[80] E. Plis, S. S. Krishna, E. P. Smith, S. Johnson, and S. Krishna, "Voltage controllable dual-band response from InAs/GaSb strained layer superlattice detectors with nBn design," *Electronics Letters*, vol. 47, no. 2, pp. 133–134, 2011.

[81] J. Huang, W. Ma, C. Cao et al., "Mid wavelength type II InAs/GaSb superlattice photodetector using SiO_xN_y passivation," *Japanese Journal of Applied Physics*, vol. 51, Article ID 074002, 2012.

[82] N. Gautam, M. Naydenkov, S. Myers et al., "Three color infrared detector using InAs/GaSb superlattices with unipolar barriers," *Applied Physics Letters*, vol. 98, no. 12, Article ID 121106, 2011.

[83] B.-M. Nguyen, D. Hoffman, P.-Y. Delaunay, and M. Razeghi, "Dark current suppression in type II InAsGaSb superlattice long wavelength infrared photodiodes with M-structure barrier," *Applied Physics Letters*, vol. 91, no. 16, Article ID 163511, 2007.

[84] E. H. Aifer, J. H. Warner, C. L. Canedy et al., "Shallow-etch mesa isolation of graded-bandgap W-structured type II superlattice photodiodes," *Journal of Electronic Materials*, vol. 39, pp. 1070–1079, 2010.

[85] O. Salihoglu, A. Muti, K. Kutluer et al., "N-structure for type-II superlattice photodetectors," *Applied Physics Letters*, vol. 101, no. 7, Article ID 073505, 2012.

[86] J. B. Rodriguez, E. Plis, G. Bishop et al., "NBn structure based on InAs/GaSb type-II strained layer superlattices," *Applied Physics Letters*, vol. 91, no. 4, Article ID 043514, 2008.

[87] H. S. Kim, E. Plis, J. B. Rodriguez et al., "Mid-IR focal plane array based on type-II InAsGaSb strain layer superlattice detector with nBn design," *Applied Physics Letters*, vol. 92, no. 18, Article ID 183502, 2008.

[88] A. Rogalski, J. Antoszewski, and L. Faraone, "Third-generation infrared photodetector arrays," *Journal of Applied Physics*, vol. 105, no. 9, Article ID 091101, 2009.

[89] A. Rogalski, "Progress in focal plane array technologies," *Progress in Quantum Electronics*, vol. 36, p. 342, 2012.

[90] M. Sundaram, A. Reisinger, R. Dennis et al., "Evolution of array format of longwave infrared Type-II SLS FPAs with high quantum efficiency," *Infrared Physics & Technology*, vol. 59, pp. 12–17, 2013.

[91] M. Razeghi, A. Haddadi, A. Hoang et al., "Advances in antimonide-based Type-II superlattices for infrared detection and imaging at center for quantum devices," *Infrared Physics & Technology*, vol. 59, pp. 41–52, 2013.

[92] D. Z. Ting, A. Soibel, S. A. Keo et al., "Development of quantum well, quantum dot, and type II superlattice infrared photodetectors," *Journal of Applied Remote Sensing*, vol. 8, Article ID 084998, 2014.

[93] D. R. Rhiger, "Performance comparison of long-wavelength infrared type II superlattice devices with HgCdTe," *Journal of Electronic Materials*, vol. 40, no. 8, pp. 1815–1822, 2011.

[94] A. Rogalski, *New Ternary Alloy Systems for Infrared Detectors*, SPIE, Bellingham, Wash, USA, 1994.

[95] S. Maimon and G. W. Wicks, "nBn detector, an infrared detector with reduced dark current and higher operating temperature," *Applied Physics Letters*, vol. 89, no. 15, Article ID 151109, 2006.

[96] W. Shockley and J. W. T. Read, "Statistics of the recombinations of holes and electrons," *Physical Review*, vol. 87, no. 5, pp. 835–842, 1952.

[97] W. Walukiewicz, "Defect reactions at metal-semiconductor and semiconductor-semiconductor interfaces," *Materials Research Society Symposium Proceedings*, vol. 148, p. 137, 1989.

[98] S. P. Svensson, D. Donetsky, D. Wang, H. Hier, F. J. Crowne, and G. Belenky, "Growth of type II strained layer superlattice, bulk InAs and GaSb materials for minority lifetime characterization," *Journal of Crystal Growth*, vol. 334, no. 1, pp. 103–107, 2011.

[99] B. C. Connelly, G. D. Metcalfe, H. Shen, and M. Wraback, "Direct minority carrier lifetime measurements and recombination mechanisms in long-wave infrared type II superlattices using time-resolved photoluminescence," *Applied Physics Letters*, vol. 97, no. 25, Article ID 251117, 2010.

[100] E. H. Steenbergen, B. C. Connelly, G. D. Metcalfe et al., "Significantly improved minority carrier lifetime observed in a long-wavelength infrared III-V type-II superlattice comprised of InAs/InAsSb," *Applied Physics Letters*, vol. 99, no. 25, Article ID 251110, 2011.

[101] G. Belenky, G. Kipshidze, D. Donetsky et al., "Effects of carrier concentration and phonon energy on carrier lifetime in Type-2 SLS and properties of $InAs_{1-x}Sb_X$ alloys," in *Infrared Technology and Applications XXXVII*, vol. 8012 of *Proceedings of SPIE*, April 2011.

[102] L. Murray, K. Lokovic, B. Olson, A. Yildirim, T. Boggess, and J. Prineas, "Effects of growth rate variations on carrier lifetime and interface structure in InAs/GaSb superlattices," *Journal of Crystal Growth*, vol. 386, pp. 194–198, 2014.

[103] Q. K. Yang, C. Pfahler, J. Schmitz, W. Pletschen, and F. Fuchs, "Trap centers and minority carrier lifetimes in InAs/(GaIn)Sb superlattice long wavelength photodetectors," in *Quantum Sensing: Evolution and Revolution from Past to Future*, vol. 4999 of *Proceedings of SPIE*, pp. 448–456, January 2003.

[104] E. C. F. Da Silva, D. Hoffman, A. Hood, B. M. Nguyen, P. Y. Delaunay, and M. Razeghi, "Influence of residual impurity background on the nonradiative recombination processes in high purity InAsGaSb superlattice photodiodes," *Applied Physics Letters*, vol. 89, no. 24, Article ID 243517, 2006.

[105] H. Mohseni, M. Razeghi, G. J. Brown, and Y. S. Park, "High-performance InAs/GaSb superlattice photodiodes for the very long wavelength infrared range," *Applied Physics Letters*, vol. 78, no. 15, pp. 2107–2109, 2001.

[106] J. V. Li, S. L. Chuang, E. M. Jackson, and E. Aifer, "Minority carrier diffusion length and lifetime for electrons in a type-II InAs/GaSb superlattice photodiode," *Applied Physics Letters*, vol. 85, no. 11, pp. 1984–1986, 2004.

[107] D. Donetsky, G. Belenky, S. Svensson, and S. Suchalkin, "Minority carrier lifetime in type-2 InAs-GaSb strained-layer superlattices and bulk HgCdTe materials," *Applied Physics Letters*, vol. 97, no. 5, Article ID 052108, 2010.

[108] B. V. Olson, E. A. Shaner, J. K. Kim et al., "Time-resolved optical measurements of minority carrier recombination in a mid-wave infrared InAsSb alloy and InAs/InAsSb superlattice," *Applied Physics Letters*, vol. 101, Article ID 092109, 2012.

[109] D. Lackner, M. Steger, M. L. W. Thewalt et al., "InAs/InAsSb strain balanced superlattices for optical detectors: material properties and energy band simulations," *Journal of Applied Physics*, vol. 111, no. 3, Article ID 034507, 2012.

[110] C. M. Ciesla, B. N. Murdin, C. R. Pidgeon et al., "Suppression of Auger recombination in arsenic-rich $InAs_{1-x}Sb_x$ strained layer super-lattices," *Journal of Applied Physics*, vol. 80, no. 5, pp. 2994–2997, 1996.

[111] T. Schuler-Sandy, S. Myers, B. Klein et al., "Gallium free type II $InAs/InAs_xSb_{1-x}$ superlattice photodetectors," *Applied Physics Letters*, vol. 101, no. 7, Article ID 071111, 2012.

[112] O. O. Cellek, Z. Y. He, Z. Y. Lin, H. S. Kim, S. Liu, and Y. H. Zhang, "InAs/InAsSb type-II superlattice infrared nBn photodetectors and their potential for operation at high temperatures," in *Quantum Sensing and Nanophotonic Devices X*, vol. 8631 of *Proceedings of SPIE*, 2013.

[113] B. V. Olson, E. A. Shaner, J. K. Kim et al., "Identification of dominant recombination mechanisms in narrow-bandgap InAs/InAsSb type-II superlattices and InAsSb alloys," *Applied Physics Letters*, vol. 103, no. 5, Article ID 052106, 2013.

[114] R. Q. Yang, Z. Tian, J. F. Klem, T. D. Mishima, M. B. Santos, and M. B. Johnson, "Interband cascade photovoltaic devices," *Applied Physics Letters*, vol. 96, no. 6, Article ID 063504, 2010.

[115] R. Q. Yang, Z. Tian, Z. Cai, J. F. Klem, M. B. Johnson, and H. C. Liu, "Interband-cascade infrared photodetectors with superlattice absorbers," *Journal of Applied Physics*, vol. 107, no. 5, Article ID 054514, 2010.

[116] A. Tian, R. T. Hinkey, R. Q. Yang et al., "Interband cascade infrared photodetectors with enhanced electron barriers and p-type superlattice absorbers," *Journal of Applied Physics*, vol. 111, no. 2, Article ID 024510, 2012.

[117] N. Gautam, S. Myers, A. V. Barve et al., "High operating temperature interband cascade midwave infrared detector based on type-II InAs/GaSb strained layer superlattice," *Applied Physics Letters*, vol. 101, no. 2, Article ID 021106, 2012.

[118] Z. B. Tian, T. Schuler-Sandy, and S. Krishna, "Electron barrier study of mid-wave infrared interband cascade photodetectors," *Applied Physics Letters*, vol. 103, Article ID 083601, 2013.

[119] K. Banerjee, S. Ghosh, S. Mallick, E. Plis, S. Krishna, and C. Grein, "Midwave infrared InAs/GaSb strained layer superlattice hole avalanche photodiode," *Applied Physics Letters*, vol. 94, no. 20, Article ID 201107, 2009.

[120] S. Mallick, K. Banerjee, S. Ghosh, J. B. Rodriguez, and S. Krishna, "Midwavelength infrared avalanche photodiode using InAs-GaSb strain layer superlattice," *IEEE Photonics Technology Letters*, vol. 19, no. 22, pp. 1843–1845, 2007.

[121] S. Mallick, K. Banerjee, S. Ghosh et al., "Ultralow noise midwave infrared InAs-GaSb strain layer superlattice avalanche photodiode," *Applied Physics Letters*, vol. 91, no. 24, Article ID 241111, 2007.

[122] A. P. Ongstad, R. Kaspi, C. E. Moeller et al., "Spectral blueshift and improved luminescent properties with increasing GaSb layer thickness in InAs-GaSb type-II superlattices," *Journal of Applied Physics*, vol. 89, no. 4, pp. 2185–2188, 2001.

[123] P.-Y. Delaunay, A. Hood, B. M. Nguyen, D. Hoffman, Y. Wei, and M. Razeghi, "Passivation of type-II InAsGaSb double heterostructure," *Applied Physics Letters*, vol. 91, no. 9, Article ID 091112, 2007.

[124] E. Plis, M. N. Kutty, and S. Krishna, "Passivation techniques for InAs/GaSb strained layer superlattice detectors," *Laser & Photonics Reviews*, vol. 7, pp. 1–15, 2012.

[125] F. M. Mohammedy and M. J. Deen, "Growth and fabrication issues of GaSb-based detectors," *Journal of Materials Science: Materials in Electronics*, vol. 20, no. 11, pp. 1039–1058, 2009.

[126] G. J. Brown, "Type-II InAs/GaInSb superlattices for infrared detection: an overview," in *Infrared Technology and Applications XXXI*, vol. 5783 of *Proceedings of SPIE*, pp. 65–77, April 2005.

[127] K. Banerjee, S. Ghosh, E. Plis, and S. Krishna, "Study of short-and long-term effectiveness of ammonium sulfide as surface passivation for InAs/GaSb superlattices using X-ray photoelectron spectroscopy," *Journal of Electronic Materials*, vol. 39, no. 10, pp. 2210–2214, 2010.

[128] T. Wada and N. Kitamura, "X-Ray photoelectron spectra of an electron-beam oxide layer on GaSb," *Japanese Journal of Applied Physics*, vol. 27, no. 4, pp. 686–687, 1988.

[129] E. Plis, M. N. Kutty, S. Myers et al., "Passivation of long-wave infrared InAs/GaSb strained layer superlattice detectors," *Infrared Physics and Technology*, vol. 54, no. 3, pp. 252–257, 2011.

[130] A. Gin, Y. Wei, J. Bae, A. Hood, J. Nah, and M. Razeghi, "Passivation of type II InAs/GaSb superlattice photodiodes," *Thin Solid Films*, vol. 447-448, pp. 489–492, 2004.

[131] E. K. Huang, B.-M. Nguyen, D. Hoffman, P.-Y. Delaunay, and M. Razeghi, "Inductively coupled plasma etching and processing techniques for type-II InAs/GaSb superlattices infrared detectors toward high fill factor focal plane arrays," in *Quantum Sensing and Nanophotonic Devices VI*, vol. 7222 of *Proceedings of SPIE*, January 2009.

[132] E. K.-W. Huang, D. Hoffman, B.-M. Nguyen, P.-Y. Delaunay, and M. Razeghi, "Surface leakage reduction in narrow band gap type-II antimonide-based superlattice photodiodes," *Applied Physics Letters*, vol. 94, no. 5, Article ID 053506, 2009.

[133] R. Rehm, M. Walther, F. Fuchs, J. Schmitz, and J. Fleissner, "Passivation of InAs/(GaIn) Sb short-period superlattice photodiodes with 10 μm cutoff wavelength by epitaxial overgrowth with $Al_xGa_{1-x}As_ySb_{1-y}$," *Applied Physics Letters*, vol. 86, no. 17, Article ID 173501, pp. 1–3, 2005.

[134] B.-M. Nguyen, D. Hoffman, E. K.-W. Huang, P.-Y. Delaunay, and M. Razeghi, "Background limited long wavelength infrared type-II InAs/GaSb superlattice photodiodes operating at 110 K," *Applied Physics Letters*, vol. 93, no. 12, Article ID 123502, 2008.

[135] R. Chaghi, C. Cervera, H. Aït-Kaci, P. Grech, J. B. Rodriguez, and P. Christol, "Wet etching and chemical polishing of InAs/GaSb superlattice photodiodes," *Semiconductor Science and Technology*, vol. 24, no. 6, Article ID 065010, 2009.

[136] C. Cervera, J. B. Rodriguez, R. Chaghi, H. At-Kaci, and P. Christol, "Characterization of midwave infrared InAs/GaSb superlattice photodiode," *Journal of Applied Physics*, vol. 106, no. 2, Article ID 024501, 2009.

[137] Y. Chen, A. Moy, S. Xin, K. Mi, and P. P. Chow, "Improvement of R0A product of type-II InAs/GaSb superlattice MWIR/LWIR photodiodes," *Infrared Physics and Technology*, vol. 52, no. 6, pp. 340–343, 2009.

[138] S. dip Das, S. L. Tan, S. Zhang, Y. L. Goh, C. H. Ting, and J. David, "Development of LWIR photodiodes based on InAs/GaSb type-II strained layer superlattices," in *Proceedings of the 6th EMRS DTC Technical Conference*, vol. B7, Edinburgh, UK, 2009.

[139] M. N. Kutty, E. Plis, A. Khoshakhlagh et al., "Study of surface treatments on InAs/GaSb superlattice lwir detectors," *Journal of Electronic Materials*, vol. 39, no. 10, pp. 2203–2209, 2010.

[140] M. Sundaram, A. Reisinger, R. Dennis et al., "Longwave infrared focal plane arrays from type-II strained layer superlattices," *Infrared Physics and Technology*, vol. 54, no. 3, pp. 243–246, 2011.

[141] Y. T. Kim, D. S. Kim, and D. H. Yoon, "PECVD SiO_2 and SiON films dependant on the rf bias power for low-loss silica waveguide," *Thin Solid Films*, vol. 475, no. 1-2, pp. 271–274, 2005.

[142] J. A. Nolde, R. Stine, E. M. Jackson et al., "Effect of the oxide-semiconductor interface on the passivation of hybrid type-II superlattice long-wave infrared photodiodes," in *Quantum Sensing and Nanophotonic Devices VIII*, vol. 7945 of *Proceedings of SPIE*, January 2011.

[143] G. Chen, B.-M. Nguyen, A. M. Hoang, E. K. Huang, S. R. Darvish, and M. Razeghi, "Elimination of surface leakage in gate controlled type-II InAs/GaSb mid-infrared photodetectors," *Applied Physics Letters*, vol. 99, no. 18, Article ID 183503, 2011.

[144] T. Tansel, K. Kutluer, Ö. Salihoglu et al., "Effect of the passivation layer on the noise characteristics of mid-wave-infrared InAs/GaSb superlattice photodiodes," *IEEE Photonics Technology Letters*, vol. 24, no. 9, pp. 790–792, 2012.

[145] A. Hood, P.-Y. Delaunay, D. Hoffman et al., "Near bulk-limited R0A of long-wavelength infrared type-II InAs/GaSb superlattice photodiodes with polyimide surface passivation," *Applied Physics Letters*, vol. 90, no. 23, Article ID 233513, 2007.

[146] H. S. Kim, E. Plis, A. Khoshakhlagh et al., "Performance improvement of InAs/GaSb strained layer superlattice detectors by reducing surface leakage currents with SU-8 passivation," *Applied Physics Letters*, vol. 96, no. 3, Article ID 033502, 2010.

[147] E. A. DeCuir Jr., J. W. Little, and N. Baril, "Addressing surface leakage in type-II InAs/GaSb superlattice materials using novel approaches to surface passivation," in *Infrared Sensors, Devices, and Applications; and Single Photon Imaging II*, vol. 8155 of *Proceedings of SPIE*, August 2011.

[148] H. S. Kim, E. Plis, N. Gautam et al., "Reduction of surface leakage current in InAs/GaSb strained layer long wavelength superlattice detectors using SU-8 passivation," *Applied Physics Letters*, vol. 97, no. 14, Article ID 143512, 2010.

[149] D. Sanchez, L. Cerutti, and E. Tournie, "Selective lateral etching of InAs/GaSb tunnel junctions for mid-infrared photonics," *Semiconductor Science and Technology*, vol. 27, no. 8, Article ID 085011, 2012.

[150] G. A. Umana-Membreno, H. Kalaa, B. Klein et al., "Electronic transport in InAs/GaSb type-II superlattices for long wavelength infrared focal plane array applications," in *Infrared Technology and Applications XXXVIII*, vol. 8353 of *Proceedings of SPIE*, May 2012.

[151] C. J. Sandroff, M. S. Hegde, and C. C. Chang, "Structure and stability of passivating arsenic sulfide phases on GaAs surfaces," *Journal of Vacuum Science & Technology B: Microelectronics and Nanometer Structures*, vol. 7, pp. 841–844, 1989.

[152] C. J. Spindt, D. Liu, K. Miyano et al., "Vacuum ultraviolet photoelectron spectroscopy of $(NH_4)_2$S-treated GaAs (100) surfaces," *Applied Physics Letters*, vol. 55, no. 9, pp. 861–863, 1989.

[153] V. N. Bessolov, E. V. Konenkova, and M. V. Lebedev, "Sulfidization of GaAs in alcoholic solutions: a method having an impact on efficiency and stability of passivation," *Materials Science and Engineering B*, vol. 44, no. 1-3, pp. 376–379, 1997.

[154] H. Oigawa, J.-F. Fan, Y. Nannichi, H. Sugahara, and M. Oshima, "Universal passivation effect of $(NH_4)_2S_x$ treatment on the surface of III-V compound semiconductors," *Japanese Journal of Applied Physics*, vol. 30, no. 3, pp. 322–325, 1991.

[155] D. Y. Petrovykh, M. J. Yang, and L. J. Whitman, "Chemical and electronic properties of sulfur-passivated InAs surfaces," *Surface Science*, vol. 523, no. 3, pp. 231–240, 2003.

[156] M. Pérotin, P. Coudray, L. Gouskov et al., "Passivation of GaSb by sulfur treatment," *Journal of Electronic Materials*, vol. 23, no. 1, pp. 7–12, 1994.

[157] P. S. Dutta, K. S. Sangunni, H. L. Bhat, and V. Kumar, "Sulphur passivation of gallium antimonide surfaces," *Applied Physics Letters*, vol. 65, no. 13, pp. 1695–1697, 1994.

[158] S. Basu and P. Barman, "Chemical modification and characterization of Te-doped n-GaSb (111) single crystals for device application," *Journal of Vacuum Science & Technology B: Microelectronics and Nanometer Structures*, vol. 10, no. 3, pp. 1078–1080, 1992.

[159] V. N. Bessolov, M. V. Lebedev, E. B. Novikov, and B. V. Tsarenkov, "Sulfide passivation of III-V semiconductors: kinetics of the photoelectrochemical reaction," *Journal of Vacuum Science & Technology B: Microelectronics and Nanometer Structures*, vol. 11, pp. 10–14, 1993.

[160] V. N. Bessolov, Y. V. Zhilyaev, E. V. Konenkova, and M. V. Lebedev, "Sulfide passivation of III-V semiconductor surfaces: role of the sulfur ionic charge and of the reaction potential of the solution," *Journal of Technical Physics*, vol. 43, no. 8, pp. 983–985, 1998.

[161] T. Ohno, "Passivation of GaAs(001) surfaces by chalcogen atoms (S, Se and Te)," *Surface Science*, vol. 255, no. 3, pp. 229–236, 1991.

[162] T. Ohno and K. Shiraishi, "First-principles study of sulfur passivation of GaAs(001) surfaces," *Physical Review B*, vol. 42, no. 17, pp. 11194–11197, 1990.

[163] A. Gin, Y. Wei, A. Hood et al., "Ammonium sulfide passivation of Type-II InAs/GaSb superlattice photodiodes," *Applied Physics Letters*, vol. 84, no. 12, pp. 2037–2039, 2004.

[164] K. Banerjee, J. Huang, S. Ghosh et al., "Surface study of thioacetamide and zinc sulfide passivated long wavelength infrared type-II strained layer superlattice," in *Infrared Technology and Applications XXXVII*, vol. 8012 of *Proceedings of SPIE*, April 2011.

[165] E. Plis, J.-B. Rodriguez, S. J. Lee, and S. Krishna, "Electrochemical sulphur passivation of InAs/GaSb strain layer superlattice detectors," *Electronics Letters*, vol. 42, no. 21, pp. 1248–1249, 2006.

[166] R. Xu and C. G. Takoudis, "Chemical passivation of GaSb-based surfaces by atomic layer deposited ZnS using diethylzinc and hydrogen sulfide," *Journal of Vacuum Science and Technology A: Vacuum, Surfaces and Films*, vol. 30, no. 1, Article ID 01A145, 2012.

[167] L. Zhang, L. X. Zhang, X. W. Shen et al., "Research on passivation of type II InAs/GaSb superlattice photodiodes," in *International Symposium on Photoelectronic Detection and Imaging 2013: Infrared Imaging and Applications*, vol. 890731 of *Proceedings of SPIE*, 2013.

[168] S. C. Lee, E. Plis, S. Krishna, and S. R. J. Brueck, "Mid-infrared transmission enhancement through sub-wavelength metal hole array using impedance-matching dielectric layer," *Electronics Letters*, vol. 45, no. 12, pp. 643–645, 2009.

[169] M. Zamiri, E. Plis, J. O. Kim et al., "MWIR superlattice detectors integrated with back side illuminated plasmonic coupler," Unpablished.

[170] F. Szmulowicz and G. J. Brown, "GaSb for passivating type-II InAs/GaSb superlattice mesas," *Infrared Physics and Technology*, vol. 53, no. 5, pp. 305–307, 2010.

[171] E. Plis, B. Klein, S. Myers et al., "Type-II InAs/GaSb strained layer superlattices grown on GaSb (111)B substrate," *Journal of Vacuum Science and Technology B*, vol. 31, no. 1, 2013.

[172] F. Szmulowicz, H. J. Haugan, and G. J. Brown, "Proposal for (110) InAs/GaSb superlattices for infrared detection," in *Quantum Sensing and Nanophotonic Devices V*, vol. 6900 of *Proceedings of SPIE*, January 2008.

[173] C. Zhou, B. M. Nguyen, M. Razeghi, and M. Grayson, "Thermal conductivity of InAs/GaSb type-II superlattice," *Journal of Electronic Materials*, vol. 41, pp. 2322–2325, 2012.

[174] V. V. R. Kishore, B. Partoens, and F. M. Peeters, "Electronic structure of InAs/GaSb core-shell nanowires," *Physical Review B*, vol. 86, Article ID 165439, 2012.

[175] B. Ganjipour, M. Ek, B. M. Borg et al., "Carrier control and transport modulation in GaSb/InAsSb core/shell nanowires," *Applied Physics Letters*, vol. 101, no. 10, Article ID 103501, 2012.

[176] R. Li, Y. Lu, S. D. Chae et al., "InAs/AlGaSb heterojunction tunnel field-effect transistor with tunnelling in-line with the gate field," *Physica Status Solidi C*, vol. 9, no. 2, pp. 389–392, 2012.

[177] W. Pan, J. F. Klem, J. K. Kim, M. Thalakulam, M. J. Cich, and S. K. Lyo, "Chaotic quantum transport near the charge neutrality point in inverted type-II InAs/GaSb field-effect transistors," *Applied Physics Letters*, vol. 102, no. 3, Article ID 033504, 2013.

[178] H. Lotfi, R. T. Hinkey, L. Li, R. Q. Yang, J. F. Klem, and M. B. Johnson, "Narrow-bandgap photovoltaic devices operating at room temperature and above with high open-circuit voltage," *Applied Physics Letters*, vol. 102, no. 21, Article ID 211103, 2013.

3

All Pass Network Based MSO Using OTRA

Rajeshwari Pandey,[1] **Neeta Pandey,**[1] **Romita Mullick,**[1]
Sarjana Yadav,[1] **and Rashika Anurag**[2]

[1]*Department of Electronics and Communication Engineering, Delhi Technological University, Delhi 110042, India*
[2]*Department of Electronics and Communication Engineering, JSS Academy of Technical Education, C-20/1,
Sector 62, Noida, Uttar Pradesh 201301, India*

Correspondence should be addressed to Neeta Pandey; n66pandey@rediffmail.com

Academic Editor: Liwen Sang

This paper presents multiphase sinusoidal oscillators (MSOs) using operational transresistance amplifier (OTRA) based all pass networks. Both even and odd phase oscillations of equal amplitudes which are equally spaced in phase can be produced using single all pass section per phase. The proposed MSOs provide voltage output and can readily be used for driving voltage input circuits without increasing component count. The effect of nonideality of OTRA on the circuit performance is also analysed. The functionality of the proposed circuit is verified through PSPICE simulations.

1. Introduction

Multiphase sinusoidal oscillators (MSO) provide multiple outputs of the same frequency, equally spaced in phase, and find extensive application in the field of communications, instrumentation, and power electronics. In communications MSO circuits are used in single-sideband generators, phase modulators, and quadrature mixers [1]. Selective voltmeters and vector generator are common applications of MSOs in the field of instrumentation [2]. In power electronics three-phase MSOs are frequently utilized in PWM converters [3] and inverters [4].

A large number of MSO realizations using various analog building blocks (ABB) [2, 5–25] are available in literature. These MSOs are based on a basic design philosophy of forming closed loop using n ($n \geq 3$) cascaded phase shifting networks thereby producing n equally spaced phases. For phase shifting either first-order low pass networks (LPNs) [5–7, 9–21, 24, 25] or first-order all pass networks (APNs) [2, 8, 22, 23] are used. These reported structures provide either voltage [5–16] or current [2, 17–24] outputs.

The MSOs of [5–8] are realized using operational amplifiers (op-amps). However due to constant gain-bandwidth product and lower slew rate of the op-amps, their high frequency operations are limited. Additionally the active R implementations of [5, 6] lack tunability as these structures make use of the op-amp parasitic capacitance. The current feedback operational amplifier (CFOA) based MSO structure [9] is capable of producing high frequencies but requires an accessible compensation terminal of a CFOA. The MSOs of [10, 11] are OTA based electronically tunable structures; however, they provide voltage output at high impedance making a buffer necessary to drive the voltage input circuits. In addition, for both the structures the output voltage is temperature sensitive too. The MSO configurations presented in [12–15] are CC based designs and provide voltage output at high impedance. Further the MSO of [14], being active R structure, lacks tunability. Three topologies of OTRA based MSOs are proposed in [16] and are designed using LPNs. The structures proposed in [2, 17–25] provide current outputs which need to be converted back to voltage for circuits requiring voltage inputs, which would considerably increase the component count. A detailed comparison of these structures is given in Table 1 which suggests that OTRA based MSO is the most suitable choice for voltage output.

In this paper authors aim at presenting OTRA based MSO, designed using APNs. The proposed circuit utilizes n ($n \geq 3$) APNs to produce n phase oscillations of equal

TABLE 1: Detailed comparison of available MSOs.

Reference	ABB	Phase shift network	Output	Output impedance	Tunability
[2]	CCCDTA	APN	Current	High	Yes
[5]	Op-amp	LPN	Voltage	Low	No
[6]	Op-amp	LPN	Voltage	Low	No
[7]	Op-amp	LPN	Voltage	Low	Yes
[8]	Op-amp	APN	Voltage	Yes	Yes
[9]	CFOA	LPN	Voltage	No	Yes
[10]	OTA	LPN	Voltage	High	Yes
[11]	OTA	LPN	Voltage	High	Yes
[12]	CCII	LPN	Voltage	High	Yes
[13]	CCII	LPN	Voltage	High	Yes
[14]	CCII	LPN	Voltage	High	No
[15]	CCII	LPN	Voltage	High	Yes
[16]	OTRA	LPN	Voltage	Low	Yes
[17]	Current follower	LPN	Current	High	Yes
[18]	CCII	LPN	Current	High	Yes
[19]	CCII	LPN	Current	High	Yes
[20]	CCII	LPN	Current	High	Yes
[21]	CDTA	LPN	Current	High	Yes
[22]	CDTA	APN	Current	High	Yes
[23]	CDTA	APN	Current	High	Yes
[24]	CC	LPN	Current	High	Yes
[25]	CFTA	LPN	Current	High	Yes
Proposed work	OTRA	APN	Voltage	Low	Yes

amplitudes with a phase difference of $(360/n)°$. The APN can produce a phase shift up to $180°$ as against a maximum of $90°$ produced by LPN. As a result the APN based scheme can be used to implement an even phase system for $n = 4$ also whereas $n = 6$ is the minimum for the systems designed using LPNs [8].

2. Circuit Description

In this section the generalized APN based MSO scheme [8] is described first which is followed by the design adaption using OTRA.

2.1. The APN Based MSO Design Scheme. The APN based MSO structure consists of n cascaded first-order APN blocks. The output of the nth stage is fed back to the input of the first stage thus forming a closed loop as shown in Figure 1(a). The output of nth stage is fed back directly for odd phase system whereas for even phase system it is inverted, for sustained oscillations. In case of even phased system the structure can be modified for obtaining n even phased oscillations by replacing $n/2$ APN with $n/2$ inverters as shown in Figure 1(b).

The transfer function $G(s)$ of each APN block is given by

$$G(s) = K\frac{(1 - s\tau)}{(1 + s\tau)}, \qquad (1)$$

where K represents the gain and time constant τ determines the corner frequency of the APN. The phase (ϕ) of each APN block is computed to be

$$\phi = -2\tan^{-1}(\omega\tau). \qquad (2)$$

Thus an APN can introduce signal phase lag from 0 to $180°$ as frequency ω varies from zero to infinity. From Figure 1 the open loop gain $L(s)$ can be expressed as

$$L(s) = K^n\left(\frac{(1 - s\tau)}{(1 + s\tau)}\right)^n. \qquad (3)$$

The Barkhausen criterion [26] for sustained oscillations at a frequency ω_0 can be expressed by

$$|L(j\omega_0)| = K^n\left(\frac{1 - j\omega_0\tau}{1 + j\omega_0\tau}\right)^n = 1, \qquad (4)$$

$$\angle|L(j\omega_0)| = n\phi = -2\pi. \qquad (5)$$

Since $((1 - j\omega_0\tau)/(1 + j\omega_0\tau))^n = 1$, (4) results in condition of oscillation (CO) as

$$CO : K = 1. \qquad (6)$$

Substituting from (2) in (5) total phase shift of the loop can be computed as

$$n\left(-2\tan^{-1}(\omega_0\tau)\right) = -2\pi. \qquad (7)$$

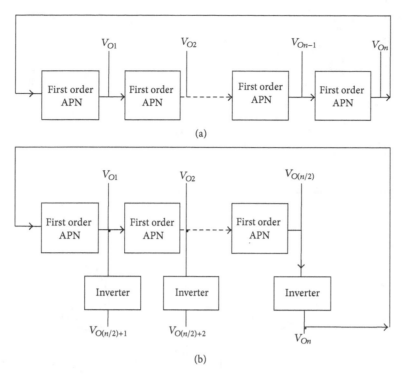

(a)

(b)

FIGURE 1: Generalized APN based MSO structure for producing n phase oscillations. (a) Odd phased oscillations. (b) Even phased oscillations.

Equation (7) will converge only for values of n such that $n \geq 3$. This results in frequency of oscillation (FO) as

$$\text{FO}: f_0 = \frac{1}{2\pi\tau} \tan\left(\frac{\pi}{n}\right). \tag{8}$$

The spacing between different phases is given by

$$\phi = \frac{2\pi}{n}. \tag{9}$$

Thus the circuit gives rise to equally spaced oscillations having a phase difference of $(360/n)°$.

2.2. The OTRA Based MSO Implementation.

The OTRA is a high gain, current input voltage output ABB. The circuit symbol of OTRA is shown in Figure 2 and the port characteristics in matrix form are given by (10), where R_m is transresistance gain of OTRA. For ideal operations the R_m of OTRA approaches infinity and forces the input currents to be equal. Thus OTRA must be used in a negative feedback configuration [25, 27]:

$$\begin{bmatrix} V_p \\ V_n \\ V_O \end{bmatrix} = \begin{bmatrix} 0 & 0 & 0 \\ 0 & 0 & 0 \\ R_m & -R_m & 0 \end{bmatrix} \begin{bmatrix} I_p \\ I_n \\ I_O \end{bmatrix}. \tag{10}$$

The MSO scheme outlined in Section 2.1 can be implemented using OTRA based first-order all pass sections. The OTRA based APN proposed in [27] and shown in Figure 3 is used for MSO implementation.

The transfer function of the APN can be written as

$$G(s) = \frac{V_O(s)}{V_i(s)} = K \frac{(1 - s\tau)}{(1 + s\tau)}, \tag{11}$$

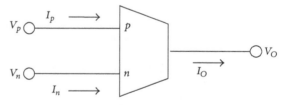

FIGURE 2: OTRA circuit symbol.

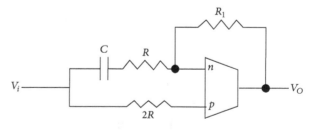

FIGURE 3: The OTRA based APN [27].

where

$$K = \frac{R_1}{2R}, \qquad \tau = \frac{1}{CR}, \tag{12}$$

and the phase relation is expressed as

$$\phi = -2\tan^{-1}(\omega CR). \tag{13}$$

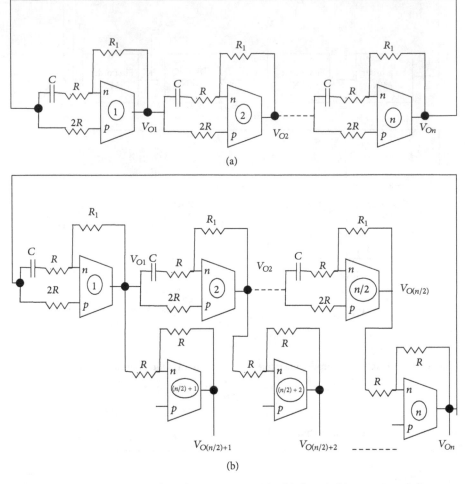

FIGURE 4: The OTRA based MSO circuits: (a) odd phased, (b) even phased.

The OTRA based odd and even phased MSO structures are shown in Figures 4(a) and 4(b), respectively. The loop gain can be written as

$$L(s) = \left(\frac{R_1}{2R}\right)^n \left(\frac{(1 - s/RC)}{(1 + s/RC)}\right)^n. \tag{14}$$

CO and FO can be expressed as

$$\text{CO}: R_1 = 2R,$$
$$\text{FO}: f_0 = \frac{1}{2\pi RC} \tan\left(\frac{\pi}{n}\right). \tag{15}$$

3. Nonideality Analysis

Ideally the transresistance gain R_m is assumed to approach infinity. However, practically R_m is a frequency dependent finite value. Considering a single pole model for the transresistance gain, R_m can be expressed as

$$R_m(s) = \left(\frac{R_0}{1 + s/\omega_0}\right). \tag{16}$$

For high frequency applications the transresistance gain $R_m(s)$ reduces to

$$R_m(s) = \left(\frac{1}{sC_p}\right) \quad \text{where } C_p = \left(\frac{1}{R_0\omega_0}\right). \tag{17}$$

C_p is known as parasitic capacitance of OTRA. Taking the effect of C_p into account (11) modifies to

$$G(s) = \frac{R_1}{2R} \left(\frac{(1 - sCR)}{(1 + sCR)(1 + sC_pR_1)}\right). \tag{18}$$

Thus the C_p of OTRA results in introduction of another pole having pole frequency $\omega_p = 1/R_1C_p$. However with the value of C_p being small (typically 5 pF) the parasitic pole frequency would be far off from the operating frequency of the APN and would not affect the MSO operation.

4. Simulation Results

The proposed circuits are simulated using PSPICE to validate the theoretical predictions. The CMOS realization of OTRA presented in [28] and reproduced in Figure 5 is used for

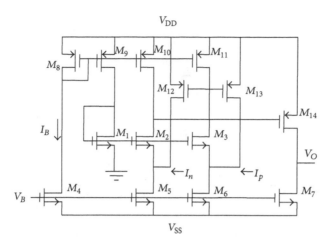

FIGURE 5: The CMOS implementation of OTRA [28].

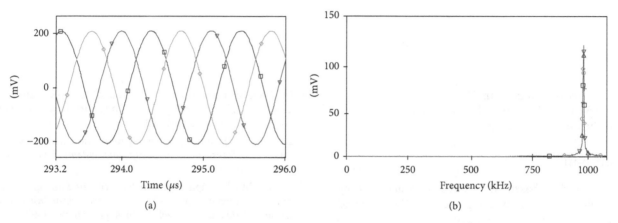

FIGURE 6: Odd phased MSO output for $n = 3$: (a) steady state output; (b) frequency spectrum.

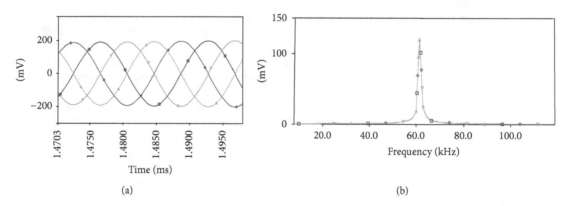

FIGURE 7: Even phased MSO output for $n = 4$: (a) steady state output; (b) frequency spectrum.

simulation. The output of the MSO of Figure 4(a) for $n = 3$, with component values $R = 2.5\,\mathrm{K\Omega}$, $R_1 = 5\,\mathrm{K\Omega}$, and $C = 0.1\,\mathrm{nF}$, is depicted in Figure 6. The steady state output is shown in Figure 6(a) while the frequency spectrum is depicted in Figure 6(b). The simulated frequency of oscillations is observed to be 1 MHz against the calculated value of 1.1 MHz.

Simulation results for $n = 4$, with component values $R = 2.5\,\mathrm{K\Omega}$, $R_1 = 5\,\mathrm{K\Omega}$, and $C = 1\,\mathrm{nF}$, are depicted in Figure 7. The simulated frequency is found to be 61.69 KHz while the theoretical calculation yields an FO of 63.69 KHz.

It may be observed from (8) that the FO can be varied either through R or by changing C. Variation of FO with respect to R while keeping $C = 1\,\mathrm{nF}$ has been depicted in

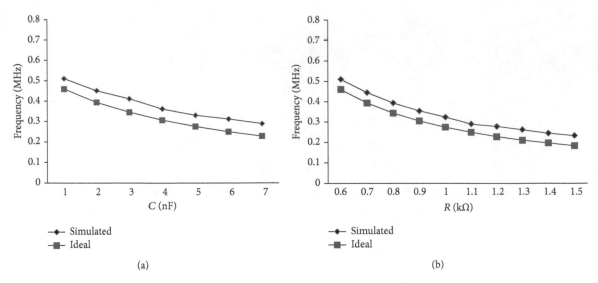

FIGURE 8: Frequency tuning of MSO (a) with R and (b) with C.

Figure 8(a) whereas tuning with C is shown in Figure 8(b) with $R = 0.5$ KΩ. It shows that the simulated values closely follow the theoretically calculated values.

5. Conclusion

In this work OTRA based MSO circuits, designed using first-order all pass networks, are presented. The proposed structures produce "n" phase oscillations of equal amplitudes which are equally spaced in phase. These circuits provide voltage output at low impedance and thus can readily be used to drive voltage input circuits without increasing component count. The proposed circuits are very accurate in providing the desired phase shift. The workability of the circuit has been demonstrated through PSPICE simulations.

Conflict of Interests

The authors declare that there is no conflict of interests regarding the publication of this paper.

References

[1] W. Tomasi, *Electronic Communications System*, Prentice-Hall, Upper Saddle River, NJ, USA, 1998.

[2] W. Jaikla and P. Prommee, "Electronically tunable current-mode multiphase sinusoidal oscillator employing CCDTA-based allpass filters with only grounded passive elements," *Radioengineering*, vol. 20, no. 3, pp. 594–599, 2011.

[3] D. C. Lee and Y. S. Kim, "Control of single-phase-to-three-phase AC/DC/AC PWM converters for induction motor drives," *IEEE Transactions on Industrial Electronics*, vol. 54, no. 2, pp. 797–804, 2007.

[4] V. P. Ramamurthi and B. Ramaswami, "A novel three-phase reference sine wave generator for PWM invertors," *IEEE Transactions on Industrial Electronics*, vol. 1, no. 3, pp. 235–240, 1982.

[5] M. T. Abuelma'atti and W. A. Almansoury, "Active-R multiphase oscillators," *IEE Proceedings G: Electronic Circuits and Systems*, vol. 134, no. 6, pp. 292–294, 1987.

[6] D. Stiurca, "On the multiphase symmetrical active-R oscillators," *IEEE Transactions on Circuits and Systems II*, vol. 41, no. 2, pp. 156–158, 1994.

[7] S. J. G. Gift, "Multiphase sinusoidal oscillator using inverting-mode operational amplifiers," *IEEE Transactions on Instrumentation and Measurement*, vol. 47, no. 4, pp. 986–991, 1998.

[8] S. J. G. Gift, "The application of all-pass filters in the design of multiphase sinusoidal systems," *Microelectronics Journal*, vol. 31, no. 1, pp. 9–13, 2000.

[9] D.-S. Wu, S.-I. Liu, Y.-S. Hwang, and Y.-P. Wu, "Multiphase sinusoidal oscillator using the CFOA pole," *IEE Proceedings: Circuits, Devices and Systems*, vol. 142, no. 1, pp. 37–40, 1995.

[10] C.-L. Hou, J.-S. Wu, J. Hwang, and H.-C. Lin, "OTA-based even-phase sinusoidal oscillators," *Microelectronics Journal*, vol. 28, no. 1, pp. 49–54, 1997.

[11] I. A. Khan, M. T. Ahmed, and N. Minhaj, "Tunable OTA-based multiphase sinusoidal oscillators," *International Journal of Electronics*, vol. 72, no. 3, pp. 443–450, 1992.

[12] D.-S. Wu, S.-I. Liu, Y.-S. Hwang, and Y.-P. Wu, "Multiphase sinusoidal oscillator using second-generation current convey-ors," *International Journal of Electronics*, vol. 78, no. 4, pp. 645–651, 1995.

[13] C.-L. Hou and B. Shen, "Second-generation current conveyor-based multiphase sinusoidal oscillators," *International Journal of Electronics*, vol. 78, no. 2, pp. 317–325, 1995.

[14] M. T. Abuelma'atti and M. A. Al-Qahtani, "A grounded-resistor current conveyor-based active-R multiphase sinusoidal oscilla-tor," *Analog Integrated Circuits and Signal Processing*, vol. 16, no. 1, pp. 29–34, 1998.

[15] G. D. Skotis and C. Psychalinos, "Multiphase sinusoidal oscil-lators using second generation current conveyors," *AEU—International Journal of Electronics and Communications*, vol. 64, no. 12, pp. 1178–1181, 2010.

[16] R. Pandey, N. Pandey, M. Bothra, and S. K. Paul, "Operational transresistance amplifier-based multiphase sinusoidal oscillators," *Journal of Electrical and Computer Engineering*, vol. 2011, Article ID 586853, 8 pages, 2011.

[17] M. T. Abuelma'atti, "Current-mode multiphase oscillator using current followers," *Microelectronics Journal*, vol. 25, no. 6, pp. 457–461, 1994.

[18] M. T. Abuelma'Atti and M. A. Al-Qahtani, "A new current-controlled multiphase sinusoidal oscillator using translinear current conveyors," *IEEE Transactions on Circuits and Systems II: Analog and Digital Signal Processing*, vol. 45, no. 7, pp. 881–885, 1998.

[19] M. T. Abuelma'atti and M. A. Al-Qahtani, "Low component second-generation current conveyor-based multiphase sinusoidal oscillator," *International Journal of Electronics*, vol. 84, no. 1, pp. 45–52, 1998.

[20] C. Loescharataramdee, W. Kiranon, W. Sangpisit, and W. Yadum, "Multiphase sinusoidal oscillators using translinear current conveyors and only grounded passive components," in *Proceedings of the 33rd Southeastern Symposium on System Theory*, pp. 59–63, IEEE, Athens, Ga, USA, March 2001.

[21] W. Tangsrirat and W. Tanjaroen, "Current-mode multiphase sinusoidal oscillator using current differencing transconductance amplifiers," *Circuits, Systems, and Signal Processing*, vol. 27, no. 1, pp. 81–93, 2008.

[22] W. Tangsrirat, W. Tanjaroen, and T. Pukkalanun, "Current-mode multiphase sinusoidal oscillator using CDTA-based all-pass sections," *AEU—International Journal of Electronics and Communications*, vol. 63, no. 7, pp. 616–622, 2009.

[23] W. Jaikla, M. Siripruchyanun, D. Biolek, and V. Biolkova, "High-output-impedance current-mode multiphase sinusoidal oscillator employing current differencing transconductance amplifier-based allpass filters," *International Journal of Electronics*, vol. 97, no. 7, pp. 811–826, 2010.

[24] M. Kumngern, J. Chanwutitum, and K. Dejhan, "Electronically tunable multiphase sinusoidal oscillator using translinear current conveyors," *Analog Integrated Circuits and Signal Processing*, vol. 65, no. 2, pp. 327–334, 2010.

[25] P. Uttaphut, "Realization of electronically tunable current mode multiphase sinusoidal oscillators using CFTAs," *World Academy of Science, Engineering and Technology*, vol. 6, pp. 657–660, 2012.

[26] A. S. Sedra and K. C. Smith, *Microelectronic Circuits*, Oxford University Press, New York, NY, USA, 2004.

[27] S. Kilinç and U. Çam, "Cascadable allpass and notch filters employing single operational transresistance amplifier," *Computers and Electrical Engineering*, vol. 31, no. 6, pp. 391–401, 2005.

[28] H. Mostafa and A. M. Soliman, "A modified realization of the OTRA," *Frequenz*, vol. 60, no. 3-4, pp. 70–76, 2006.

4

Reconfigurable CPLAG and Modified PFAL Adiabatic Logic Circuits

Manoj Sharma[1,2] and Arti Noor[3]

[1]Department of ECE, Mewar University, Rajasthan 312901, India
[2]Department of ECE, BVCOE, Paschim Vihar, New Delhi 110063, India
[3]SoE, CDAC Noida, Ministry of Communications and IT, Government of India, Noida, Uttar Pradesh 201307, India

Correspondence should be addressed to Manoj Sharma; manojs110281@gmail.com

Academic Editor: Meiyong Liao

Previously, authors have proposed CPLAG and MCPLAG circuits extracting benefits of CPL family implemented based upon semiadiabatic logic for low power VLSI circuit design along with gating concept. Also authors have communicated RCPLAG circuits adding another dimension of reconfigurability into CPLAG/MCPLAG circuits. Moving ahead, in this paper, authors have implemented/reconfigured RCPLAG universal Nand/And gate and universal Nor/Or gate for extracting behavior of dynamic positive edge triggered DFF. Authors have also implemented Adder/Subtractor circuit using different techniques. Authors have also reported modification in PFAL semiadiabatic circuit family to further reduce the power dissipation. Functionality of these is verified and found to be satisfactory. Further these are examined rigorously with voltage, C_{load}, temperature, and transistor size variation. Performance of these is examined with these variations with power dissipation, delays, rise, and fall times associated. From the analysis it is found that best operating condition for DFF based upon RCPLAG universal gate can be achieved at supply voltage lower than 3 V which can be used for different transistor size up to 36 μm. Average power dissipation is 0.2 μW at 1 V and 30 μW at 2 V at 100 ff C_{load} 25°C approximately. Average power dissipated by CPLAG Adder/Subtractot is 58 μW. Modified PFAL circuit reduces average power by 9% approximately.

1. Introduction

Previously electronic products released by the companies used to drive the electronic market needs. But with the advances in technologies, competition, and conscious customer behavior now people/clients compel companies to design products according to their desired wish list. The ever increase in demand of more and more functionalities in lesser area is well accomplished with technological upgradation in different VLSI circuit design stages. But with reducing chip area, power dissipation per unit area has increased enormously. Heat dissipation has become comparable to the temperature ranges, unit area in nuclear reactors, turbines, and even sun's surface temperature, in many high end complex integrated circuit chips. Removal of this volcanic heat is essential for correct functionality, which, otherwise, would damage the chip. Because of different constraints in present technologies and scientific theories, the power equations are reaching their boundaries. This has forced researchers, designers, globally, to conceive some alternative circuit implementation methodologies targeting high performance with low power dissipation. Adoptions of adiabatic logic concept in VLSI circuit design mechanisms have shown lucrative options. Two primary rules need to be taken care of while implementing adiabatic circuits. These rules are as follows.

(i) Never "Switch On" the transistor when there is potential difference across its two terminals.

(ii) Never "Turn Off" the switch when a current is flowing through it.

These factors decide the adiabatic and nonadiabatic power loss in the circuit. Adiabatic power loss is inherent energy loss, even after above-stated prerequisites are completely followed. On the other hand nonadiabatic power loss involves some degree of violation in these rules which lead to power loss in access to the adiabatic loss. Depending upon these adiabatic circuits can be broadly classified into full-adiabatic circuits and semi-adiabatic circuits, respectively. Full-adiabatic circuits are ideal cases which are very difficult to comply with [1–6].

The fundamental concept behind adiabatic circuit is to reduce the charge transfer rate leading to reduction in power dissipated by the circuit. Equation (1) shows the power dissipation relationship for adiabatic circuits. It is evident from this that energy can be reduced by slowing down the charge transfer rate. Hence adiabatic circuits are inherently slower in their operating ranges as compared to traditional SCMOS circuits. This proves to be a major drawback for adiabatic circuits in the era of high computing. The usage of CPL based circuit implementation, embedded with adiabatic principles, has been reported in prior art [7, 8]. Authors have also proposed use of CPL based circuit implementation based upon the adiabatic concept [9–11]. The inherent speed advantage in CPL compensates the operating speed tradeoff with the power involved in favor of adiabatic circuits. Authors also proposed the use of gated signal in said CPL adiabatic circuit topologies. The gated signals enable smooth inter-stages and intra-stages integration reducing the efforts involved in meeting the timing constraints involved. In previous work authors have shown good results with integration of said concepts. This is further appreciated with the use of reconfigurability in implementing adiabatic circuits based upon CPL functionality with gated signals:

$$E = \int_0^T R \frac{C^2 V_{DD}^2}{T^2} dt = \frac{RC}{T} C V_{DD}^2. \tag{1}$$

Positive Feedback Adiabatic Logic (PFAL) was introduced by Vetali in 1996 which uses dual rail logic implementation strategy. It involves realization of function (F) and complement-function (F_{bar}) needing prime and unprime inputs simultaneously. It consists of cross-coupled inverter stages maintaining the output terminals. The logic blocks are implemented with NMOS transistors only. Authors have proposed addition of extra transistor in Vetali's PFAL structure trading the area in favor of power diminution.

Section 2 discusses the circuit implementation for DFF, Adder/Subtractor with different techniques. Modified PFAL circuit is also discussed here. Results are discussed in Section 3. Finally the work is concluded in the Conclusion section.

2. Circuit Implementation

A time varying power source is used having 4 phases as depicted in Figure 1, having different circuit operation in these 4 phases. During "Evaluation phase," the circuit functionality is evaluated depending upon the supplied input

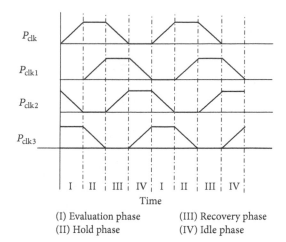

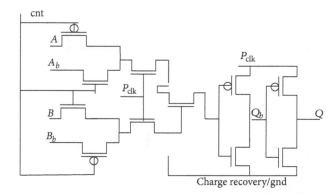

FIGURE 1: Four-phase power clock P_{clk}.

(I) Evaluation phase   (III) Recovery phase
(II) Hold phase   (IV) Idle phase

FIGURE 2: Proposed RCPLAG Nand/Nor universal gate.

signals. In "Hold phase," the computed functional value levels are retained at the output terminals for proper latching in the subsequent stages. The charge stored in the circuit nodes is recovered back during "Recovery phase." In "Idle phase," the circuit remains idle and next input signal permutation may be applied into the circuit. The circuit implementation parameters are listed in Table 1.

2.1. RCPLAG Universal Gate Reconfiguration for Dynamic PET D FF. The circuit topology for the RCPLAG Nand/Nor universal gate implementation is shown in Figure 2 [12]. Control signal reconfigures the circuit to work as Nand/And or Nor/Or functionality. The circuit takes both primed and unprimed inputs and evaluates functionality. It produces primed and unprimed output with the help of traditional inverter pair driven by the power clock P_{clk}. The output signal integrity is well maintained with proper levels. This enables the circuit to drive the next stage logic with ease. This RCPLAG circuit can also be reconfigured to implement the functionality of dynamic positive edge triggered D FF. The power source P_{clk} is also used as clock signal required. The PMOS transistor at input latching may be replaced with NMOS transistor with primed signal levels accordingly.

TABLE 1: Simulation parameters.

Simulation parameters			
Technology	Value	Simulation	Value
Channel length	0.180 μm	Power clock	Pulse type with Trise and Tfall
Min. width	0.180 μm	Input signal	Bit type
Max. width	36 microns	Delay calculation	50% points
|Vton|	0.3932664	Data sequence	8 cycles
TOX	4.10E − 09	Power clock time period	40 μs
MOS Gate Capacitance Model:			
capmod = 0			
Conditions:			
Voltage		1 V to 5 V (+0.5 V)	
Temperature		25, 30, 40, 70, 100, 200	

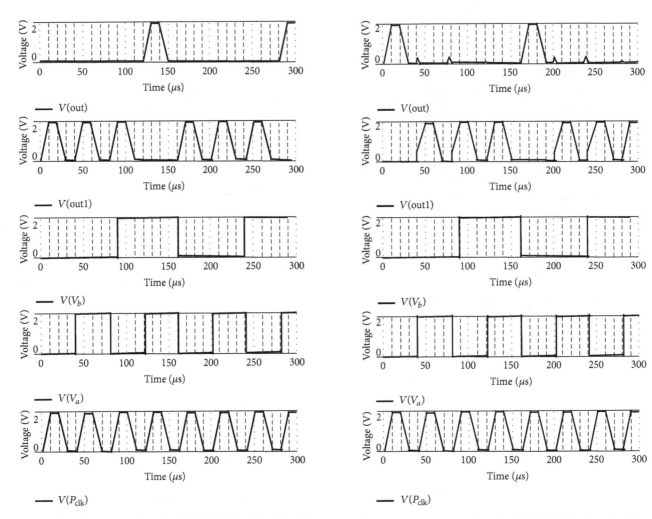

FIGURE 3: Simulation waveform for RCPLAG-Nand functionality.   FIGURE 4: Output waveform for RCPLAG-Nor functionality.

This can further optimize the associated area and speed equations. The simulation waveform for Nand, Nor, and DFF functionalities using RCPLAG universal gate implementation strategy is shown in Figures 3, 4, 5, and 6. The respective power comparison values for the implemented circuits are tabulated in Tables 2, 3, and 4.

TABLE 2: Power utilization for And/Nand.

AND CKT. IMP.	Pavg_Vpuls (W)
PFAL	13.58 μ
ECRL	6.91 μ
RCPLAG	12 n

TABLE 3: Power utilization for Or/Nor.

NOR CKT. IMP.	Pavg_Vpuls (W)
PFAL	13.46 μ
ECRL	7.007 μ
RCPLAG	2.5 μ

TABLE 4: Power utilization for DFF.

AND CKT. IMP.	Pavg_Vpuls (W)
PFAL	74.7 μ
ECRL	59.1 μ
RCPLAG	39.8 μ

TABLE 5: Power utilization for Adder/Subtractor.

ADDER CKT. IMP.	Pavg_Vpuls (W)
CPL	$6.11E-04$
CPL_PCLK	$2.22E-04$
CPLAG PCLK	$5.87E-05$
CPLAG VDD	$6.18E-05$

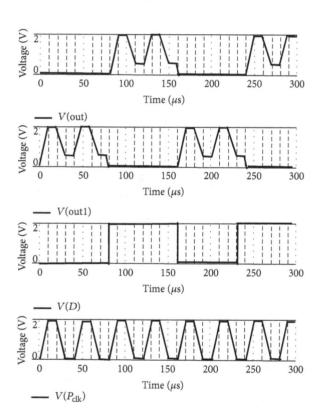

FIGURE 5: Output waveform for DFF @ RCPLAG universal Nand gate.

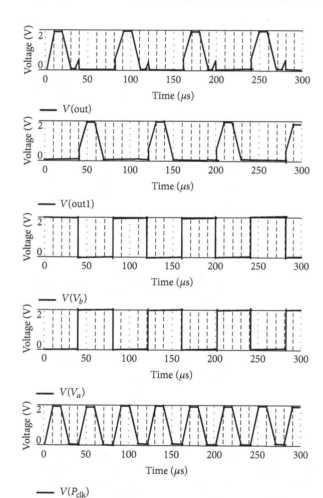

FIGURE 6: Output waveform for DFF @ RCPLAG Nor universal gate.

2.2. Adder/Subtractor Circuit. Addition is a major functionality in system data-path used in various components like multiplication, division filters, and so forth. Adder circuit is implemented with CPL techniques, with four different methodologies, namely, (i) CPL technique [13–15], (ii) CPL Adder circuit with power clock as driver, (iii) CPLAG technique [10], and (iv) CPLAG with constant power supply. These circuits are shown in Figures 7 and 8. The circuits are verified with different input combinations and permutations. Simulation waveform for CPLAG Adder with P_{clk} is shown in Figure 9. The power results are tabulated in Table 5.

2.3. Modified PFAL Circuit. The circuit diagram for modified PFAL adiabatic circuit is shown in Figure 10. It implements basic Inverter/Buffer functionality. It uses an additional drain gate connected NMOS transistor, in between the source and ground terminal of PFAL cross-coupled inverters. Figure 11 shows simulation result for modified PFAL based inverter.

TABLE 6: Power utilization for PFAL versus modified PFAL.

(W)	Modified PFAL inverter	PFAL inverter	% reduction	Total reduction
Pavg_Vpuls	$1.58E - 11$	$6.60E - 10$	$9.76E + 01$	
Pavg_Vin	$4.59E - 11$	$4.57E - 11$	$-4.44E - 01$	$9.80E + 01$
Pavg_Vinb	$2.29E - 10$	$2.31E - 10$	$8.76E - 01$	

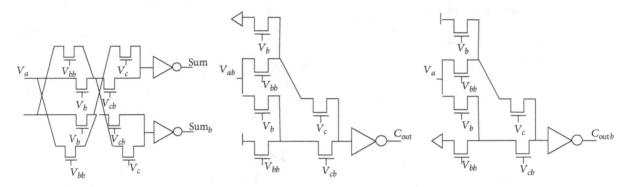

FIGURE 7: CPL Adder circuit implementation.

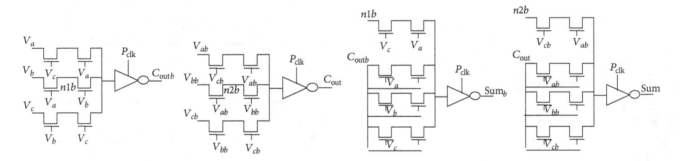

FIGURE 8: CPLAG Adder circuit implementation.

3. Result Discussion

Table 6 tabulates the power comparison for PFAL and modified PFAL inverter circuit implementation.

The average power consumption variation with respect to supply voltage variation for RCPLAG Nand based D FF is shown in Figure 12. As expected the power dissipation with respect to power source P_{clk}/V_{puls} increases with increase in voltage level of the source. For the range of 1 V to 2.5 V the variation can be linearly approximated. For voltage levels above 2.5 V the rate of increment in the power dissipation is varying exponentially. Hence it can be inferred that, for best case power equations, the circuit should be operated with voltage levels less than 2.5 V. The power dissipation at 3 V is 0.2 mW approximately. For extreme cases one may also consider operating 3 V supply.

The input signal strength required for driving the circuit and evaluating its functionality varies linearly approximately with respect to voltage level as shown in Figure 13. For a supply voltage range up to 3 V the power required is less than 3 nW approximately. Hence a small rating input source for V_b and V_a would work fine to actuate and evaluate the circuit functionality. For signal V_a the signal power requirement is around 10 pW as shown in Figure 14, for 2 V voltage supply. On the other hand, the input signal power requirement for V_{bb} varies parabolically with respect to voltage. At its base lines up to 3 V power voltage supply the variation may be interpolated linearly but beyond this the rate of increase in power with respect to V_{bb} increases sharply with voltage levels, but still this rate is less as compared to signals V_a, V_b.

Circuit is operated for a maximum load of 1 nF with average power dissipation of 0.1 mW at 2 V trapezoidal power supply. The average power dissipation with respect to C_{load} is shown in Figure 15. Till C_{load} of 0.1 nf the power dissipation from the circuit is constant, approximately equal to 40 μW. As the C_{load} is increased beyond 0.1 nf the circuit dissipates comparatively very large power, because of the energy required to charge the large capacitance which is

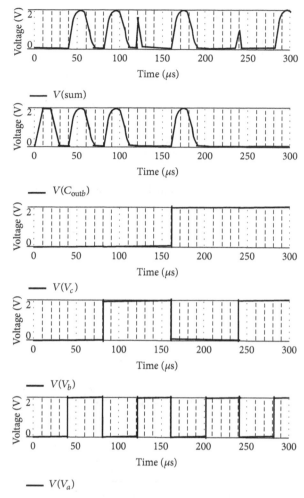

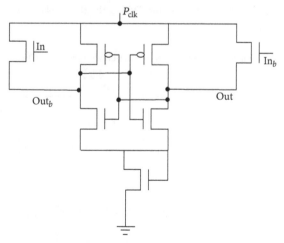

FIGURE 9: Output waveform for CPLAG Adder with P_{clk}.

FIGURE 10: Modified PFAL circuit implementation @ inverter.

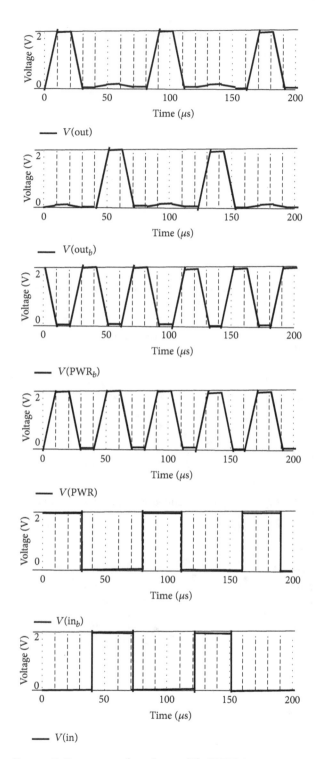

FIGURE 11: Output waveform for modified PFAL inverter gate.

directly proportional to incremental power dissipation. The quantum of effort required to charge the C_{load} up to 0.1 nf is more or less the same, dissipating approximately same power from the power source. The average power variation with power clock source, for DFF using RCPLAG Nand with respect to C_{load}, is shown in Figure 16.

The signal strength required for input signal is quite independent with the variation of C_{load} as depicted in Figure 17. The variant capacitive loading has no effect on the signal which is approximately 5.8 nW and 1.44 nW for V_{bb} and V_b signal approximately.

With rise in temperature the input resistance for input terminal V_{bb} decreases with constant slope proportional to

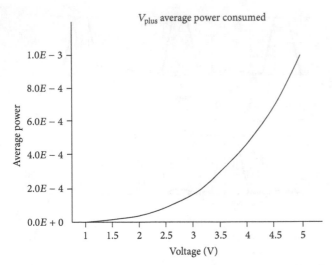

FIGURE 12: V_{plus} P_{avg} variation for DFF @ RCPLAG Nand with respect to supplied voltage level.

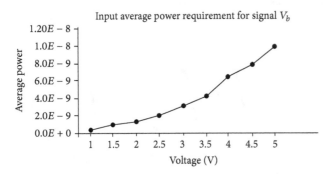

FIGURE 13: Avg. input power required for V_b for DFF @RCPLAG Nand.

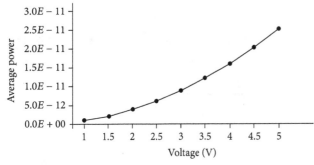

FIGURE 14: Avg. input power required for V_a for DFF @RCPLAG Nand.

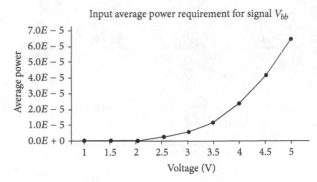

FIGURE 15: Avg. input power required for V_{bb} for DFF @RCPLAG Nand.

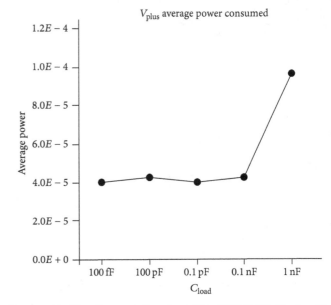

FIGURE 16: V_{plus} P_{avg} variation for DFF @RCPLAG Nand with respect to C_{load}.

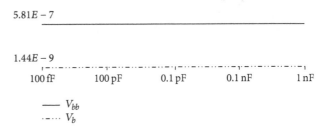

FIGURE 17: Signal power requirement with respect to C_{load}.

temperature coefficient concerned as shown in Figure 18. The input signal V_b has approximately constant input resistance, with very small rate of decrease, with respect to temperature increase.

Similarly as shown in Figure 19 the output resistance decreases with increase in temperature. This is because of more availability of charge carriers at high temperature giving rise in current associated and hence reduction in resistance associated. At 25°C the Output resistance is 3.5 × 10^9 Ω which reduces to 2.5 × 10^8 Ω approximately at

100°C. The average power dissipation reduces linearly with a constant slope with temperature variation as shown in Figure 20. The power dissipation varies from 45 nW to 30 nW from room temperature 25°C to 100°C.

The average power dissipation by D FF circuit using Nor RCPLAG universal gate is shown in Figure 21 with respect to C_{load}. As with Nand based implementation the average power is constant approximately up to a loading of 0.1 nf

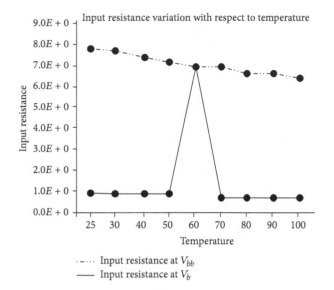

FIGURE 18: Input resistance variation wrt temperature.

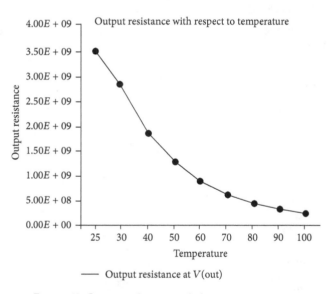

FIGURE 19: Output resistance variation wrt temperature.

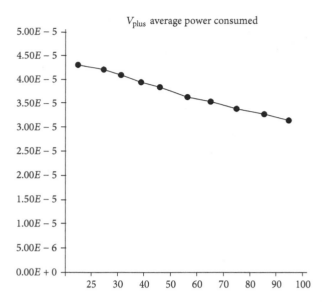

FIGURE 20: V_{puls} avg. power consumed with respect to temperature variation.

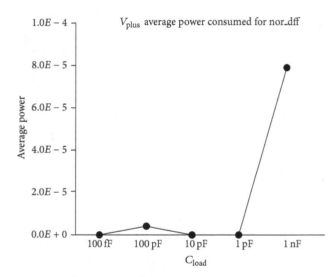

FIGURE 21: Avg. power dissipation for DFF @RCPLAG Nor gate with respect to C_{load}.

and increases greatly for further increase in loading. Also the average power required for input signal for successfully actuating the circuit is independent of loading as shown in Figure 22.

The PDP distribution for 20 different transistor sizes is shown in Figure 23. The transistor size is varied from 1.8 μm to 36 μm. This PDP is examined for 9 different voltage levels as shown. Up to a voltage level of 3 V and 12.6 μm transistor size, PDP shows a linear behavior. For higher transistor size the PDP variation increases parabolically with supply voltage. Similarly for high voltage the rate of increase in PDP with respect to transistor size varies parabolically, but the quantum rate is comparatively less. For the permutation of high transistor size and voltage level the PDP factor increases rapidly indicating inefficient functional operation for the circuit. It can be inferred that for a voltage level of 3 V

the circuit can be operated with constant PDP with different transistor sizes till 36 μm.

Hence with the voltage range of up to 3 V with 23.4 μm transistor size the circuit can be operated in best PDP scenario. The average power distribution with voltage and transistor size is shown in Figure 24. Normalized power drawn and feedback to the source are shown in Figure 25. Up to 2.5 V voltage supply the power feedback shown negative axes are approximately constant. As circuit is based upon CPL family, a good amount of circuit transistors is driven by the input signals as shown in positive axes in the figure. Majority of the power drawn from the source is feedback. The variation in the power drawn and feedback is quite similar for different transistor sizes as shown in Figure 25. The cone point behavior is shown whose base point reduces a small

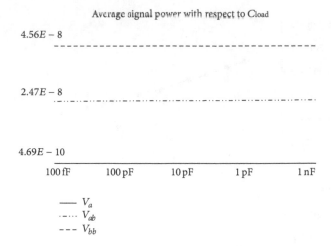

FIGURE 22: Input signal power requirement with respect to C_{load}.

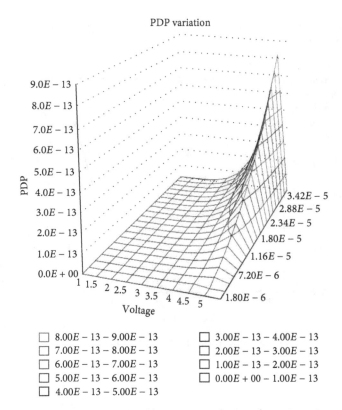

FIGURE 23: PDP variation with respect to voltage and transistor size.

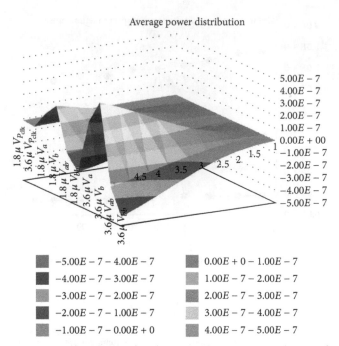

FIGURE 24: Avg. power distribution with respect to voltage and transistor size.

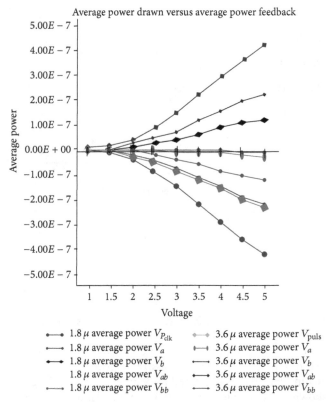

FIGURE 25: Normalized power drawn and feedback to power source.

value with increase in transistor size retaining the behavior variation.

4. Conclusion

In this paper authors have implemented dynamic positive edge triggered D FF configured from reconfigurable complementary pass transistor adiabatic logic gated (RCPLAG) Nand/And and Nor/or universal gates. Authors have also implemented Adder/Subtractor circuit with 4 different techniques. Authors have also reported modification in PFAL

adiabatic techniques for further optimizing of the power equations. These circuits implemented are functionally verified and found to work to a good level of satisfaction. These circuits are examined for power dissipation, timings, and

PDP for 20 different transistor sizes and 9 different voltage levels and 9 different temperature values.

From the analysis it is found that the best operating condition for the D FF based upon RCPLAG universal gate can be achieved at supply voltage lower than 3 V which can be used for different transistor size up to 36 μm. The average power dissipation is 0.2 μW at 1 V and 30 μW at 2 V at 100 ff C_{load} 25°C approximately. For C_{load} less than 0.1 nf the average power dissipation remains constant. Also smaller strength input signal can successfully actuate the circuit and evaluate its functionality. The input resistance is independent of the temperature variation. The circuits work fine for a wider range of temperature. Majority of the power drawn from the power source is feedback after circuit evaluation in the Recovery phase of power clock to the voltage source. This assists in reutilization of the energy and helps in reducing the amount of power drawn from the source. The average power dissipated by CPLAG Adder/Subtractor is the least as compared to other implementations, which is 58 μW at 2 V trapezoidal power source. The modified PFAL circuit successfully reduces the average power by 9% approximately as compared to PFAL circuit.

Conflict of Interests

The authors declare that there is no conflict of interests regarding the publication of this paper.

References

[1] R. Landauer, "Irreversibility and heat generation in the computing process," *IBM Journal of Research and Development*, vol. 5, no. 3, pp. 183–191, 1961.

[2] A. Vetuli, S. D. Pascoli, and L. M. Reyneri, "Positive feedback in adiabatic logic," *Electronics Letters*, vol. 32, no. 20, pp. 1867–1869, 1996.

[3] A. Blotti and R. Saletti, "Ultralow-power adiabatic circuit semi-custom design," *IEEE Transactions on Very Large Scale Integration (VLSI) Systems*, vol. 12, no. 11, pp. 1248–1253, 2004.

[4] M. P. Frank, "Common mistakes in adiabatic logic design and how to avoid them," in *Proceedings of the International Conference on Embedded Systems and Applications (ESA'03)*, pp. 216–222, Las Vegas, Nev, USA, June 2003.

[5] M. P. Frank, Realistic Cost-Efficiency Advantages for Reversible Computing in Coming Decades, UF Reversible Computing Project Memo #M16, 2002, http://www.cise.ufl.edu/research/revcomp/writing.html.

[6] V. S. Kanchana Bhaaskaran, "Asymmetrical positive feedback adiabatic logic for low power and higher frequency," in *Proceedings of the International Conference on Advances in Recent Technologies in Communication and Computing*, pp. 5–9, October 2010.

[7] M. Kumari, M. L. Keote, and N. Aman, "Low power adiabatic complementary pass transistor logic for sequential circuit," *International Journal of Research in Computer and Communication Technology*, vol. 3, no. 1, 2014.

[8] C. P. Kumar, S. K. Tripathy, and R. Tripathi, "High performance sequential circuits with adiabatic complementary pass-transistor logic (ACPL)," in *Proceedings of the IEEE Region 10 Conference (TENCON '09)*, pp. 1–4, November 2009.

[9] M. Sharma and A. Noor, "Modified CPL adiabatic gated logic—MCPLAG based DPET DFF with XOR," *International Journal of Computer Applications*, vol. 89, no. 19, pp. 35–41, 2014.

[10] M. Sharma and A. Noor, "CPL-adiabatic gated logic (CPLAG) XOR gate," in *Proceedings of the International Conference on Advances in Computing, Communications and Informatics (ICACCI '13)*, pp. 575–579, August 2013.

[11] M. Sharma and A. Noor, "Positive feed back adiabatic logic: PFAL single edge triggered semi-adiabatic D flip flop," *African Journal of Basic & Applied Sciences*, vol. 5, no. 1, pp. 42–46, 2013.

[12] M. Sharma and A. Noor, "Reconfigurable CPL adiabatic gated logic-RCPLAG based universal NAND/NOR gate," *International Journal of Computer Application*, vol. 95, no. 26, pp. 27–32, 2015.

[13] N. H. E. Weste and D. M. Harris, *CMOS VLSI Design: A Circuits and Systems Perspective*, chapter 6, section 6.2.5.2, Pearson Education, 2006.

[14] S. M. Kang and Y. Leblebici, *CMOS Digital Integrated Circuits Analysis and Design*, chapter 7, Tata McGraw Hill Education Private, 3rd edition, 2003.

[15] J. M. Rabaey, A. Chandrakasan, and B. Nikolic, *Digital Integrated Circuits: A Design Perspective*, chapter 3, Prentice Hall, 2nd edition, 2003.

5

Shifting the Frontiers of Analog and Mixed-Signal Electronics

Arthur H. M. van Roermund

Mixed-Signal Microelectronics Group, Department of Electrical Engineering, Eindhoven University of Technology, Den Dolech 2, P.O. Box 513, 5600 MB Eindhoven, The Netherlands

Correspondence should be addressed to Arthur H. M. van Roermund; a.h.m.v.roermund@tue.nl

Academic Editor: Frederick Mailly

Nowadays, analog and mixed-signal (AMS) IC designs, mainly found in the frontends of large ICs, are highly dedicated, complex, and costly. They form a bottleneck in the communication with the outside world, determine an upper bound in quality, yield, and flexibility for the IC, and require a significant part of the power dissipation. Operating very close to physical limits, serious boundaries are faced. This paper relates, from a high-level point of view, these boundaries to the Shannon channel capacity and shows how the AMS circuitry forms a matching link in transforming the external analog signals, optimized for the communication medium, to the optimal on-chip signal representation, the digital one, for the IC medium. The signals in the AMS part itself are consequently not optimally matched to the IC medium. To further shift the frontiers of AMS design, a matching-driven design approach is crucial for AMS. Four levels will be addressed: technology-driven, states-driven, redundancy-driven, and nature-driven design. This is done based on an analysis of the various classes of AMS signals and their specific properties, seen from the angle of redundancy. This generic, but abstract way of looking at the design process will be substantiated with many specific examples.

1. Introduction

The performance of analog and mixed-signal (AMS) electronics is characterized by a multitude of function and resource related parameters, like speed, bandwidth, accuracy, linearity, resolution, phase noise, dynamic range, SN(D)R, power dissipation and efficiency, robustness, chip area, yield, and design time. Moreover, most of them are dependent on each other. This leads to a high order of complexity and highly dedicated designs. In future technologies this complexity will further increase. The definition of "figures-of-merit" (performance parameter combinations) helps, but lack of insight slows down the shift of frontiers and even leads to wrongful "boundaries" and intuitive design approaches that are difficult to carry over to other designers.

The purpose of this paper is, in line with the invitation, to show the scientific view and vision on AMS that has been built up in my research group in the past 15 years, a vision that we see as crucial for further shifting the frontiers in AMS, and to elucidate with examples how it drives our research.

In contradiction to the hundreds of papers we published on all kinds of specific AMS IC designs for various applications, this paper tries to formulate our high-level view on designing analog mixed-signal ICs. In this generalized view we leave out the (important) details of specific designs and focus on the fundamental properties, including all kinds of boundaries and restrictions of the various AMS signal representations, of the various IC technologies (the physical aspects) in which we want to implement the system functions, and of the design environments that provide us with the resources to do the design. As such, we deliberately make an abstraction by leaving out specific design properties and focus on the underlying fundamental and generic properties of signals and hardware. This makes it a scientific analysis that can be subsequently used in the synthesis of a plurality of existing and new applications and can support the designer to make scientifically founded choices and to structure his design approach, of course within the constraints of the specific system specifications at hand. Many references to practical designs from our group, which comprise all practical details that go with these specific designs, support the reader to link this high abstraction view to the lower implementation levels of specific designs.

Following this approach, we will argument that the primary role for AMS circuitry in general will be to provide a technology-matching function, and, in practice, this takes

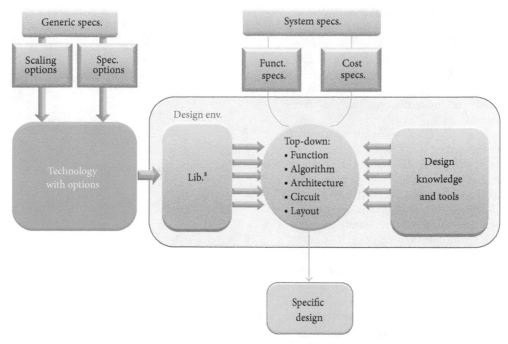

FIGURE 1: Generic design process.

place in the so-called frontend. As such, the word frontend is defined here as any interface between an analog communication channel, sensor, or actuator, on the one hand, and the digital processing chip, on the other hand. The reasoning behind that is that the information processing that is required by the system function (not to be confused with the abovementioned signal-conditioning functions in the frontend) is usually best done in the digital domain, after having the signal domain and signal carrier optimally matched, in the frontend, to the technology.

We will also introduce the fundamental concept of redundancy and explain that this plays a crucial role in this matching function. We will use for this reasoning the high-level concept of channel capacity, as introduced long ago by Shannon [1]. In this paper we will show that this concept also applies to IC design.

Based on all of this, we propose four complementary "matching-driven" design procedures for AMS frontend design that differ fundamentally from the top-down approaches used for the implementation of the system function in the digital domain:

(i) *technology-driven matching;*

(ii) *states-driven matching;*

(iii) *redundancy-driven matching;*

(iv) *nature-driven matching.*

The paper also includes a forward-looking analysis of the issues that we believe will drive future research in this field.

The paper starts with a generic view on IC-design procedures and discusses how often promoted function-driven top-down approach leads to wrongful boundaries for AMS design (Section 2). To find out how the design approach should be changed in order to shift the frontiers across these "boundaries," we discuss and classify in Section 3 the various AMS signal representations and explain the crucial roles of abstraction and redundancy and relate that to Shannon's definition of channel capacity [1]. In Section 4, we describe the primary AMS function, matching the communication channel to the digital core via the frontend, and explain that the inevitable mismatch occurring inside the frontend itself requires a smooth and stepwise matching based on transformations in and between the various AMS domains. In Section 5 we elaborate on the proposed four complementary matching-based AMS design procedures: technology-, states-, redundancy-, and nature-driven matching. Section 6 addresses shortly the future of AMS and in Section 7 we will draw conclusions. For reasons of clarity and paper size, we restrict ourselves in this paper to intuitive discussions, rather than to exact formulations.

2. IC-Design and Boundaries

The generic IC-design process is visualized in Figure 1. System specifications, both functional and cost-related, are directing the design process towards a specific design. To reduce complexity, various abstraction layers are defined, each decoupling further details from lower levels. At all levels, libraries providing generic technology-related information (like clearance rules, transistor models, and IP blocks) are available, and general design knowledge and tools. Technology development is a decoupled process, driven by generic design-independent specifications; scaling plays an important role here, and special process options might be provided for certain performance domains, like low power or high speed.

The commonly preferred design procedure follows a top-down approach, allowing iterations where necessary, going step by step from high functional level towards final layout level. Intrinsically, higher level requirements are given preference over lower-level details, to further reduce design complexity. This *"function-driven approach"* is, for sure, the recommended approach for implementing complex functionality in digital systems.

For AMS design too, this *function-driven* approach is usually followed, but this paper will dispute its preference. Indeed, in non-state-of-the-art AMS designs, far enough away from the boundaries, you might follow a function-driven approach, but for such non-state-of-the-art designs implementation in AMS instead of digital is questionable at all, as we will explain. On the other hand, for state-of-the-art AMS designs, the appearance of a "boundary" is *not* due to technological limitations per se, as is often thought, but rather to a *nonoptimal match* between the functional requirements and the (limited) properties of the hardware. Proper transformations can provide an optimal match (For example, the "Walden FoM" for AD converters [2] is not technology constrained, as often assumed, but design constrained: a consequence of nonoptimal matching to the technology.) The design process must therefore principally be *matching driven*. By the end, performance is always resource limited, how much power, chip area, design time, and so forth do we want to spend.

To find an optimal match between functional requirements and technological properties, we first need to know what are the characteristic properties of AMS signals and how these enable signal-domain transformations.

3. AMS Subdomains, Shannon Capacity, and Redundancy

The double-log curves in Figure 2 show relative costs (here in terms of power dissipation) versus accuracy, for analog and digital. For analog, costs increase with (at least) a factor 4 for every factor of 2 in accuracy, whereas for digital relative cost increase diminishes with word size: every factor 2 in accuracy requires just 1 bit extra. This suggests that for accurate functions we better use digital and for simple functions analog. However, this is only partly true. First, all physics-related accuracy problems have been shifted to the (required, but here neglected) converter. Second, the difference is more fundamentally related to redundancy; digital is just an example of (low-level) redundancy.

Redundancy can be applied at all design layers (Figure 1), as will be discussed in Section 5, but already at the (abstract) signal level we can distinguish four classes of signal representations, leading to different types of hardware redundancy, depending on whether or not the time domain and/or the amplitude domain of the (abstract) signal is discrete; see Figure 3(a), with red for continuous and blue for discrete. For simplicity we will use in this paper the short names as mentioned within brackets. The three classes that have at least one continuous subdomain together belong to the AMS domain (Figure 3(b)).

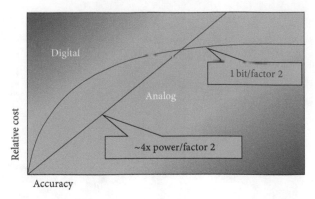

FIGURE 2: Relative cost versus accuracy for analog and digital.

Note that Shannon [1] already explained that every "propagation channel," generalised here to every "physical processing medium," and hence every IC, is limited in information-handling capacity, defined by its SNR and bandwidth: $C = BW \log_2(1 + SNR)$. Note that BW and SNR are resources (particularly power dissipation) related and thus expensive, especially when working at the frontiers. The difference between this *available* channel capacity and the amount of the capacity that is actually *used* to represent the signal information makes up the hardware *redundancy*. This redundancy provides a margin that makes the capacity-limited channel robust, not only to unwanted signals that "eat up" part of the capacity in an actual implementation but also to uncertainties in hardware capacity due to limited model accuracy at design time, technology variations at processing time, and environmental, system, application, and user variations at runtime.

Note that any *physical* carrier, independent of the *abstract* signal domain chosen to represent the information, is purely analog by nature and that the capacity of any hardware system is given by the Shannon formula. In case of discrete-time or discrete-amplitude representation, a large part of the time, respectively, amplitude domain of the physical carrier will not be used by the abstract signal, which leaves extra redundancy in the hardware that can be translated to robustness in those specific domains. Digital signals are in general so robust, in both domains, that they allow full abstraction from carriers and hardware; see (Figure 3(c)). Staircase and asd signals allow abstraction in their discrete domain.

It should be mentioned that, besides the well-known discrete-time and discrete-amplitude domain signals, there are many more signal representations that use only restricted subparts of time and/or amplitude domain. These representations can also be used to provide robustness and also belong to the class of AMS domains. Examples are pulse width or period modulation and amplitude multiplex (e.g., power amp per subrange). Visualizing this by more than just four AMS domains seems more correct, but impractical in this case, as it will fuzzify the message of this paper. Therefore we will stick here, without loss of generality, to the four mentioned domains and consider the other domains as inside subdomains.

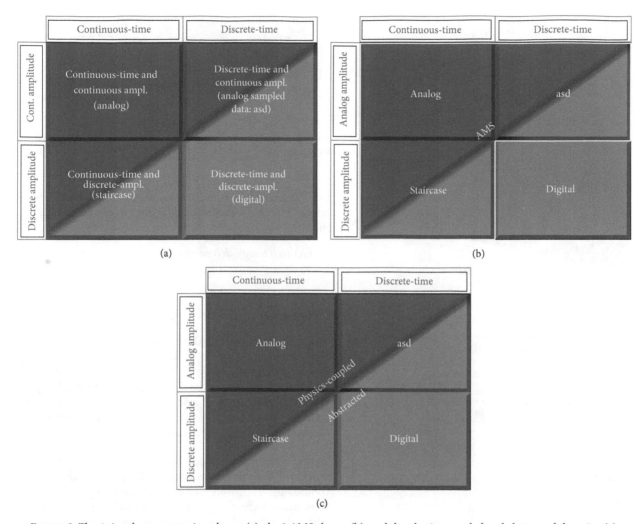

FIGURE 3: The 4 signal-representation classes (a), the 3 AMS classes (b), and the physics-coupled and abstracted domains (c).

4. The Matching Function as Primary Task for AMS

It is out of the question that digital is by far the best implementation for nowadays' complex functions on a chip, as it provides decoupling from hardware impairments and from hardware architectures, primitives, and structural properties. This provides process insensitivity (thus allowing generic CMOS processes), robustness, scalability, data bandwidth scaling, portability, flexibility, adaptivity, high yield, low design cost, high design reuse, and so on. It shows the power of the redundancy, and therefore the abstraction, in the digital signal waveforms.

However, most digital systems communicate with the outside world via communication channels, sensors, and/or actuators; see Figure 4. Therefore, data conversion is required. The benefits of digital usually justify this overhead, except for a small class of low complexity systems, where a full AMS implementation might outperform a digital one. However, the problem is far more than "just" data conversion: the signals from outside are optimized for their own medium (sensor, communication channel) and the on-chip digital

signals for the IC-technology. (For simplicity reasons, we will neglect in this paper the third medium, the antenna. If on-chip, it should be taken into account with the frontend [3, 4]). As these media are different, the signal properties at both sides do not match (in terms of frequency, bandwidth, signal strength, SNR, interferences, etc.), and a "matching network" is required to perform signal transformations in time, amplitude, or frequency domain.

Figure 5 shows this problem suggestively: the information "package" should be transformed to fit the "straitjacket" of the silicon or vice versa, to fit the propagation medium. This is where AMS, with its *intra- and interdomain* transformations, plays its primary and very crucial role.

Note, however, that this matching network, the frontend (including data converters), is also implemented on-chip; see Figure 6, so *the signals in this frontend by definition show a strong mismatch to the (silicon) medium.* The figure visualizes this in "Shannon language": we have three channel sections, but only two media; the left and right sections are optimized; the frontend is not.

The mismatch is the strongest at the input of the IC and decreases along the frontend. Brute-force direct digitizing

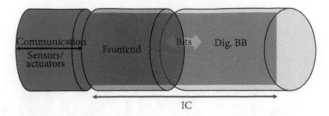

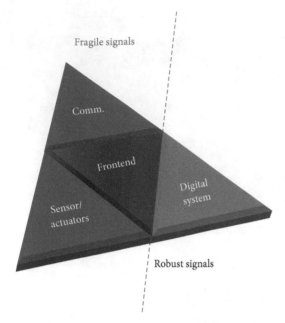

FIGURE 4: The AMS frontend in between digital system and sensor/actuator/communication channels.

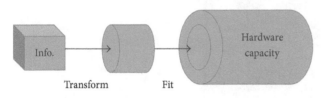

FIGURE 5: Signal transforms to fit the information to the hardware.

would create an abrupt mismatch in time and/or amplitude within the frontend, forming a bottleneck that can only be counteracted by increasing power dissipation. As a chain is not stronger than its weakest link, this is clearly not the optimal solution. We have to carefully *(re)distribute redundancy* over amplitude, time, and frequency, providing a better balance of robustness over the domains "channel coding," to achieve higher overall robustness.

5. Matching-Based Design

As discussed, we need to match the signals in the AMS domain gradually between communication medium and digital processing medium (silicon), taking into account that these matching steps themselves take place inside the (as yet unmatched) silicon medium. The question is now: *how do we design this AMS matching system, the frontend, such that we have an optimal smooth match between signal properties on the one hand and technology properties (the medium) on the other hand, via appropriate signal transformations and hardware changes and adaptations*? To answer this, we go back to the design process and look more in detail to it, Figure 7. The function-driven approach is exchanged now for a "matching process," as the primary goal should be to match smoothly the functional specifications with the technology properties.

FIGURE 6: Three channel sections in the overall chain, implemented in two media.

This matching procedure can still be driven from different angles. Here we will discuss four design approaches (2A–2D in the figure), each complementary to the others:

(A) *technology-driven matching;*

(B) *states-driven matching;*

(C) *redundancy-driven matching;*

(D) *nature-driven matching.*

Besides adapting our hardware by proper design, we can also adapt the technology, to achieve a better match, by adding specific options (1A in Figure 7) or even by using a technology with special options for the frontend (1B in Figure 7). The last approach would require a separate IC for the frontend or a technology overhead for the overall IC. This paper focuses, however, on the design process.

(A) Technology-Driven Matching. Technologies are defined by foundries; the only freedom for the designer is in the choice of technology. "Hardware" is based on technology and therefore is subject to technology limitations but also implies all circuit and architecture choices, defined by the designer. By proper hardware design and signal-processing choices an optimal matching with technology can be achieved. The technology properties form therefore the hard constraints that should be taken into account directly from the beginning of the design. These properties (e.g., signal ranges and noise levels) should then be compared to the system-level specifications (e.g., required signal resolution). Architecture, circuit design, and signal-domain transformations should provide the optimal match to the medium in the different (physical) domain and at all intermediate steps in the frontend.

As capacity limitations related to technology (and resources) are a fundamental problem, it seems in the first instance paradoxical to further restrict ourselves to go over to AMS signal domains that use only subsets of time and amplitude. However, as mentioned, these restrictions provide robustness. This allows decoupling from imperfections, which creates options for reducing, for example, the power dissipation.

Moreover, the abstraction it allows enables decoupling of individual algorithmic operations, creating a lot of freedom in assigning operations to hardware and in scheduling in time, as we very well know from digital. This in turn creates options to balance the time-domain and amplitude-domain redundancies to provide a better match. Finally, it enables using more hardware to increase the overall capacity.

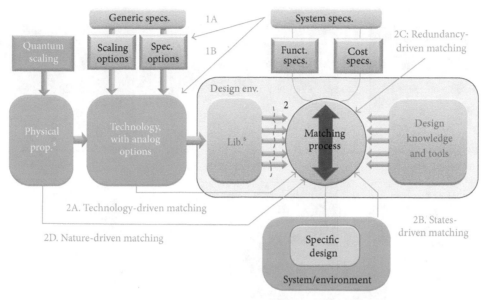

FIGURE 7: The four design approaches shown in relation to the generic design process.

So, redundancy and abstraction enable far better matching between functional requirements and technology properties, which more than compensates the loss in used capacity "per amount of hardware." Note that technology capacity is limited, but hardware capacity is not: it is finally only subject to cost, as expressed in Section 2: "final performance is not technology but is resource limited." For example, we can cross the "bandwidth boundary" of the hardware by switching over to discrete-time and parallelize operations in time interleave or the "resolution boundary" by sequencing the algorithm in lower-resolution suboperations that better match hardware resolution, like in pipelined data conversion. Redundancy created by restrictions in the signal domains thus generates many new options for transformations. Here we will discuss very shortly, with a helicopter view, a number of intra- and interdomain transformations; to elucidate this, see Figure 8. For details we refer to the references.

Figure 8(a) refers to all transformations within the purely analog domain. Most of them are well known, for example, offset, gain, compression, and expansion in the amplitude domain; frequency offset (mixing) and bandwidth compression or expansion in the frequency domain; amplitude-frequency transformations (AM-to-FM modulation); and so forth. (Note that intra-time-domain transformations are severely restricted in the purely analog domain, as information is continuously present and we cannot go back in time. Redundancy can therefore only be applied in time to come for example, by positive time shifts or time stretching. Transformations to discrete-time signals and hence to the second column of Figure 8 create time redundancy and enable many intra-time-domain transformations.) For power amplifiers the voltage, current and temperature boundaries of a technology can be circumvented by using distributed amplification [5, 6].

Unconventional technologies may ask for unconventional approaches. For example, in organic technologies, used for large-area, flexible-foil, and ultra-low cost electronics, hard faults (functional defects) and soft faults (variability in performance) completely dominate the design. This (still) leads to a preference for p-only technologies and circuits with a minimum number of transistors to reduce the hard faults. The severe p-only restriction leads to quite unconventional circuit architectures to provide an optimal match to the specific organic-technology properties, while reducing the vulnerability to the high technology spreads and the lack of spatial correlation (bad matching). Examples can be found in the references: in [7] the transistor output impedance is applied for filtering purposes, in [8] a specific "positive-feedback level shifter logic" is proposed, and in [9] a parametric way of amplification is used.

The p-only circuit limitation also makes discrete-time operation not realistic, which means that transformations are restricted to the two left quadrants in Figure 8. Complementary organic technologies (or hybrid technologies, with inorganic amorphous metal oxides), using evaporated or printed materials, face even more hard faults but can provide higher resilience to variability at circuit level, as all established complementary circuit options are available now to provide optimal resilience [10, 11]. Moreover, discrete-time operation becomes available now, enabling AD and DA, and more generally designs in the asd and staircase quadrants of AMS [12, 13].

Nonlinear transformations form a specific class of (unconventional) transformations. They can, for example, be used to exceed the frequency limitations of (already high-speed) technologies. Figure 9 shows how signals, generated at regular "electronics frequencies," matching the technology bandwidth, are frequency shifted in a transmitter with passive nonlinear transmission lines (NLTL), to the THz domain, so above the technology bandwidth. Subsequently, this THz signal is transmitted to and reflected by a device under test. In the receiver, another nonlinear transmission line (NLTL)

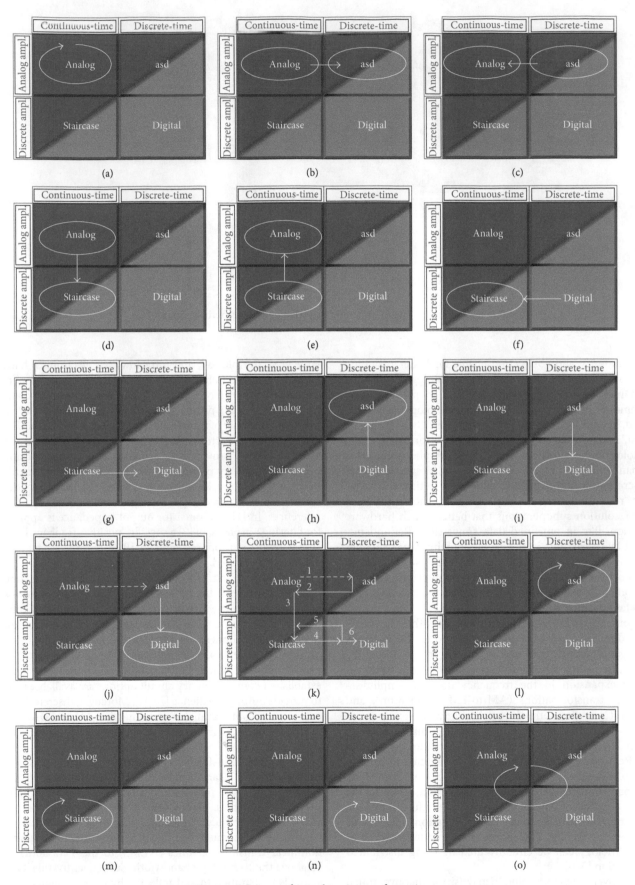

FIGURE 8: Intra- and interdomain transformations.

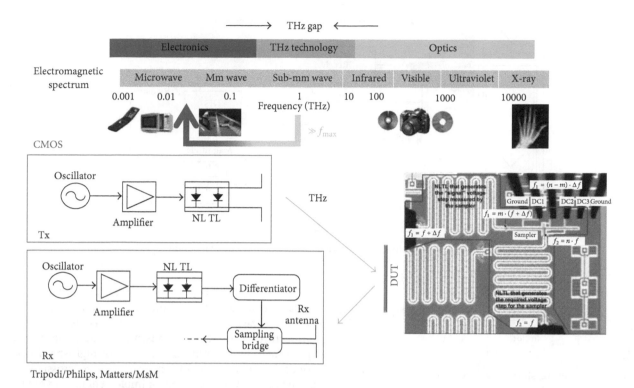

FIGURE 9: Passive nonlinear frequency shifting to match THz frequencies to IC technology frequencies.

mixes back these THz frequencies to "normal" frequencies that fit again in the bounded frequency domain of the technology. By this way we have created an on-chip THz spectrometer with frequencies far above the "boundary f_{max}" of the technology [14].

Another application of nonlinear transformations is shown in Figure 10. It shows how a strongly nonlinear low-noise input amplifier, here called "nonlinear interference suppressor" (NIS), is used, in a very unconventional and counterintuitive way in a transceiver like a smartphone. The deliberately introduced strong nonlinearity at the input of the receiver is very well controlled as a function of the known signal (transmitted by the same frontend) that forms a close-by interferer for the receiver. The non-linearity is controlled in such a way that the fundamental component of the interferer is maximally suppressed (also called "compression"), and "replaced" by one or more if its harmonics outside the channel bandwidth [15, 16]. These harmonics are further away from the received signal and can therefore easily be suppressed by a filter.

Figure 8(b) refers to the well-known sampling, leaving only signal information at discrete-time moments, shifting a surplus of redundancy in the frequency domain (the surplus of bandwidth the hardware can handle) to the time domain without information loss. The redundant time in the physical carrier creates new freedoms for the designer. Examples are: time-multiplexing of the system functions, by clock-phase-controlled switches (time-variant operation); rescheduling of operations in time; positioning of the discrete-time samples of your signal in such a way that they are shifted in time with respect to the periodic glitches that are caused by clock

generation or by digital blocks, so that the signal is not interfered by these glitches; making the signal processing insensitive to absolute errors (by making them only dependent on device ratios, like in switched-capacitor circuits); the use of scalar integration within a single domain (e.g., voltage to voltage, like in switched-capacitor integrators); the use of hardware sharing (time-multiplex use of hardware); and so forth. In fact, it enables within the analog domain a plurality of transformations that are already well known and heavily used in the digital domain; indeed, these properties are fundamentally related to discrete-time (and not to discrete amplitude) and as such not reserved for digital as most designers assume.

Figure 8(c) shows the well-known reconstruction to continuous-time signals, possibly done stepwise via oversampling and analog filtering. Information is then shifted from the amplitude to the time domain, resulting in less time-domain redundancy and more redundancy in frequency domain (less spectrum occupied) and amplitude domain.

Figure 8(d) shows a less conventional domain, here called the staircase domain. The amplitude domain is limited here to discrete levels, providing signal redundancy in the amplitude domain. An example is an asynchronous sigma-delta modulator that we designed to control a power amplifier with only discrete-amplitude levels, enabling efficient switched-mode (discrete-level) amplification; see Figure 11. With the number of quantization levels we can balance the redundancy in time and amplitude domain. The asynchronous "data-driven" sampling provides the time redundancy without introducing quantization errors: there is no loss of analog information; the information is only shifted partly from the amplitude domain

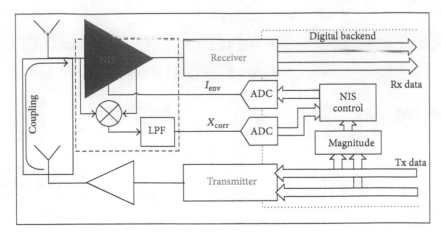

FIGURE 10: A nonlinear transform to shift a known close-by interferer away from the signal band.

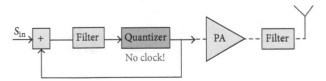

FIGURE 11: A staircase transformation to enable efficient switched mode power amplification.

to the time domain, where it is present in the position in time of the transitions in the square wave multilevel (possibly 2 level) signal, at the cost of a larger occupied bandwidth. More detailed information can be found in [17–21].

The so-called LINC amplifier uses an alternative staircase transformation to achieve efficient power amplification: a continuous-amplitude signal is split into 2 level signals, with different phases (information is thus now transferred to relative time domain). Both are subsequently efficiently amplified by two switched-mode amplifiers. Adding the outputs recreates the original continuous-amplitude signal but now amplified with respect to the original signal [22].

Yet another example for this class of transformations is the folding operation in an AD converter. The amplitude range is split into subranges and all of them are mapped to one and the same subrange [23].

Aside from hard splits in the amplitude domain, also smooth changeovers are possible, like in, for example, Doherty amplifiers, where for the larger amplitudes the amplification is gradually taken over from a main amplifier to a peak amplifier, via load pulling, an interaction between the amplifiers (with outputs connected via transmission lines) via their output impedances [24, 25].

In envelope-elimination-and-restoration (EER) amplifiers, a signal that originally has all its information in the amplitude domain is split into a signal in (again) continuous-amplitude domain (the envelope) and a two-level signal in the staircase domain (comprising the carrier phase information), such that a transistor (drain-modulated by the amplitude-continuous first signal) can be operated very efficiently in switched mode by switching it with the second (two-level) signal [26].

Yet another example is that of a VCO-based ADC. In that case, a signal with amplitude information is translated to a square wave signal of which the zero crossings contain the information. Subsequent translation to the digital domain is performed by counting the number of high-speed clock periods (of another clock) between the zero crossings. In [27] such an ADC is described for implementation in organic electronics, to provide robustness to the high spreads in such a technology. (Note that, in case of amplitude-to-phase transformation, the information is put on a continuous-time basis in the phase; in that case the transformation belongs to the situation described in Figure 8(a).)

In Figure 8(e), the way back is performed by spectral filtering; see, for example, the filter before the antenna in Figure 11. Such a filter reduces the bandwidth in this example to keep the transmitted signal within the boundaries of the allowed communication channel, without loss of information, as its output signal comprises more information in the amplitude domain than its input signal. This increases spectral redundancy (spectral purity) at the cost of amplitude redundancy.

In Figure 8(f), the transition from digital to staircase implies a transition from discrete-time to continuous-time and as such there is a reduction of the time-domain redundancy. This is used, for example, in asynchronous event-driven converters that exploit maximum technology speed: a next algorithmic step is done instantaneously after the hardware is ready with the previous step, instead of letting it wait till a next clock phase (in synchronous operation). See, for example, [28], where a "slope AD converter" (which is intrinsically slow) is geared to maximum speed: the slope automatically steps up at the maximum speed the hardware can provide (like a domino effect), until a comparator detects that it has reached the value of the analog input signal. The amount of steps is then the digital countervalue of the analog input.

In Figure 8(g), a transformation from staircase to digital implies (synchronous) sampling, or saying otherwise: a change from asynchronous to synchronous, but this sampling process experiences reduced constraints compared to the transition from purely analog (continuous both in amplitude and time) to discrete-time analog (analog sampled data),

Figure 8(b), as the amplitude is constant during the sampling process, thanks to the applied redundancy in the amplitude domain [18]. A quantization error is introduced (a loss of information, as both domains have now made discrete), and some hardware capacity lost (SNR deteriorated), but the robustness to timing errors has improved.

In Figure 8(h), from digital to asd means filtering with an infinite impulse response (IIR) filter, which translates the digital multilevel signal to a continuous-amplitude signal. Note that a finite impulse response (FIR) filter also increases the resolution in the amplitude domain (to more discrete levels), but the output remains multilevel, which implies a transformation within the digital domain (Note that binary is often mixed up with digital; however, binary is only a subset of digitals (multilevel).).

In Figure 8(i), the reverse direction reflects the well-known quantization, performed in all AD converters that start from the asd domain, introducing a quantization error with associated loss in used capacity in the amplitude domain. However, the gain in redundancy and robustness has formed the basis for many AD architectures [29, 30], thanks to the freedom that the increased redundancy provides in mapping over time and amplitude domains, like pipelining, time interleaving, recycling, sigma-delta modulation, and all kinds of combinations of them.

In Figure 8(j), introducing a sampling preceding the previously discussed quantization step reflects all conventional AD converters that have continuous-time input and perform the sampling before the quantization (which is not strictly necessary, as this paper shows).

In Figure 8(k), more complex AD conversion is possible, exploiting even more the various subdomains. A time-interleaved asynchronous SAR AD converter, for example, the one discussed in [31] and shown in Figure 12, first performs the sampling; see number "1" in both Figures 8(k) and 12. This sampling creates a redundancy in time which paves the way for time-multiplexing of the following iterative AD suboperations. A hold function translates the sampled signal back to a continuous-time hold signal (2), without losing the redundancy in time, enabling subsequent fast asynchronous operation of comparator and logic (3 + 4). Simultaneously, an approximating quantized signal is generated iteratively, each iteration with more accuracy, by the DA converter that translates the asynchronous digital words, in each iteration step, back to the staircase domain (5). After all asynchronous iteration slots, when the comparator output flips, meaning the analog input has been approached optimally, the final digital output is delivered. In a similar way, redundancy in time is created and used in [32–35].

In Figure 8(l), also within the asd domain, we can perform transformations. Sample-rate transformations with interpolating or decimating filters redistribute redundancy between amplitude and time domain. Many other asd functions are possible in this domain and are known from, for example, switched capacitor literature, like the ones discussed already above in the discussion of Figure 8(b).

In Figure 8(m), within the staircase domain, we find all kinds of period-modulation transformations, with information (partly) in continuous-time domain [19]. Any

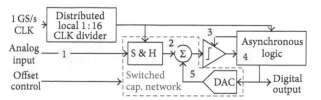

FIGURE 12: Example of time-multiplexed AD suboperations, enabled by first creating time redundancy by overall synchronous sampling, with transforms in all 4 subdomains of Figure 8.

transformation from purely analog to this domain needs ("data-driven" or asynchronous) sampling, to achieve the required time redundancy for time-domain operations; this asynchronous sampling was already discussed in Figure 8(d). The so-called time-domain AD converters first perform this transition from purely analog to staircase (two-level signals in this case, with the information now in the pulse duration) and subsequently use synchronous sampling to create the time redundancy required to do the suboperations of the AD algorithm in the "empty" (redundant) time periods. Synchronously counting the output periods introduces now the quantization, transferring the information to the digital domain [28].

In Figure 8(n), the digital transformations fall outside the AMS discussion; they reflect all processing in the digital backend of the IC.

In Figure 8(o), this figure emphasises that in any frontend with feedback loops we can have continuously transitions between domains. A continuous-time sigma-delta converter is an example.

(B) States-Driven Matching. On top of the above discussed technology-driven matching a states-driven matching is required to shift the boundaries; see Figure 7. As mentioned before, the actual situation, represented by the states, will differ from the system you wanted to design, as unwanted signals will "eat up" part of the capacity and because of uncertainties in the hardware capacity due to limited model accuracy at design time, technology variations at processing time, and environmental, system, application, and user variations at runtime. If we can do both the measurement and the correction at runtime, we can come far closer to the ideal situation [3, 4, 36, 37].

Figure 13 shows the generic situation for an AMS frontend [37]. First, the frontend must be made adaptive (reconfigurable architecture, adaptable parameters, test signals, etc.) which requires redundant hardware. For example, for a DA we can put redundancy in the decoder [38] or in the core, by using unary code, or even several sets of DAs, each with unary/binary cores [39–41].

Second, any information from outside or inside states, obtained if necessary with optional extra hardware (to detect, actuate, or generate test signals, to analyse the situation and to control feedback and learning algorithms), will help and should be used. Both core hardware and pre- and postprocessing can be adapted. Note that a states-driven approach also needs redundant hardware. Next, some examples will be shortly addressed to illustrate the states-driven matching.

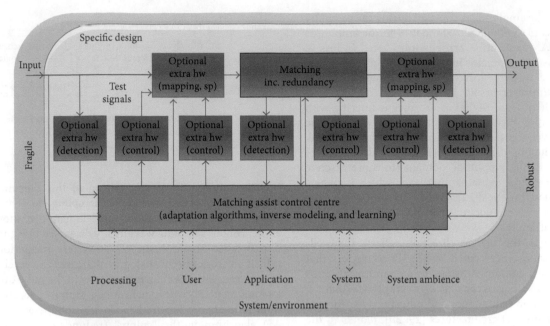

FIGURE 13: Generic diagram for a states-driven frontend system.

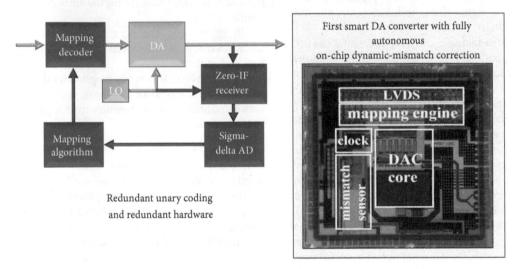

FIGURE 14: Dynamic-mismatch correction in a DA, based on code redundancy.

Figure 14 shows, as an example, how the algorithmic redundancy in the unary code of a DA converter in combination with an on-chip dynamic-error measurement and a feedback-controlled mapping decoder can be used to improve performance without excessive power consumption. The unary code provides many code options that, for an ideal DA, give the same output value. However, the errors in an actual DA lead to different results for these codes. The mapping algorithm that controls the decoder finds the best code options for the specific situation (with specific errors). The criterion for "best codes" is in this example the accuracy at high frequencies. This is detected by first downmixing that part of the spectrum in the "zero-IF receiver" and subsequently measuring the signal at baseband with a sigma-delta AD converter [42–44].

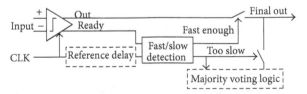

FIGURE 15: Improved SAR-ADC with comparator-state adaptive algorithm.

Another example is the "fast/slow detection circuit" in the SAR ADC in Figure 15 [33–35], which detects if the actual (processing and signal dependent) comparator is ready in time and adapts the SAR algorithm if necessary (more approximations in critical situations, followed by a

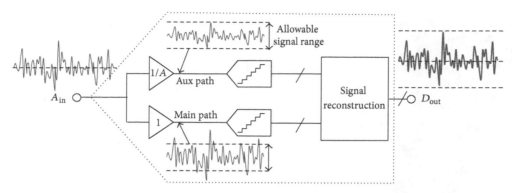

FIGURE 16: AD converter that uses amplitude multiplex based on signal statistics.

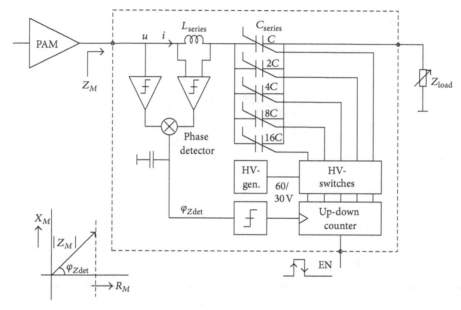

FIGURE 17: Adaptive antenna matching, dependent on reflection status.

majority voting). This is a favorable solution compared to overdesigning the power-consuming comparator such that it is fast enough for all potential situations.

Figure 16 shows an AD that splits the amplitude domain into subranges [45–47]. The performance of each AD is made proportional to the probability of the input signal in its subrange, which saves resources like power consumption. This is favorable for communication signals where the input signal shows a quite large probability for the lower amplitude range and a low probability for the higher amplitude range.

Figure 17 shows an example of a situation-dependent mismatch between the power amplifier and the transmit antenna (e.g., if a user holds his hand close to the antenna of his smart phone). This mismatch induces reflections, which can be measured and used to improve the antenna matching. This is a far more efficient solution than brute-force overdesigning for all potential situations [48–52].

Figure 18 shows a pregnancy monitoring system with smart sensing patches on the belly of the mother-to-be that can be made adaptive to the actual status of the mother and the baby, to the phase and to the history of the pregnancy, to

interferers (by choosing the optimum set of patches), and so forth. Details can be found in [53, 54].

(C) Redundancy-Driven Matching. We have already discussed extensively the use of redundancy at the signal and hardware level to enable optimal matching transformations. We also discussed that redundancy is required to enable states-driven matching. This redundancy-driven matching approach can (and should) be extended to all higher levels of a system (algorithmic, circuit, and application level) and can as such relax considerably the requirements at the hardware (AMS) level. Many examples of this use of redundancy at the higher system levels can be found in literature, from error correction in AD, up to redundant parallel systems.

(D) Nature-Driven Matching. In the previous examples the hardware itself was calibrated based on redundancy and/or state information. Nature learns that redundancy is used abundantly, enabling very small and ultra-low-power primitive hardware cells. This leads to an efficiency in the use of hardware that exceeds the efficiency in our electronics with

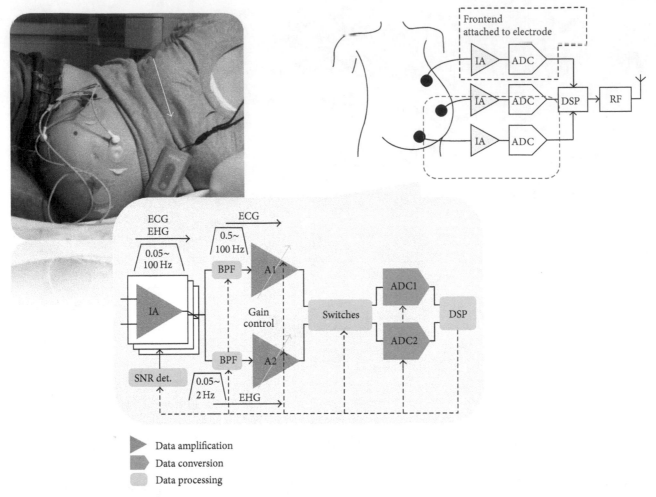

FIGURE 18: Pregnancy monitoring system based on status of baby and mother, pregnancy status, signal conditions, and history.

orders of magnitude. Indeed, these natural primitive cells are intrinsically very inaccurate, but that is corrected at the higher system levels, like ear and eye corrections in the brain. This correction is enabled by the abundant redundancy that is applied at the lower levels. Future IC-design should find an optimal balance also in that aspect. Further, nature relies on other paradigms, not yet used in IC design, as will be discussed next.

6. Future Prospects and Changing Paradigms

We are entering currently nanoscale electronics, clearly, but this is just a first step in a major transition that will be physics-driven and nature-inspired.

   (i) Next to nanoelectronics come quantum electronics, with distinction between individual electron energy levels; with new, nanoscale, and quantum-scale devices at quantum-energy levels; with increased uncertainties in modelling and variability in processing; and with statistics-based design approaches.

   (ii) A reduction in signal-handling capacity per device will be overcompensated by a capacity increase based on massively parallel hardware and signal processing and massive amounts of redundancy.

   (iii) The digital paradigm, based on redundancy and correct primitive operation at the lowest level, might make place for analog mixed-signal processing (like in our brains!) in combination with redundancy at higher levels and learning mechanisms.

   (iv) Alternative and hybrid technologies will arise, such as MEMS, organic and flexible electronics, optoelectronics, and molecular electronics.

   (v) Alternative manufacturing technologies in terms of materials, processing, and lithography will be followed by self-organising and self-assembling hardware.

   (vi) Self-organisation, self-learning, and self-assembling will enable autonomous short-term matching; self-growing, evolution, and inheritance will further lead to autonomous long-term matching.

7. Conclusions

Nowadays, system functionality is done digital, whereas AMS should perform a matching function between two media, while this AMS matching function itself is performed in an unmatched situation. This requires a shift in design approach, technology-driven, states-driven, redundancy-driven, and nature-driven matching, to enable crossing "boundaries" that are not fundamental boundaries but are consequences of a nonoptimal design approach. A high-level vision and accordingly a high-level design approach as discussed in this paper are necessary to be able to shift the frontiers in AMS design further.

We just face the very start of a new age of electronics. The changes we foresee will lead to an increased role for AMS in the future, even for system functionality implementation, combined with massive amounts of redundancy at various system levels, autonomous operation, self-organisation, and self-assembling. Like living cells.

Conflict of Interests

The author declares that there is no conflict of interests regarding the publication of this paper.

Acknowledgments

The author would like to acknowledge all staff members and partners from business the author cooperated with, for all the actual design work and for the examples mentioned above. Cooperating with these highly qualified persons in an open atmosphere gave the author the opportunity to build up the vision presented in this paper.

References

[1] C. E. Shannon, "A mathematical theory of communication," *The Bell System Technical Journal*, vol. 27, pp. 379–423, 623–656, 1948.

[2] R. H. Walden, "Analog-to-digital converter survey and analysis," *IEEE Journal on Selected Areas in Communications*, vol. 17, no. 4, pp. 539–550, 1999.

[3] A. H. M. van Roermund, "An integral Shannon-based view on smart front-ends," in *Proceedings of the European Conference on Wireless Technology (EuWiT '08)*, Amsterdam, The Netherlands, October 2008.

[4] H. M. van Roermund, P. Baltus, A. van Bezooijen et al., "Smart front-ends, from vision to design," *IEICE Transactions on Electronics*, vol. 92, no. 6, pp. 747–756, 2009.

[5] J. Essing, R. Mahmoudi, Y. Pei, and A. Van Roermund, "A fully integrated 60GHz distributed transformer power amplifier in bulky CMOS 45nm," in *Proceedings of the IEEE Radio Frequency Integrated Circuits Symposium (RFIC '11)*, pp. 1–4, Baltimore, Md, USA, June 2011.

[6] Y. Pei, R. Mahmoudi, J. Essing, and A. Van Roermund, "A 60GHz fully integrated power amplifier using a distributed ring transformer in CMOS 65nm," in *Proceedings of the 19th IEEE Symposium on Communications and Vehicular Technology in the Benelux (SCVT '12)*, pp. 1–4, Eindhoven, The Netherlands, November 2012.

[7] D. Raiteri, F. Torricelli, E. Cantatore, and A. H. M. van Roermund, "A tunable transconductor for analog amplification and filtering based on double-gate organic TFTs," in *Proceedings of the 37th European Solid-State Circuits Conference (ESSCIRC '11)*, pp. 415–418, Helsinki, Finland, September 2011.

[8] D. Raiteri, P. van Lieshout, A. van Roermund, and E. Cantatore, "Positive-feedback level shifter logic for large-area electronics," *IEEE Journal of Solid-State Circuits*, vol. 49, no. 2, pp. 524–535, 2014.

[9] D. Raiteri, A. H. M. van Roermund, and E. Cantatore, "A discrete-time amplifier based on Thin-Film Trans-Capacitors for sensor systems on foil," *Microelectronics Journal*, 2014.

[10] S. Abdinia, M. Benwadih, E. Cantatore et al., "Design of analog and digital building blocks in a fully printed complementary organic technology," in *Proceedings of the European Solid State Circuits Conference (ESSCIRC '12)*, pp. 145–148, Bordeaux, France, September 2012.

[11] S. Abdinia, F. Torricelli, G. Maiellaro et al., "Variation-based design of an AM demodulator in a printed complementary organic technology," *Organic Electronics*, vol. 15, no. 4, pp. 904–912, 2014.

[12] S. Abdinia, M. Benwadih, R. Coppard et al., "A 4b ADC manufactured in a fully-printed organic complementary technology including resistors," in *Proceedings of the 60th IEEE International Solid-State Circuits Conference (ISSCC '13)*, pp. 106–107, February 2013.

[13] S. Abdinia, T. H. Ke, M. Ameys et al., "Organic CMOS line drivers," *Journal of Display Technology*. In press.

[14] L. Tripodi, X. Hu, R. Götzen et al., "Broadband CMOS millimeter-wave frequency multiplier with vivaldi antenna in 3-D chip-scale packaging," *IEEE Transactions on Microwave Theory and Techniques*, vol. 60, no. 12, pp. 3761–3768, 2012.

[15] E. J. G. Janssen, D. Milosevic, P. G. M. Baltus, A. H. M. van Roermund, and H. Habibi, "Frequency-independent smart interference suppression for multi-standard transceivers," in *Proceedings of the 42nd European Microwave Integrated Circuits Conference (EuMIC '12)*, pp. 909–912, Amsterdam, The Netherlands, October 2012.

[16] E. J. G. Janssen, H. Habibi, D. Milosevic, P. G. M. Baltus, and A. H. M. van Roermund, "Smart self-interference suppression by exploiting a nonlinearity," in *Frequency References, Power Management for SoC, and Smart Wireless Interfaces: Advances in Analog Circuit Design*, A. Baschirotto, K. A. A. Makinwa, and P. J. A. Harpe, Eds., pp. 249–263, Springer, Dordrecht, The Netherlands, 2013.

[17] S. Ouzounov, E. Roza, J. A. Hegt, G. van der Weide, and A. H. M. Van Roermund, "Analysis and design of high-performance asynchronous sigma-delta modulators with a binary quantizer," *IEEE Journal of Solid-State Circuits*, vol. 41, no. 3, pp. 588–596, 2006.

[18] S. Ouzounov, H. Hegt, and A. van Roermund, "Sigma-delta modulators operating at a limit cycle," *IEEE Transactions on Circuits and Systems II: Express Briefs*, vol. 53, no. 5, pp. 399–403, 2006.

[19] A. H. M. van Roermund, F. A. Malekzadeh, M. Sarkeshi, and R. Mahmoudi, "Extended modelling for time-encoding converters," in *Proceedings of the IEEE International Symposium on Circuits and Systems (ISCAS '10)*, pp. 1077–1080, Paris, France, May 2010.

[20] F. A. Malekzadeh, R. Mahmoudi, M. Sarkeshi, and A. Roermund, "Fine tuning of switching frequency for minimal distortion in high frequency PWM systems," in *Proceedings of the*

IEEE MTT-S International Microwave Symposium (IMS '11), pp. 1–4, Baltimore Md, USA, June 2011.

[21] F. A. Malekzadeh, R. Mahmoudi, and A. H. M. van Roermund, "A new approach for nonlinear metric estimation of limit cycle amplifiers," in *European Microwave Conference (EuMC '09)*, pp. 1804–1807, Rome, Italy, September–October 2009.

[22] H. Chireix, "High power outphasing modulation," *Proceedings of the Institute of Radio Engineers (IRE)*, vol. 23, no. 11, pp. 1370–1392, 1935.

[23] R. van de Plassche and P. Baltus, "An 8b 100mhz folding Adc," in *Proceedings of the IEEE International Solid-State Circuits Conference, Digest of Technical Papers (ISSCC '88)*, San Francisco, Calif, USA, February 1988.

[24] W. H. Doherty, "A new high efficiency power amplifier for modulated waves," *Proceedings of the Institute of Radio Engineers*, vol. 24, no. 9, pp. 1163–1182, 1936.

[25] M. Sarkeshi, L. Ooi Bang, and A. H. M. van Roermund, "A novel Doherty amplifier for enhanced load modulation and higher bandwidth," in *Proceedings of the IEEE MTT-S International Microwave Symposium Digest*, Atlanta, Ga, USA, June 2008.

[26] L. Kahn, "Single-sideband transmission by envelope elimination and restoration," *Proceedings of the IRE-IEEE RFIC Virtual Journal*, vol. 40, no. 7, pp. 803–806, 1952.

[27] D. Raiteri, P. V. Lieshout, A. V. Roermund, and E. Cantatore, "An organic VCO-based ADC for quasi-static signals achieving 1LSB INL at 6b resolution," in *Proceedings of the 60th IEEE International Solid-State Circuits Conference (ISSCC '13)*, pp. 108–109, February 2013.

[28] M. Ding, P. Harpe, H. Hegt, K. Philips, H. de Groot, and A. van Roermund, "A 5bit 1GS/s 2.7 mW 0.05 mm^2 asynchronous digital slope ADC in 90 nm CMOS for IR UWB radio," in *Proceedings of the Radio Frequency Integrated Circuits Symposium (RFIC '12)*, pp. 487–490, Montreal , Canada, June 2012.

[29] R. van de Plassche, *CMOS Integrated Aanlog-to-Digital and Digital-to-Analog Converters*, Kluwer Academic Publishers, 2nd edition, 2003.

[30] F. Maloberti, *Data Converters*, Springer, 2007.

[31] P. Harpe, B. Busze, K. Philips, and H. de Groot, "A 0.47–1.6mW 5bit 0.51GS/s time-interleaved SAR ADC for low-power UWB radios," ESSCIRC, 2011.

[32] P. J. A. Harpe, C. Zhou, K. Philips, and H. de Groot, "A 0.8-mW 5-bit 250-MS/s time-interleaved asynchronous digital slope ADC," *IEEE Journal of Solid-State Circuits*, vol. 46, no. 11, pp. 2450–2457, 2011.

[33] P. Harpe, E. Cantatore, and A. Van Roermund, "A 2.2/2.7fJ/conversion-step 10/12b 40kS/s SAR ADC with data-driven noise reduction," in *Proceedings of the 60th IEEE International Solid-State Circuits Conference (ISSCC '13)*, pp. 270–271, San Francisco, Calif, USA, February 2013.

[34] P. Harpe, E. Cantatore, and A. van Roermund, "A 10b/12b 40 kS/s SAR ADC with data-driven noise reduction achieving up to 10.1b ENOB at 2.2 fJ/conversion-step," *IEEE Journal of Solid-State Circuits*, vol. 48, no. 12, pp. 3011–3018, 2013.

[35] P. Harpe, E. Cantatore, and A. van Roermund, "An oversampled 12/14b SAR ADC with noise reduction and linearity enhancements achieving up to 79.1dB SNDR," in *Proceedings of the 61st IEEE International Solid-State Circuits Conference (ISSCC '14)*, pp. 194–195, San Francisco, Calif, USA, February 2014.

[36] A. V. Roermund, H. Hegt, P. Harpe et al., "Smart AD and DA converters," in *Proceedings of the IEEE International Symposium on Circuits and Systems (ISCAS '05)*, vol. 4, pp. 4062–4065, May 2005.

[37] A. H. M. van Roermund, "Smart, flexible, and future-proof data converters," in *Proceedings of the European Conference on Circuit Theory and Design (ECCTD '07)*, pp. 308–319, Sevilla, Spain, August 2007.

[38] G. I. Radulov, P. J. Quinn, P. C. W. Van Beek, J. A. Hegt, and A. H. M. Van Roermund, "A binary-to-thermometer decoder with built-in redundancy for improved DAC yield," in *Proceedings of the IEEE International Symposium on Circuits and Systems (ISCAS '06)*, pp. 1414–1417, May 2006.

[39] G. I. Radulov, P. J. Quinn, H. Hegt, and A. Van Roermund, "A flexible 12-bit self-calibrated quad-core current-steering DAC," in *Proceedings of the IEEE Asia Pacific Conference on Circuits and Systems*, pp. 25–28, Macau, China, December 2008.

[40] G. I. Radulov, P. J. Quinn, J. A. Hegt, and A. H. M. van Roermund, *Smart and Flexible Digital-to-Analog Converters*, Analog Circuits and Signal Processing Series, Springer, Dordrecht, The Netherlands, 2011.

[41] G. I. Radulov, P. J. Quinn, and A. H. M. van Roermund, "A 28-nm CMOS 1 V 3.5 GS/s 6-bit DAC with signal-independent delta-i noise DfT scheme," *IEEE Transactions on Very Large Scale Integration (VLSI) Systems*, 2014.

[42] . Tang Yongjian, J. Briaire, K. Doris et al., "A 14b 200MS/s DAC with SFDR>78dBc, IM3<-83dBc and NSD<-163dBm/Hz across the whole Nyquist band enabled by dynamic-mismatch mapping," in *In Proceeding of IEEE Symposium on VLSI Circuits*, pp. 151–152, Honolulu, Hawaii, USA, June 2010.

[43] Y. Tang, J. Briaire, K. Doris et al., "A 14 bit 200 MS/s DAC with SFDR >78 dBc, IM3 < −83 dBc and NSD < −163 dBm/Hz across the whole Nyquist band enabled by dynamic-mismatch mapping," *IEEE Journal of Solid-State Circuits*, vol. 46, no. 6, pp. 1371–1381, 2011.

[44] Y. Tang, H. Hegt, and A. van Roermund, *Dynamic-Mismatch Mapping for Digitally-Assisted DACs*, vol. 92 of *Analog Circuits and Signal Processing*, Springer, Dordrecht, The Netherlands, 2013.

[45] Y. Lin, K. Doris, H. Hegt, and A. van Roermund, "An 11b pipeline ADC with dual sampling technique for converting multi-carrier signals," in *Proceedings of the IEEE International Symposium of Circuits and Systems (ISCAS '11)*, pp. 257–260, Rio de Janeiro, Brasil, May 2011.

[46] Y. Lin, K. Doris, E. Janssen et al., "An 11b 1GS/s ADC with parallel sampling architecture to enhance SNDR for multi-carrier signals," in *Proceedings of the 39th European Solid-State Circuits Conference (ESSCIRC '13)*, pp. 121–124, Bucharest, Romania, September 2013.

[47] Y. Lin, A. Zanikopoulos, K. Doris, J. A. Hegt, and A. H. M. van Roermund, "A power-optimized high-speed and hig-resolution pipeline ADC with a parallel sampling first stage for broadband multi-carrier systems," in *Design, Modeling and Testing Data Convertors*, P. Carbone, S. Kiaei, and F. Xu, Eds., pp. 3–28, Springer, Berlin, Germany, 2014.

[48] A. van Bezooijen, R. Mahmoudi, and A. H. M. van Roermund, "Adaptive methods to preserve power amplifier linearity under antenna mismatch conditions," *IEEE Transactions on Circuits and Systems I: Regular Papers*, vol. 52, no. 10, pp. 2101–2108, 2005.

[49] A. van Bezooijen, M. A. de Jongh, C. Chanlo et al., "A GSM/EDGE/WCDMA adaptive series-LC matching network using RF-MEMS switches," *IEEE Journal of Solid-State Circuits*, vol. 43, no. 10, pp. 2259–2268, 2008.

[50] A. Van Bezooijen, M. A. de Jongh, F. van Straten, R. Mahmoudi, and A. H. M. van Roermund, "Adaptive impedance-matching techniques for controlling L networks," *IEEE Transactions on*

Circuits and Systems I: Regular Papers, vol. 57, no. 2, pp. 495–505, 2010.

[51] A. van Bezooijen, R. Mahmoudi, and A. H. M. van Roermund, "Adaptively controlled RF-MEMS antenna tuners for hand-held applications," in *Proceedings of the IMS/RFIC Symposium*, p. WMK (IMS)-4-1/24, Anaheim, Calif, USA, May 2010.

[52] A. van Bezooijen, R. Mahmoudi, and A. H. M. van Roermund, *Adaptive RF Front-Ends for Hand-Held Applications*, Analog Circuits and Signal Processing, Springer, Dordrecht, The Netherlands, 2011.

[53] S. Song, M. J. Rooijakkers, C. Rabotti, M. Mischi, A. H. M. van Roermund, and E. Cantatore, "A low-power noise scalable instrumentation amplifier for fetal monitoring applications," in *Proceedings of the IEEE International Symposium on Circuits and Systems (ISCAS '13)*, pp. 1926–1929, Beijing, China, May 2013.

[54] S. Song, M. J. Rooijakkers, M. Mischi, C. Rabotti, A. H. M. van Roermund, and E. Cantatore, "Analysis and architecture design of a novel home-based fetal monitoring system," in *Proceedings of the ICT.OPEN*, Veldhoven, The Netherlands, November 2011.

6

Blood Glucose Measurement Using Bioimpedance Technique

D. K. Kamat,[1,2] **Dhanashri Bagul,**[2] **and P. M. Patil**[3]

[1]*SCOE, Pune 411041, India*
[2]*Department of E & TC, Sinhgad Academy of Engineering, Pune 411048, India*
[3]*KJ's Educational Institutes, Pune 411048, India*

Correspondence should be addressed to D. K. Kamat; dkkamat@gmail.com and Dhanashri Bagul; dhanashribagul@gmail.com

Academic Editor: Meiyong Liao

Bioimpedance measurement is gaining importance in wide field of bioresearch and biomedical systems due to its noninvasive nature. Noninvasive measurement method is very important to decrease infection and physical injuries which result due to invasive measurement. This paper presents basic principle of bioimpedance along with its application for blood glucose analysis and effect of frequency on impedance measurement. Input from bioimpedance sensor is given to amplifier and signal conditioner AD5933. AD5933 is then interfaced with microcontroller LPC1768 using I2C bus for displaying reading on LCD. Results can also be stored in database using UART interface of LPC1768.

1. Introduction

Impedance of any material can be defined as the opposition offered by material to the electric current flowing through it. It can be formulated as the frequency domain ratio of voltage to current. Impedance can also be represented using resistance and reactance. Every material shows property to dissipate energy and to store energy. Reactance (Xc) indicates energy storage in material whereas resistance (R) is indicator of energy dissipation [1].

When electricity is passed through body, two types of resistances that is capacitive R (reactive) and resistive R (resistance) are offered by body, where capacitance arises due to cellular membrane and resistance arises due to body water (intracellular or extracellular water). Cell membrane consists of a layer of nonconductive lipid material sandwiched between two layers of conductive protein molecules. High reactance value indicates good health and cell membrane integrity. Cell membrane structure makes them behave as capacitors when alternating current is applied to it. Hence impedance of tissue varies with frequency. At high frequency, current can flow through both intra- and extracellular water which means that it can penetrate the cellular membrane while at low frequency current cannot penetrate cellular membrane so it flows only through extracellular fluid. As

a result, at low frequency, impedance is resistive in nature and at high frequency it has a resistive as well as a reactive component [1] (see Figure 1).

The incidence of diabetes is increasing worldwide every year [2]. Therefore, it is important to control as well as to treat diabetes. There are various invasive and noninvasive methods available for blood glucose measurement. Glucometer which depends on radio wave transmission uses continuously transmitting and receiving antenna. The transmitting antenna sends a signal of frequency in a range from 5 GHz to 12 GHz while receiving antenna monitors signal attenuation to determine the blood sugar level. The main drawback of radio wave transmission is the requirement of high frequencies which helps to minimize influence of the skin and to improve the accuracy of measurement results [3].

Another glucometer based on photoplethysmography method uses principle of infrared absorption measurement. This method uses the concept that the blood with increased sugar level has higher absorption rate of infrared radiation than human skin. Requirement of additional sensor for detecting heart rhythm is the main drawback of such measurement methods [4]. The change in glucose level can be detected using electrode sensor by measuring changes in conductivity and permittivity of measuring component [5]. But the main drawback of this type of measurement

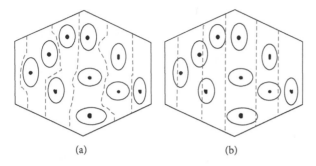

FIGURE 1: Current flow through body: (a) low frequency and (b) high frequency.

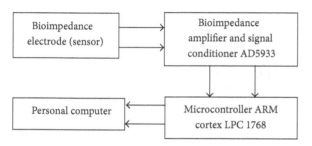

FIGURE 2: Block diagram of system.

FIGURE 3: LPC1768 ARM Cortex board.

system is the design of interdigital electrode sensor which is complicated and very expensive; also the sensor impedance depends on frequency so it is not convenient to use this parameter for blood glucose level estimation. In this paper, we have discussed blood glucose measurement system which is more accurate and less expensive as it is using AgCl electrodes [2].

2. Methodology

The measurement environment for continuous and non-invasive monitoring of the impedance for blood glucose measurement has been developed. The block diagram of the proposed measurement system, shown in Figure 2, consists of the bioimpedance electrodes, integrated circuit AD5933, microcontroller LPC1768, and a personal computer. Each block of system is discussed in detail in the following sections.

The integrated circuit AD5933 is the core of the proposed measurement system (Figure 4). Bioimpedance electrodes are used for measurement of the impedance. Impedance is calculated by the microcontroller LPC1768 through I2C interface. The microcontroller sends this measured data to a personal computer, where we can store data, using a serial interface UART. The microcontroller also provides an initial configuration of the integrated circuit AD5933 which includes mainly setting the frequency and amplitude of the input signal used for measurement of unknown impedance. The ARM Cortex LPC1768 microcontroller also controls time slots during which the measurements are performed. When microcontroller is done with measurement in respective time slot, it reads the data from AD5933 by using I2C interface and

sends measured data to PC by using UART interface where data is stored and further processed.

2.1. Microcontroller Cortex LPC1768 M3. The LPC1768 is an ARM Cortex-M3 32 bit microcontroller which is designed for embedded applications which require a high level of integration as well as low power dissipation. Here, UART is used for downloading the program and for PC interface. The communication between AD5933 and controller is through I2C bus. LPC1768 works at a maximum operating frequency of 100 MHz. Figure 3 shows hardware of LPC 1768. The ARM Cortex-M3 CPU has a 3-stage pipeline and uses Harvard architecture with separate local instruction and data buses as well as a third bus for peripherals. The I2C interfaces of LPC1768 I2C are byte oriented and have four operating modes: master transmitter mode, master receiver mode, slave transmitter mode, and slave receiver mode [6].

2.2. Bioimpedance Amplifier. Bioimpedance amplifier is heart of impedance measurement system. Impedance converter and network analyzer AD5933 functions as signal conditioner for bioimpedance signal. Low noise voltage reference IC AD820 and low power amplifier IC ADR 423 act as supporting blocks for signal amplification and noise reduction. Figure 5 shows hardware for bioimpedance amplifier and signal conditioner AD5933. A precision, low power FET input op amp AD 820 which can operate from a single supply of 5 V to 36 V, or dual supplies of ±2.5 V to ±18 V. In the AD820, N-channel JFETs are mainly used for providing a low offset, low noise, high impedance input stage which is required by most of the embedded applications. It also keeps low noise performance to low frequencies. Low noise performance, low input current, and current noise are features of the AD820 which contributes negligible noise for applications [7, 8].

Integrated circuit AD5933 consists of the various blocks such as an input signal generator, a 12-bit A/D converter, a DFT (discrete Fourier transform) circuit, a thermal sensor, and I2C interface. The function of generator is to supply a sine wave input signal of certain frequency and amplitude at the output VOUT. Unknown impedance to be calculated is connected between VOUT and VIN terminals. Therefore, magnitude and phase of the current flowing through a load depend on its impedance. The current is then transformed

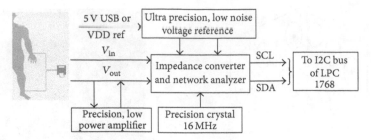

FIGURE 4: Block diagram of bioimpedance amplifier.

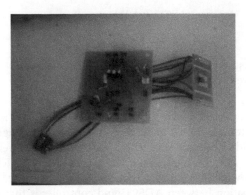

FIGURE 5: PCB of the realized monitoring system bioimpedance amplifier with signal conditioning circuit.

to voltage that is converted into a digital signal by the digital to analog converter. The DFT circuit provides discrete Fourier transform of the converted impedance signal. This will lead to measurement of real and imaginary parts which are measured. Functional block diagram of AD5933 is shown in Figure 6 [9].

The obtained response signal from the impedance is then sampled by the on-board discrete Fourier transform (DFT) and ADC. This operation returns a real (R) and imaginary (I) data-word at each output frequency. Impedance magnitude and phase are then easily calculated using the following equations:

$$\text{Magnitude} = \sqrt{R^2 + I^2} \qquad (1)$$

$$\text{Phase} = \tan^{-1}\frac{I}{R}. \qquad (2)$$

Once calibration is done, the magnitude of the impedance and relative phase of the impedance at each frequency point along the sweep can be easily calculated. This is done off chip using content of real and imaginary register, which can be read from the serial I2C interface [9].

There are two stages of operation of AD5933, namely, transmit stage and receive stage. The excitation signal required for transmit stage is given by DDS technique. AD5933 has an in built 27-bit accumulator DDS core in the transmit stage. At a particular frequency this DDS core provides on-chip output excitation signal. Input current signal is provided from unknown impedance to receive stage. The current to voltage amplifier, a programmable gain amplifier

(PGA), antialiasing filter, and ADC are main constituents of receiving stage of AD5933. This receive stage obtains input current signal from the impedance which is unknown then performs signal processing followed by digitization of the result. An external reference clock or internal oscillator provides clock for DDS [9].

The DFT operation is as follows.

A DFT is estimated for each frequency point in the sweep. The AD5933 DFT algorithm is expressed using following equation:

$$X(f) = \sum_{n=0}^{1023} \left(x(n) \left(\cos(n) - j \sin(n) \right) \right), \qquad (3)$$

where $X(f)$ is the power in the signal at the frequency point f. $X(n)$ is the ADC output. $\cos(n)$ and $\sin(n)$ are the sampled test vectors provided by DDS core at the frequency point f [9].

3. Impedance Measurement

The readings for impedance measurement are taken in the measurement frequency range of 10 kHz to 100 kHz. Electrical contact with the body was made using silver electrodes. In order to increase the accuracy and to minimize noise in measurement, high precision impedance converter system, AD5933, is used. Then the body part, which is connected between input and output ports of electrodes, is excited with different frequencies. The current which is the response from the body is then converted into voltage using a transimpedance amplifier. The output voltage of this transimpedance amplifier is then sampled and processed by a DSP engine of AD5933 at each frequency of excitation. Both the real and imaginary components were stored in two 16-bit registers of AD5933. The stored data in each of these registers must be read after each ADC conversion to get impedance reading. These results which contain real and imaginary parts are then read by the microcontroller using I2C and are further processed and displayed on a personal computer [1].

4. Result

Normal glucose level for healthy human being is 4.4 to 6.1 mmol/L. To study low as well as high blood glucose level, graph is plotted in the range of 4 to 6.8 mmol/L. The dependency of impedance on blood glucose level is initially

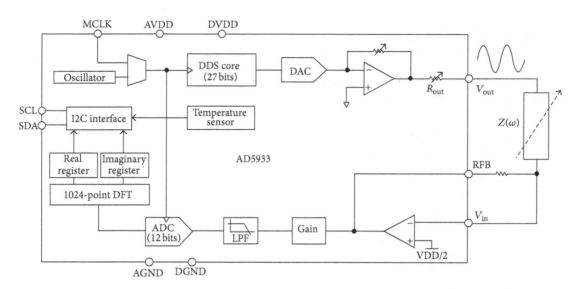

FIGURE 6: Functional block diagram of AD5933.

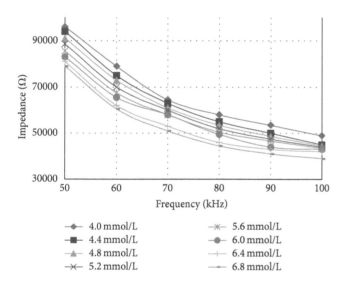

FIGURE 7: Impedance dependence on frequency and blood glucose level in a range from 50 kHz to 100 kHz.

determined using readings obtained from ACCU-CHEK glucometer that is invasive measurement of glucose level. The results obtained from blood glucose measurement system are compared with readings obtained from invasive glucometer measurements. Measurements of blood impedance at various frequencies starting from low frequency to high frequency are carried out. Impedance measurement of several subjects at various frequencies is done to study possibility of change of blood glucose level.

Impedance measurement of eight subjects is carried out at different frequencies. The results of impedance at different blood glucose levels are presented in Figure 7. It can be seen that impedance decreases when blood glucose level increases and impedance module decreases when blood glucose level decreases. Frequency is limited to the range of 10 kHz to

100 because the highest scattering of points of measurement results is observed in this range of frequencies. So frequencies outside this range are not used in further measurement.

5. Conclusion

Study of impedance blood glucose level results and its graphical representation shows that noninvasive method can be used for accurate measurement. For accurate measurement, a system is designed using AD5933. Use of noninvasive method for blood glucose level estimation eliminates continual finger pricking and risk of infection. It is determined from the results that impedance depends on blood glucose level. The achieved results have proven that the accuracy of results, input signal voltage, and frequency ranges are suitable for biomedical monitoring applications.

Additionally, the proposed measurement system is versatile, flexible, and easy to be used for different measurement approaches like heartbeat, ECG, blood pressure, skin impedance for cancer detection, and so forth. This developed measurement system will offer new approaches and opportunities to noninvasive biomedical systems in the field of medicine and other useful areas.

Conflict of Interests

The authors declare that there is no conflict of interests regarding the publication of this paper.

References

[1] E. R. Jasmine Rose, D. Pamela, and K. Rajasekaran, "Apple vitality detection by impedance measurement," *International Journal of Advanced Research in Computer Science and Software Engineering*, vol. 9, pp. 144–148, 2013.

[2] V. Pockevicius, V. Markevicius, M. Cepenas, D. Andriukaitis, and D. Navikas, "Blood glucose level estimation using interdigital electrodes," *Elektronika IR Elektrotechnika*, vol. 19, no. 6, pp. 71–74, 2013.

[3] M. Hofmann, T. Fersch, R. Weigel, G. Fischer, and D. Kissinger, "A novel approach to non-invasive blood glucose measurement based on RF transmission," in *Proceedings of the IEEE International Symposium on Medical Measurements and Applications (MeMeA '11)*, pp. 39–42, Institute of Electrical and Electronics Engineers, University of Erlangen-Nuremberg, Erlangen, Germany, May 2011.

[4] P. Brince, M. Melvin, and C. A. Zachariah, "Design and development of non-invasive glucose measurement system," in *Proceedings of the 1st International Symposium on Physics and Technology of Sensors (ISPTS '12)*, pp. 43–46, Pune , India, 2012.

[5] S. Abdalla, S. S. Al-ameer, and S. H. Al-Magaishi, "Electrical properties with relaxation through human blood," *Biomicrofluidics*, vol. 4, no. 3, Article ID 034101, 2010.

[6] Datasheet, "Analog Devices," http://www.analog.com/static/imported-files/data_sheets/LPC1768.pdf.

[7] Datasheet, "Analog Devices," http://www.analog.com/static/imported-files/data_sheets/AD820.pdf.

[8] Datasheet, "Analog Devices," http://www.analog.com/static/imported-files/data_sheets/ADR420_421_423_425.pdf.

[9] Datasheet, "Analog Devices," http://www.analog.com/en/rfif-components/direct-digital-synthesis-dds/ad5933/products/product.html.

Ultra-Low-Voltage Low-Power Bulk-Driven Quasi-Floating-Gate Operational Transconductance Amplifier

Ziad Alsibai and Salma Bay Abo Dabbous

Department of Microelectronics, Brno University of Technology, 61600 Brno, Czech Republic

Correspondence should be addressed to Ziad Alsibai; xalsib00@stud.feec.vutbr.cz

Academic Editor: Bo K. Choi

A new ultra-low-voltage (LV) low-power (LP) bulk-driven quasi-floating-gate (BD-QFG) operational transconductance amplifier (OTA) is presented in this paper. The proposed circuit is designed using 0.18 μm CMOS technology. A supply voltage of ±0.3 V and a quiescent bias current of 5 μA are used. The PSpice simulation result shows that the power consumption of the proposed BD-QFG OTA is 13.4 μW. Thus, the circuit is suitable for low-power applications. In order to confirm that the proposed BD-QFG OTA can be used in analog signal processing, a BD-QFG OTA-based diodeless precision rectifier is designed as an example application. This rectifier employs only two BD-QFG OTAs and consumes only 26.8 μW.

1. Introduction

In the late sixties, the Radio Corporation of America (RCA) and then General Electric (GE) came out with the operational transconductance amplifier, hereafter called OTA. The name means essentially a controllable resistance amplifier. OTA is a key functional block used in many analog and mixed-mode circuits. It is a special case of an ideal active element, and its implementation in IC form makes it indispensable today in discrete and fully integrated analog network design. The ideal OTA as shown in Figure 1 can be considered as a differential voltage-controlled current source (DVCCS); its transconductance "g_m" represents the ratio of the output current to the differential input voltage, that is, $I_{out}/(V_1 - V_2)$. This transconductance is used as a design parameter and it is usually adjustable by the amplifier bias current (I_{bias}). The benefit of this adjusting possibility is acquiring the ability of electronic orthogonal tunability to circuit parameters. It could be noted that tunability has a main role in integrated circuits, especially to satisfy a variety of design specifications. Thus, OTA has been implemented widely in CMOS and bipolar and also in BiCMOS and GaAs technologies [1].

The OTA is similar to the standard operation amplifier (OPA) in the sense of infinite input impedances, but its output impedance is much higher and that makes OTA more

desirable than any ordinary amplifier. Recently, the multiple-output-OTA (MO-OTA) has been introduced and used, on par with the ordinary operation amplifier, as a basic block in many applications, particularly for realizing universal filters which are able to implement several second-order transfer functions with a minimum of adjustments. The literature provides numerous examples of OTA-based biquad structures, as well as active elements such as current conveyor (CC), current differencing transconductance amplifier (CDTA), current-through transconductance amplifier (CTTA), and current-conveyor transconductance amplifier (CCTA).

The symbol and the equivalent circuit of ideal OTA are shown in Figures 1(a) and 1(b), respectively [1]. Simple applications of the OTA include voltage amplification, voltage-variable resistor (VVR), voltage summation, integration, gyrator realization, practical OTAs, current conveyor, and active RC filters. In addition, one of OTA's principal uses is in implementing electronically controlled applications such as variable frequency oscillators and variable gain amplifier stages which are more difficult to implement with standard OPAs.

Recently, Low power analog circuit design is undergoing a very considerable boom. However, circuit designers encounter difficulties to preserve reliable performance of the analog circuits with scaling down their supply voltage,

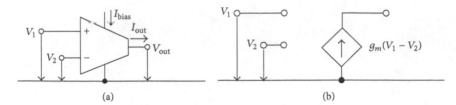

FIGURE 1: Ideal operational transconductance amplifier, (a) symbol and (b) equivalent circuit.

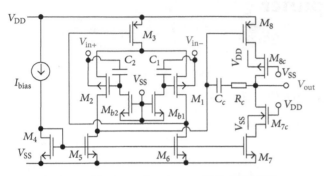

FIGURE 2: The internal structure of BD-QFG OTA.

owing to the fact that the threshold voltage and supply voltage are not decreased proportionally. Hence, various techniques based on CMOS technology have emerged to overcome the rather high threshold voltage problem of MOS transistors, such as unconventional MOS techniques, that is, floating-gate (FG), quasi-floating-gate (QFG), bulk-driven (BD), bulk-driven floating-gate (BD-FG), and bulk-driven quasi-floating-gate (BD-QFG) MOS transistor [2, 3]. Utilizing these techniques offers circuit simplicity, high functionality, extended input voltage range, and ultra-LV LP operation capability. Thus, they are very suitable for ultra-LV LP applications as battery-powered implantable and wearable medical devices.

In this paper, The BD-QFG technique has been chosen to be utilized to build LP LV OTA, since it enjoys higher transconductance value, higher bandwidth, and smaller input referred noise in comparison with other unconventional techniques. To verify the functionality of the proposed OTA voltage mode diodeless precision rectifier is introduced in this work.

The organization of this paper is as follows: in Section 2 the CMOS internal structure of the BD-QFG OTA is described. In Section 3 the principle of BD-QFG OTA-based diodeless precision rectifier is presented. The simulation results are provided in Section 4; eventually, Section 5 is the conclusions.

2. Bulk-Driven Quasi-Floating-Gate Operational Transconductance Amplifier (BD-QFG OTA)

The circuit, which is shown in Figure 2, consisted of two stages. The first stage consists of BD-QFG differential input M_1 and M_2. The gates of these transistors are tied to the

negative supply voltage V_{SS} through extremely high value resistors constructed by transistors M_{b1} and M_{b2} which are operating in cutoff region. The input terminals are connected to M_1 and M_2 from two sides: capacitively coupled to the quasi-floating-gates via C_1 and C_2 from one side and connected to bulk terminals from the other side.

Transistors M_4, M_5, M_6, and M_7 act as a multiple output current mirror applying the constant current source I_{bias} to each branch of the circuit. Transistors M_5 and M_6 form the active load and transistor M_3 acts as tail current source for the differential input stage. The input voltage terminals are connected to the bulk terminals of M_1 and M_2; therefore, high input impedance is achieved.

The use of bulk-driven quasi-floating-gate flipped voltage follower for the differential input stage makes the minimum power supply voltage $V_{DD(min)}$. The supply voltage is given by [4]

$$V_{DD(min)} = V_{GS(M_3)} + V_{DS(M_6)}. \tag{1}$$

Equation (1) shows the capability of the proposed BD-QFG OTA structure for operation under lower supply voltage.

The second stage consists of M_7, M_{7c} and M_8, M_{8c}. Cascode structure is used to implement the gain stage in order to provide significantly high-value output impedance, consequently to achieve high voltage gain. Output impedance can be calculated from the following equation:

$$
\begin{aligned}
r_o = 1 \\
\times \left(\left(\left(g_{o,M7} g_{o,M7c} \right) / \left(g_{m,M7c} + g_{mb,M7c} \right) \right) \right. \\
\left. + \left(\left(g_{o,M8} g_{o,M8c} \right) / \left(g_{m,M8c} + g_{mb,M8c} \right) \right) \right)^{-1}.
\end{aligned}
\tag{2}
$$

3. BD-QFG OTA-Based Diodeless Precision Rectifier

A precision rectifier is one of important nonlinear circuits, which is extensively used in analog signal processing systems. In precision rectification, a bidirectional signal is converted to one-directional signal. Typically, a conventional rectifier could be realized by using diodes for its rectification, but diode cannot rectify signals whose amplitudes are less than the threshold voltage (approximately 0.7 V for silicon diode and approximately 0.3 V for germanium diode). As a result, diode-only rectifiers are used in only those applications in which the precision in the range of threshold voltage is insignificant, such as RF demodulators and DC voltage supply

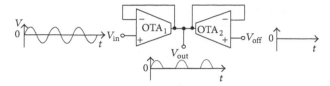

FIGURE 3: BD-QFG half-wave rectifier.

rectifiers, but for applications requiring accuracy in the range of threshold voltage the diode-only rectifier cannot be used. This can be overcome by using integrated circuit rectifiers instead.

Traditional methods of realizing precision rectifier circuits include the use of operational amplifiers, resistors, and either diodes [5–9] or alternating source-followers [10]. A number of current conveyors-based current-mode rectifier circuits were introduced in the literature [7, 11–16]. The rectifier circuits in [11–13] employ diodes and resistors in addition to second generation current conveyors (CCIIs). The circuit proposed in [7] employs bipolar current mirrors in addition to a CCII and a number of resistors. The rectifier circuit in [14] employs four current controlled conveyors (CCCIIs) and resistors. However, the use of resistor makes these circuits not ideal for integration. Therefore, precision rectifiers by using all-MOS transistors are proposed [17–26]. Authors in [27] proposed a circuit which employs two CCIIs and two MOS transistors. Authors in [28] presented a circuit which employs an amplifier and a simple voltage comparator. Author in [27] introduced a rectifier which employs two differential difference current conveyors (DDCCs). A new technique for realizing a precision half-wave voltage rectifier in CMOS technology is proposed; this technique is based on bulk-driven quasi-floating-gate operational transconductance amplifier (BD-QFG OTA).

Diodeless half wave rectifier based on bulk-driven quasi-floating-gate OTA is shown in Figure 3. This circuit is a WTA-like (winner-take-all) circuit. The principle of work is as follows: if we applied a voltage signal to V_{in} terminal and a zero to V_{off} terminal, the output voltage would equal the maximum voltage of both inputs. In other words, positive half of the signal wave is passed, while the other half is blocked. For an input voltage V_{in} the ideal half-wave rectified output V_{out} is given by

$$V_{out} = \begin{cases} V_{in} & \text{if } V_{in} > 0 \\ 0 & \text{otherwise.} \end{cases} \quad (3)$$

It is worth mentioning that the same configuration shown in Figure 3 could be used as full-wave rectifier just by applying $\overline{V_{in}}$ (an identical signal of V_{in} shifted 180°) to V_{off} terminal.

4. Simulation Results

The proposed BD-QFG OTA was designed and simulated using TSMC 0.18 μm N-well CMOS. The used PSpice model is available on [29]. The supply voltage was ±0.3 V, the biasing current was $I_{bias} = 5 \mu$A, and the power consumption was 13.4 μW. The optimal transistor aspect ratios and the values

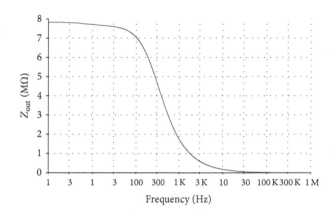

FIGURE 4: Output impedance versus frequency.

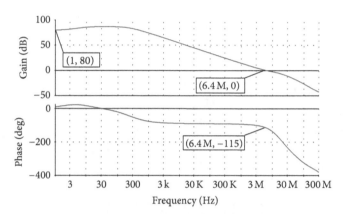

FIGURE 5: Frequency response of BD-QFG OTA.

TABLE 1: Transistors aspect ratios for Figure 2.

BD-QFG OTA	W/L [μm/μm]
M_1, M_2	20/0.3
M_{b1}, M_{b2}	30/2
M_3, M_8	100/0.3
M_4, M_5, M_6	4/0.3
M_7	8/0.3
M_{7c}	45/2
M_{8c}	100/2
$C_1 = C_2 = 0.4$ pF	
$R_c = 10$ kΩ, $C_c = 3$ pF	

of components are given in Table 1. Table 2 shows a list of measured operational amplifier benchmarks used to evaluate proposed OTA. Features of the circuit (shown in Figure 2) are listed in the first column, along with values of other works listed in other columns.

Figure 4 shows the simulated magnitude of output impedance of OTA. Z_{out} is high as expected; its value is 7.83 MΩ. The AC gain and phase responses of the BD-QFG OTA with 3 pF load capacitance are shown in Figure 5. The open-loop gain is 80 dB and the gain-bandwidth product is 6.4 MHz. The phase margin is 65° which guarantees the circuit stability. The voltage follower frequency response of the proposed

TABLE 2: BD-QFG OTA performance benchmark indicators.

Parameters	This work	Kozlel and Szczepanski [33]	Majumdar [34]	Li and Raut [35]	Zhang et al. [36]
CMOS technology [μm]	0.18	0.5	0.35	0.35	0.18
Power supply [V]	±0.3	±2.5	3.3	3.3	1.8
Power consumption [μW]	13.4	6800	3370	2330	590
Transistors number [/]	12	37	14	22	68
AC gain [dB]	51	65	80.4	65	55
Linearity range [mV]	±0.3	±0.75	±0.1	/	±0.27
Output resistance [MΩ]	0.317	3.4	/	1.2	8.3
−3 dB bandwidth [MHz]	59	100	123.2	100	200

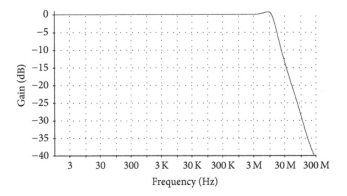

FIGURE 6: Frequency response of BD-QFG OTA as a voltage follower.

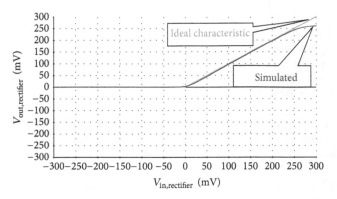

FIGURE 8: DC transfer characteristic of BD-QFG half-wave rectifier.

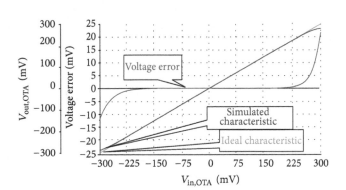

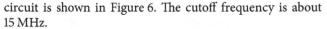

FIGURE 7: DC transfer characteristic and voltage error of the BD-QFG OTA.

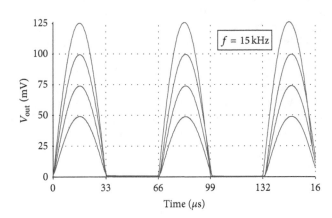

FIGURE 9: Transient analyses of output waveforms with 15 kHz and various amplitudes of the input signal.

circuit is shown in Figure 6. The cutoff frequency is about 15 MHz.

Figure 7 presents the DC transfer characteristic of the BD-QFG OTA. For input voltage range from −266 to 266 mV the voltage error is below 4 mV. Therefore, using $V_{off} = 0$ V and maximum input amplitude $V_{m \cdot max}$ of ±250 mV, the OTA is not expected to have strong impact on the overall rectifier accuracy.

The diodeless half-wave precision rectifier shown in Figure 3 was simulated using BD-QFG OTA shown in Figure 2. The supply voltage of ±0.3 V and the bias current of $I_{bias} = 5\ \mu$A for OTAs were used. The circuit consumes 26.8 μW. Figure 8 shows the DC transfer characteristic of BD-QFG half-wave rectifier in comparison with the ideal one and it confirms the precise rectification for input amplitude ranging ±250 mV. Figure 9 shows the transient response of

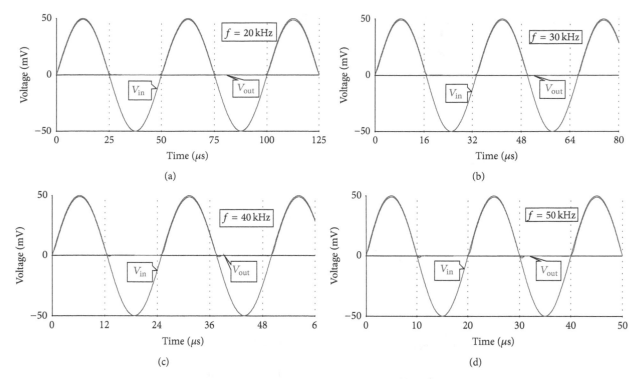

FIGURE 10: Transient analyses of input and output waveforms with $V_m = 50$ mV and (a) 20, (b) 30, (c) 40, and (d) 50 kHz.

the output waveforms for input signal of 15 kHz and amplitudes from 50 mV to 125 mV with step of 25 mV. Hence, the rectifier is capable of rectifying a wide range of amplitudes. Figure 10 shows the transient responses of the input and output waveforms with amplitude of 100 mV and frequency of 20 kHz, 30 kHz, 40 kHz, and 50 kHz. The load capacitor C_{load} for simulations done in Figures 9 and 10 was set to 5 pF.

To demonstrate the temperature performance of proposed circuit, the proposed circuit was simulated at the frequency of 10 kHz by changing temperature. Figure 11 shows the output waveforms of the proposed rectifier at temperatures of 0°C, 27°C, and 100°C. From Figure 11, it can be seen that the proposed circuit provides excellent temperature stability without any compensation technique.

To evaluate the quality of the rectification process as a function of the amplitude and the frequency of the input signal, two types of characteristics are proposed [29]. The first type is P_{AVR} (AVR: average value ratio) which is the ratio of the average value of the rectified output signal v_{out} and the average value of the sinusoidal input signal after its ideal half-wave rectification:

$$P_{AVR} = \frac{(1/T) \int_0^{T/2} v_{(out)} dt}{(1/\pi) V_m}, \tag{4}$$

where T and V_m are the period and amplitude of the sinusoidal input signal, respectively. The ideal operation of the rectifier is then characterized by the value $P_{AVR} = 1$. With increasing the frequency and decreasing the amplitude of the input signal, the deviation from the ideal operation is indicated by a change, mostly a decrease in P_{AVR} below one.

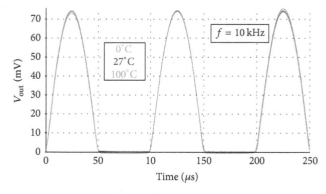

FIGURE 11: Outputs waveforms at different temperatures.

The second type of characteristic is defined more rigorously as a ratio of two root mean square "RMS" values, the RMS of the difference of the real and ideal output signals, v_{out} and v_{ideal}, and the RMS value of the ideal signal:

$$P_{RMSE} = \frac{\sqrt{(1/T) \int_0^{T/2} [v_{out}(t) - v_{ideal}(t)]^2 dt}}{(1/2) V_m}. \tag{5}$$

Here, the suffix RMSE is an abbreviation of the term "root mean square error." For ideal circuit operation, that is, $v_{out}(t) = v_{ideal}(t)$, the result is $P_{RMSE} = 0$, while in the case of total attenuation of the output signal $P_{RMSE} = 1$. For extra high distortions, when the mutual energy of signals v_{out} and v_{ideal} can be negative, one can obtain $P_{RMSE} > 1$. Figure 12 shows the P_{AVR} (a) and P_{RMSE} (b) versus frequency in range of

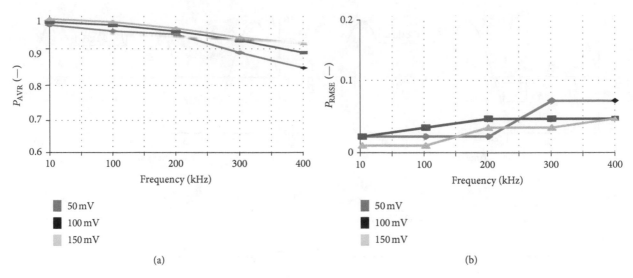

FIGURE 12: AVR (a) and RMSE (b) versus frequency for three amplitudes of the input voltage (50, 100, and 150) mV.

TABLE 3: A comparison with other reported implementations in the literature.

Parameter	Proposed rectifier	[30]	[31]	[32]
Technology used [μm]	0.18	1.5	0.18	0.35
Supply voltage [V]	±0.3	2.8	1.8	1
Power consumption [μW]	13.4	2.8	1.08	0.06
Maximum amplitude [mV]	225*	850	630	230
Maximum frequency [kHz]	400*	10	20	0.1
FOM** %	37.5	30	35	23

*$0.85 \leq P_{AVR} \leq 1, 0 < P_{RMSE} < 0.1$.
**FOM = (maximum amplitude/supply voltage) * 100.

10 kHz up to 500 kHz for three amplitudes of the input voltage (50, 100, and 150) mV. As one can notice from the figure, over the full range of frequency, the value of P_{AVR} ranges from 1 to 0.85 and the value of P_{RMSE} is below 0.1. The values of P_{AVR} and P_{RMSE} achieved confirm the quality of the rectification process of the proposed rectifier.

Table 3 provides comparison of the proposed rectifier with three half wave rectifiers introduced in [30–32]. To compare their performance figure of merit FOM is provided, which indicates the efficiency of the design regarding the maximum allowable input swing over voltage supply. Higher value of the FOM translates proportionate advantage in terms of input voltage range. It is notable that the proposed rectifier possesses the highest value of FOM. Besides, the proposed rectifier offers the highest maximum frequency with ultra-LV operation capability and simple CMOS internal structure.

5. Conclusions

In this paper, a new ultra-low-voltage supply and low-power consumption BD-QFG OTA is presented. The active building block was simulated using TSMC 0.18 μm N-well CMOS technology. It is demonstrated that the proposed BD-QFG OTA can operate at ±0.3 V supply voltage and consumes 13.4 μW of static power. The proposed BD-QFG OTA is used to realize diodeless precision rectifier as an example application. Proposed BD-QFG OTA could be used in LV and LP applications such as bioelectronics, biosensors, and biomedical systems. The simulation results confirm the workability of the active building block.

Conflict of Interests

The authors declare that there is no conflict of interests regarding the publication of this paper.

Acknowledgments

The described research was performed in laboratories supported by the SIX project, registration no. CZ. 1.05/2.1.00/03.0072, the operational program Research and Development for Innovation, and has been supported by Czech Science Foundation Project no. P102-14-07724S.

References

[1] T. Deliyannis, Y. Sun, and J. K. Fidler, *ContinuousTime Active Filter Design*, CRC Press, New York, NY, USA, 1999.

[2] F. Khateb, S. B. Abo Dabbous, and S. Vlassis, "A survey of nonconventional techniques for low-voltage low-power analog circuit design," *Radioengineering*, vol. 22, no. 2, pp. 415–427, 2013.

[3] F. Khateb, "Bulk-driven floating-gate and bulk-driven quasi-floating-gate techniques for low-voltage low-power analog circuits design," *AEU Electronics and Communications Journal*, vol. 68, no. 1, pp. 64–72, 2014.

[4] G. Raikos, S. Vlassis, and C. Psychalinos, "0.5 v bulk-driven analog building blocks," *International Journal of Electronics and Communications*, vol. 66, no. 11, pp. 920–927, 2012.

[5] J. Peyton and V. Walsh, *Analog Electronics with Op Amps: A Source Book of Practical Circuits*, Cambridge University Press, New York, NY, USA, 1993.

[6] R. G. Irvine, *Operational Amplifier Characteristics and Applications*, Prentice Hall International, Englewood Cliffs, NJ, USA, 1994.

[7] Z. Wang, "Full-wave precision rectification that is performed in current domain and very suitable for CMOS implementation," *IEEE Transactions on Circuits and Systems I*, vol. 39, no. 6, pp. 456–462, 1992.

[8] S. J. G. Gift, "A high-performance full-wave rectifier circuit," *International Journal of Electronics*, vol. 87, no. 8, pp. 925–930, 2000.

[9] P. R. Gray and R. G. Meyer, *Analysis and Design of Analog Integrated Circuits*, Wiley, New York, NY, USA, 1984.

[10] K. Yamamoto, S. Fujii, and K. Matsuoka, "A single chip FSK modem," *IEEE Journal of Solid-State Circuits*, vol. 19, no. 6, pp. 855–861, 1984.

[11] C. Toumazou, F. J. Lidgey, and S. Chattong, "High frequency current conveyor precision full-wave rectifier," *Electronics Letters*, vol. 30, no. 10, pp. 745–746, 1994.

[12] A. A. Khan, M. Abou El-Ela, and M. A. Al-Turaigi, "Current-mode precision rectification," *International Journal of Electronics*, vol. 79, no. 6, pp. 853–859, 1995.

[13] K. Hayatleh, S. Porta, and F. J. Lidgey, "Temperature independent current conveyor precision rectifier," *Electronics Letters*, vol. 30, no. 25, pp. 2091–2093, 1995.

[14] W. Surakampontorn, K. Anuntahirunrat, and V. Riewruja, "Sinusoidal frequency doubler and full-wave rectifier using translinear current conveyor," *Electronics Letters*, vol. 34, no. 22, pp. 2077–2079, 1998.

[15] A. Monpapassorn, K. Dejhan, and F. Cheevasuvit, "CMOS dual output current mode half-wave rectifier," *International Journal of Electronics*, vol. 88, no. 10, pp. 1073–1084, 2001.

[16] M. Kumngern and K. Dejhan, "Current conveyor-based versatile precision rectifier," *WSEAS Transactions on Circuits and Systems*, vol. 7, no. 12, pp. 1070–1079, 2008.

[17] E. Yuce, S. Minaei, and O. Cicekoglu, "Full-wave rectifier realization using only two CCII+s and NMOS transistors," *International Journal of Electronics*, vol. 93, no. 8, pp. 533–541, 2006.

[18] A. Monpapassorn, K. Dejhan, and F. Cheevasuvit, "A full-wave rectifier using a current conveyor and current mirrors," *International Journal of Electronics*, vol. 88, no. 7, pp. 751–758, 2001.

[19] P. D. Walker and M. M. Green, "CMOS half-wave and full-wave precision voltage rectification circuits," in *Proceedings of the 38th IEEE Midwest Symposium on Circuits and Systems*, pp. 901–904, Rio de Janeiro, Brazil, August 1995.

[20] V. Riewruja and R. Guntapong, "A low-voltage wide-band CMOS precision full-wave rectifier," *International Journal of Electronics*, vol. 89, no. 6, pp. 467–476, 2002.

[21] M. Kumngern and K. Dejhan, "High frequency and high precision CMOS full-wave rectifier," *International Journal of Electronics*, vol. 93, no. 3, pp. 185–199, 2006.

[22] M. Kumngern, P. Saengthong, and S. Junnapiya, "DDCC-based full-wave rectifier," in *Proceeding of the 5th International Colloquium on Signal Processing and Its Applications (CSPA '09)*, pp. 312–315, Kuala Lumpur, Malaysia, March 2009.

[23] M. Kumngern, B. Knobnob, and K. Dejhan, "High frequency and high precision CMOS half-wave rectifier," *Circuits, Systems, and Signal Processing*, vol. 29, no. 5, pp. 815–836, 2010.

[24] M. Kumngern, "CMOS precision full-wave rectifier using current conveyor," in *Proceeding of the IEEE International Conference of Electron Devices and Solid-State Circuits (EDSSC '10)*, Hong Kong, December 2010.

[25] A. Virattiya, B. Knobnob, and M. Kumngern, "CMOS precision full-wave and half-wave rectifier," in *Proceedings of the IEEE International Conference on Computer Science and Automation Engineering (CSAE '11)*, vol. 4, pp. 556–559, Shanghai, China, June 2011.

[26] H. Mitwong and V. Kasemsuwan, "A 0.5 V quasi-floating gate self-cascode DTMOS current-mode precision full-wave rectifier," in *Proceedings of the 9th International Conference on Electrical Engineering/Electronics, Computer, Telecommunications and Information Technology (ECTI-CON '12)*, pp. 1–4, May 2012.

[27] M. Kumngern, "Precision full-wave rectifier using two DDCCs," *Circuits and Systems*, vol. 2, pp. 127–132, 2011.

[28] Wafer Electrical Test Data and SPICE Model Parameters, http://www.mosis.com/pages/Technical/Testdata/tsmc-018-prm.

[29] F. Khateb, J. Vávra, and D. Biolek, "A novel current-mode full-wave rectifier based on one CDTA and two diodes," *Radioengineering*, vol. 19, no. 3, pp. 437–445, 2010.

[30] S. M. Zhak, M. W. Baker, and R. Sarpeshkar, "A low-power wide dynamic range envelope detector," *IEEE Journal of Solid-State Circuits*, vol. 38, no. 10, pp. 1750–1753, 2003.

[31] B. Rumberg and D. W. Graham, "A low-power magnitude detector for analysis of transient-rich signals," *IEEE Journal of Solid-State Circuits*, vol. 47, no. 3, pp. 676–685, 2012.

[32] E. Rodriguez-Villegas, P. Corbishley, C. Lujan-Martinez, and T. Sanchez-Rodriguez, "An ultra-low-power precision rectifier for biomedical sensors interfacing," *Sensors and Actuators, A: Physical*, vol. 153, no. 2, pp. 222–229, 2009.

[33] S. Koziel and S. Szczepanski, "Design of highly linear tunable CMOS OTA for continuous-time filters," *IEEE Transactions on Circuits and Systems II: Analog and Digital Signal Processing*, vol. 49, no. 2, pp. 110–122, 2002.

[34] D. Majumdar, "Comparative study of low voltage OTA designs," in *Proceedings of the 17th International Conference on VLSI Design, Concurrently with the 3rd International Conference on Embedded Systems Design*, pp. 47–51, January 2004.

[35] R. Li and R. Raut, "A very wideband OTA-C filter in CMOS VLSI technology," in *Proceedings of the 7th World Multiconference on Systemics, Cyberetics and Informatics*, pp. 1–6, 2003.

[36] L. Zhang, X. Zhang, and E. El-Masry, "A highly linear bulk-driven CMOS OTA for continuous-time filters," *Analog Integrated Circuits and Signal Processing*, vol. 54, no. 3, pp. 229–236, 2008.

8

Low Power Data Acquisition System for Bioimplantable Devices

Sadeque Reza Khan and M. S. Bhat

Department of Electronics and Communication Engineering, National Institute of Technology Karnataka,
Surathkal, Mangalore 575025, India

Correspondence should be addressed to Sadeque Reza Khan; sadeque_008@yahoo.com

Academic Editor: Sebastian Hoyos

Signal acquisition represents the most important block in biomedical devices, because of its responsibilities to retrieve precise data from the biological tissues. In this paper an energy efficient data acquisition unit is presented which includes low power high bandwidth front-end amplifier and a 10-bit fully differential successive approximation ADC. The proposed system is designed with 0.18 μm CMOS technology and the simulation results show that the bioamplifier maintains a wide bandwidth versus low noise trade-off and the proposed SAR-ADC consumes 450 nW power under 1.8 V supply and retain the effective number of bit 9.55 in 100 KS/s sampling rate.

1. Introduction

In the past few years, the rapid developments in the field of microelectronics and VLSI have driven forward the advent of implantable medical sensors and devices. Multichannel devices are emerging due to the fact of recording numerous number of biological tissue activities collectively [1]. Such multichannel sensors first collect the extracellular signals from a micromachined array including several electrodes and process them through embedded microelectronic circuits for conditioning, multiplexing, and digitization. A fully implantable recording device would then wirelessly transfer the digital data through an inductive link to an external controller. As the capability to integrate more recording channels is growing, suitable data acquisition systems are needed to meet smaller silicon area and lower power dissipation requirements [2].

Biopotential signals, such as electrooculogram (EOG), electroencephalogram (EEG), electromyogram (EMG), and electrocardiogram (ECG), cover a wide range of spectrum and signal bandwidth ranging from few hertz to 10 kHz [3] and the acquired signals through dense microelectrode arrays are very low in amplitude and susceptible to environment noises [4]. Proper processing of these signals requires amplification and noise cancellation, digitization, and digital signal processing before being considered for analysis. Figure 1 shows the block diagram of the proposed system architecture.

Bioamplifiers are the primary building blocks in biomedical sensing devices [5]. The most common characteristics of bioamplifiers are band pass characteristics, DC offset, low noise or noise reduction, and reduced power consumption. For designing bioamplifiers the power dissipation should be restricted to several orders of below 80 mWcm^{-2} [6] in order not to harm the tissues. Implantable bioamplifiers must dissipate little power so that surrounding tissues are not damaged by heating. For a 1000-electrode system the maximum power dissipation should be limited to some few microwatt per amplifier. Noise elimination is another primary concern of such amplifiers which brings out the band pass feature of these front-end amplifiers. The resulting amplifier in this paper passes signals from 2.52 Hz to 8.07 kHz with an input referred noise of 2.83 μV_{rms} and a power dissipation of 18 μW.

Among the signal processing blocks within a bioimplant, analog to digital converter is a critical interface to convert the amplified signal coming from the front-end amplifier for further backend processing. The SAR ADC architecture suits well the biomedical applications due to its moderate speed, moderate resolution, and very low power consumption characteristics [7]. The primary sources of power consumption in a SAR ADC are the comparator and charge/discharge of the capacitor arrays. This paper describes a fully differential asynchronous SAR ADC in 1.8 μm CMOS technology that

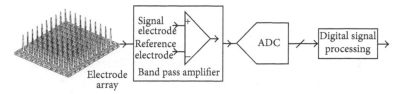

FIGURE 1: System overview.

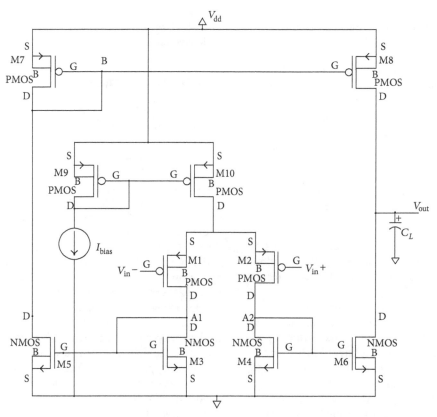

FIGURE 2: Three-mirror OTA.

uses a charge distribution differential DAC with monotonic capacitor switching architecture designed for energy efficient application. The proposed architecture occupies sample rate of 100 KS/s and a supply voltage of 1.8 V; the 10-bit SAR achieves an ENOB of 9.55 and consumes 450 nW power.

2. Front-End Amplifier

The necessity of front-end bioamplifiers in biomedical signal acquisition system is increasing for applications such as electroencephalogram (EEG), electrocardiogram (ECG), and electromyogram (EMG). A low noise amplifier is required for direct recording of signals from the dense microelectrode arrays as these signals are very weak in amplitude (typically 10–500 μV) and have a wide range (1–10 Khz) [13]. Due to electrochemical effects at the electrode-tissue interface DC offsets are common across differential electrodes. It is necessary to reject the DC offset [14] as the electrode electrolyte interface may reach few hundreds of millivolts sometimes. Amplifier required for the purpose of bioimplantable data acquisition system must pursue high gain,

acceptable bandwidth, and good stability with low power consumption and low noise. For this purpose symmetric operational transconductance amplifier (OTA) or three-current-mirror OTA [15] is used, as shown in Figure 2.

This symmetric OTA has a large transconductance, slew rate, and gain bandwidth compared to the other basic OTA's. In the given architecture input stage is a differential pair where M1, M3 and M2, M4 pairs are self-biased inverters.

Transistors M3, M5; M4, M6; M7, M8; and M9, M10 are simple current mirrors. In this symmetric OTA a dominant pole (P1) is present at the output node and two nondominant poles are present at nodes A (P2) and B (P3). Due to symmetric behavior at the input stage, the amplifier has a RHP (right hand plane) zero (Z) which can reduce the phase margin and cause instability of the system. To ensure stability P2, P3 $\gg$ P1 and Z = 2 * P2. This can be satisfied by making CL large. The poles P2 and P3 can also be increased by decreasing the size of M3, M4, and M7 transistors. Next, the reduction of noise especially the flicker noise which dominates at the low frequencies depends on the sizing of the transistors. A practical technique to reduce the flicker noise is

TABLE 1: Comparison with other amplifiers.

Topology	Technology	V_{dd}	Current (μA)	Gain (dB)	Bandwidth (Hz)	V_{rms} (μ)	C_L (pF)
[8]	CMOS $0.5\,\mu$	±1.8	8	70.4	116.9 m–1.5 K	2.5	15
[9]	CMOS $0.35\,\mu$	±1.5	2	46	13–8.9 K	5.7	—
[6]	CMOS $1.5\,\mu$	±2.5	16	40	130 m–7.5 K	3.1	17
This work	CMOS $0.18\,\mu$	1.8	10	54.57	2.52–8.07 K	2.83	1

TABLE 2: Comparison and performance evaluation.

Parameter	[10]	[11]	[12]	This work
Technology	$0.18\,\mu$	$0.18\,\mu$	$0.18\,\mu$	$0.18\,\mu$
Bits	10	12	10	10
Supply (V)	1	1	1	1.8
Power (W)	$6.7\,\mu$	$3.8\,\mu$	$42\,\mu$	450 n
Sampling rate (KS/s)	100	100	500	100
SNDR (dB)	61.2	58	58.4	59.28
SFDR (dB)	84.8	64.2	75	78.74
ENOB	9.87	9.4	9.4	9.55
FOM (fJ/conv.-step)	71	56	124	6

to use PMOS transistor as input stage with large gate areas. So flicker noise can be avoided by adjusting the (W/L) ratio of M1 and M2 transistors. The band pass amplifier architecture is shown in Figure 3.

Such an amplifier features band pass characteristics. The depicted topology consists of a low noise amplifier (A1) following an inverting miller integrator in its feedback path. The miller integrator uses a second amplifier (A2), a capacitor (C_f), and a high value resistor (Req). In this scheme the DC rejection is achieved by an active integrator located in the feedback loop. The high pass cutoff frequency is set by the small capacitor C_f and MOS bipolar equivalent resistor (Ma, Mb). So A2 OTA occupies the miller integrator configuration. The integrator's time constant τ is set by a small capacitor (C_f) and the MOS bipolar devices acting as a high equivalent pseudoresistor in a configuration of a diode connected PMOS. The mid band gain of the bioamplifier is the same as the DC gain of main OTA (A1) and its $-3\,$dB low pass cutoff is set by OTA (A1)'s dominant pole. Common mode rejection is provided through V_{ref} which senses the common mode voltage from terminal. $-3\,$dB high pass cutoff frequency can be defined as

$$f_{hp} = \left(\frac{A_{v1}}{2\pi\tau}\right), \qquad (1)$$

where $\tau = \text{Req} * C_f$.

The gain bandwidth and the input referred noise output are given in Figures 4 and 5.

The mid band gain achieved in this case is 54.57 dB for the input voltage of 1.8 V. The circuit is consuming $10\,\mu$A current here. The bandwidth is between 2.52 Hz and 8.07 KHz which is providing the opportunity of wide range of operation for the designed bioamplifier. The calculated input referred noise is 2.83 μV_{rms} which is lower than some other previously designed bioamplifiers. A comparison with the previously

published results is shown in the performance evaluation table given in Table 1.

3. ADC Architecture

Figure 6 shows the designed SAR ADC block that is based on architecture from [16] where the building blocks are the comparator, T/H circuit, DAC, and a 10-bit successive approximation register (SAR) controller. For achieving better linearity binary-weighted capacitor array is used rather than a C-$2C$ capacitor array [17] in the designed SAR ADC.

In this charge redistribution based architecture, the capacitor network serves as both a T/H circuit and a reference DAC capacitor array. Therefore, this architecture does not require a high power consuming gigantic size Track and Hold circuit. Since this ADC is fully differential, the operation of the two input sides is complementary. After fetching the input signal, first, the ADC samples those signals on the top plates of the capacitor array via the Track and Hold circuits: $V_{in}+ = V_{ip}$ and $V_{in}- = V_{in}$, and at the same time the bottom plates of the capacitor array are connected to V_{ref}, the reference voltage of the designed ADC. The comparator then directly performs the first comparison without switching any capacitor as soon as ADC turns off the Track and Hold switches. Now, the largest capacitor (C_1) on the higher voltage potential side is switched to ground according to the comparator output. At the same moment of time the other one capacitor (on the lower side) remains unchanged and the digital output D1 is generated. The ADC repeats the procedure until the LSB is decided. For each bit cycle, there is only one capacitor switch, which reduces the power dissipation by the DAC network and switch buffer. The flow chart of the proposed successive approximation procedure is shown in Figure 7. Figure 8 shows an example of how V_{in} and V_{ip} vary at every comparison.

To extract 10 bits, 10 time slots are required for comparison and 9 for DAC switching. After each switching procedure, the common mode voltage of the DAC gradually decreases towards GND or 0 V as shown in the timing diagram in Figure 8.

3.1. Track and Hold Circuit. Figure 9 is the Track and Hold circuit using a bootstrapped switch originally presented in [18]. When the CLK is low ("off" phase or "Hold" phase) [19], MN7 and MN8 discharge the gate of MN4 to ground. At the same time, V_{dd} is applied across the capacitor C by MP1 and MN6 transistor. The capacitor will act as the battery across the gate and source of MN4 during "on" phase. MP2 and MN3 isolate the bootstrapped switch from the capacitor C while it

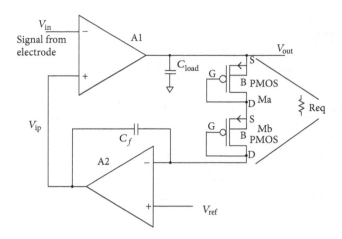

FIGURE 3: Loaded inductive link.

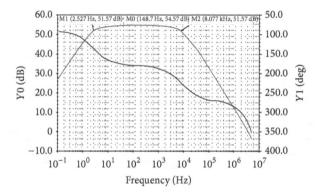

FIGURE 4: Gain bandwidth and phase diagram.

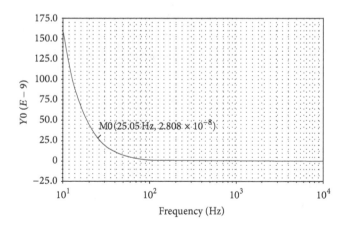

FIGURE 5: Input noise diagram.

is charging in "off" phase. When CLK is high ("on" phase or "Track" phase), MN5 pulls down the gate of MP2, allowing charge from battery capacitor C to flow on the gate of MN4. This turns both MN3 and MN4 on and give Track of the input voltage at V_{in}. Capacitor C must be sufficiently large (160 pF in this case) to supply charge to the gate of the switching device in addition to all parasitic capacitance in the charging path. The bulk of MP1 and MP2 are connected to the highest potential voltage, top plate of the bootstrapped capacitor C rather than V_{dd}.

To avoid the body effect (which can cause distortion at high frequency input signal) of the tracking switch MN4, bulk switching technique [20] is adopted. Transistors MN1 and MN2 are added to mitigate the body effect problem. During Track or "on" phase the bulk of MN4 is connected to input V_{in} via MN1 cancelling the body effect. In the Hold phase or

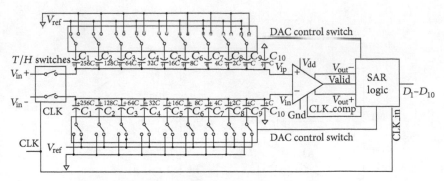

FIGURE 6: Proposed fully differential SAR ADC architecture.

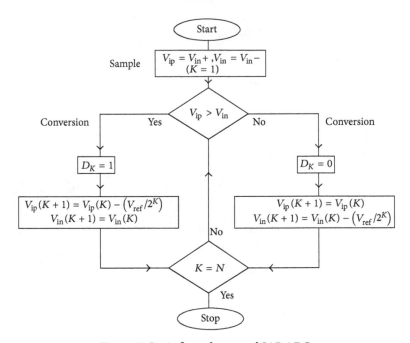

FIGURE 7: Logic flow of proposed SAR ADC.

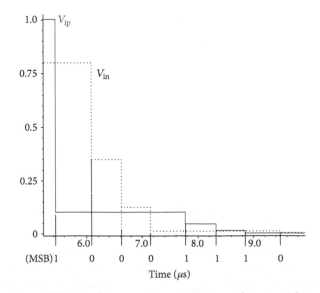

FIGURE 8: Timing diagram with monotonic switching procedure.

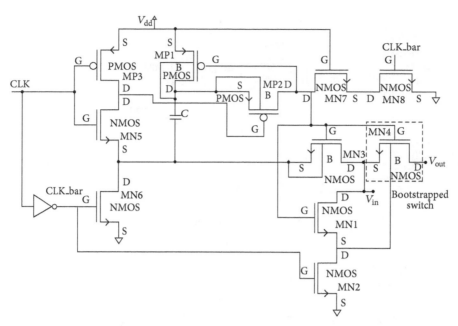

FIGURE 9: Track and Hold circuit.

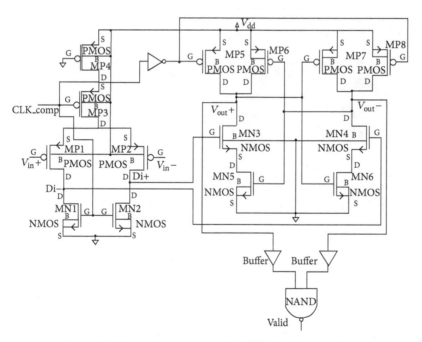

FIGURE 10: Dynamic comparator with Valid signal generation.

"off" phase the bulk is connected to GND via MN2 to prevent negative source-bulk voltage.

3.2. Dynamic Comparator.

Figure 10 is showing the proposed dynamic comparator [21] circuit. It is a two-stage comparator: first stage is the input gain stage and 2nd stage is the output latch. This architecture made this comparator operate in a lower and stable offset. It also operates in a wide range of common mode voltages and at a lower supply voltage.

In precharge phase (CLK_comp = 1), Di+ and Di− are grounded by MN1, MN2 transistors and the V_{out} are precharged by MP5, MP8 transistors. In comparison phase (CLK_comp = 0), MP3 is on, so Di+ and Di− nodes voltage start to charge from ground to V_{dd} with a different time rate proportional to each input voltage. So MP1 and MP2 generate a differential voltage at Di+ and Di− nodes. This differential voltage is now passed to output latch through MN3 and MN4 transistors.

So the cross coupled inverter in output latch regenerates the V_{out} voltages according to the difference present at the input gain stage. Figure 12 is showing the output generated by the dynamic comparator.

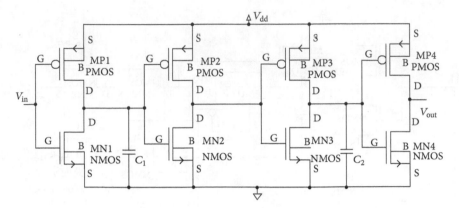

FIGURE 11: Delay buffer.

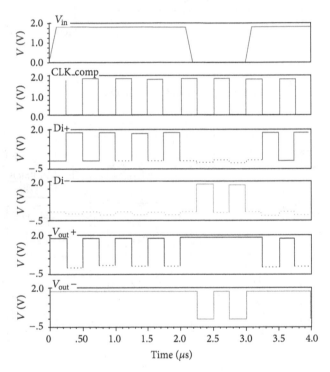

FIGURE 12: Dynamic comparator output.

The dynamic offset of the proposed comparator is minimized by adding MP4 transistor at the top of the gain stage. As MP4 is in the saturation region, the change of its drain to source voltage has a slight influence on the drain current. Hence MP4 keeps the effective voltage of the input pair near a constant value when common mode voltage changes. The dynamic offset thus has a minor influence on the conversion linearity. Again the MP1 and MP2 have large size to minimize the offset.

CLK_comp signal generation is dependent on a signal "Valid" extracted from the dynamic comparator using a two-input NAND gate. Figure 11 is showing a delay buffer of approximately 250 ns in between the NAND and the comparator output. In comparison phase (CLK_comp = 0), one of the comparator output V_{out} is going low which makes the NAND output high. This high signal passes through the

SAR logic and brings CLK_comp signal to the precharge phase (CLK_comp = 1) and this process continues; thus an asynchronous clock signal (CLK_comp) is generated. But if the Valid signal is generated concurrently with the V_{out}, the CLK_comp will not continue in the comparison phase for a proper period of time; consequently the comparison will not be done appropriately. Figure 13 is showing the generation of Valid signal.

3.3. Asynchronous SAR Control Logic.
Figure 14 shows the implemented SAR logic [16] and Figure 15 is showing the asynchronous static DFF circuit. The proposed circuit can generate necessary clock inside it, so no high frequency clock generator is required. SAR logic is asynchronous; by using the "Valid" indication from the comparator, a timing state machine is controlled.

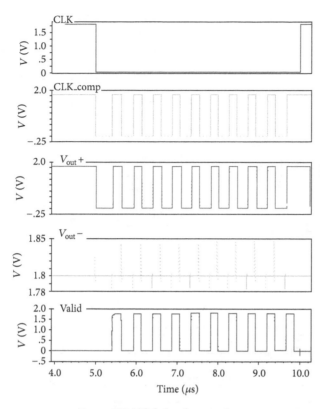

FIGURE 13: Valid signal generation.

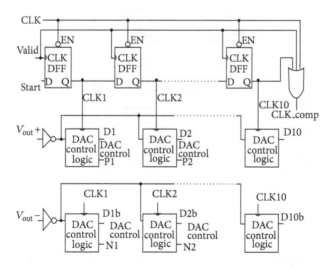

FIGURE 14: SAR logic controller with DAC control.

Figure 16 shows the timing diagram of asynchronous control logic. CLK1 to CLK10 sample the digital output codes of the comparator and serve as control signals for the capacitor arrays to perform the monotonic switching procedure via the DAC control logic. Figures 17 and 18 show a schematic and a timing diagram of the DAC control logic, respectively.

$V_{out}+$, $V_{out}-$ are precharged to VDD in precharge phase. When comparator works in comparison phase, one of the two outputs will go low. A logical NAND operation detects this high-to-low transition and generates an active-high Valid indication that will be used for the asynchronous SAR controller. The CLK1 to CLK10 also serve as control signals for the DAC capacitor arrays to perform monotonic switching procedure via 10 DAC control logic units.

As shown in Figures 14 and 15, the first 9 DAC control logic parts include a DFF, an AND gate, and a delay buffer to make sure that CLK triggers AND gate when the output of

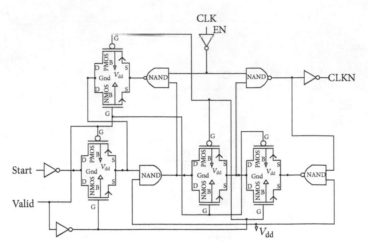

FIGURE 15: Positive edge triggered asynchronous static DFF.

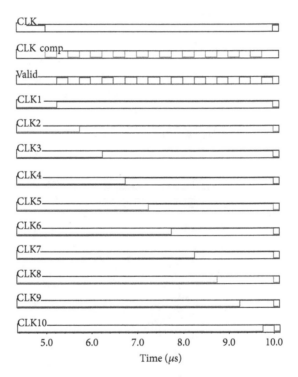

FIGURE 16: Timing diagram.

DFF generated; and the last DAC control part is only a DFF that will generate last bit, D10. Figure 18 shows example of the simulated waveforms of DAC control logic. At the rising edge of CLK1 and CLK2, DFF samples the comparator output $V_{out}+$. As $V_{out}+$ is low, the DAC control signals DAC CON P1 and DAC CON P2 are high to switch the relevant capacitor switches. Again at rising edge of CLK 3 comparator output $V_{out}-$ is high, which results in DAC CON N3 being high to control the corresponding capacitor switch.

3.4. Charge Redistribution DAC. Figure 19 shows the architecture of charge redistributing DAC. DAC control logic generates control signals DAC Control P ⟨1:9⟩ and DAC Control N ⟨1:9⟩ controlling the DAC capacitor arrays. The

inverters can be used as switches to generate V_{ref} (= 1.8 V) and GND (= 0 V) for bottom plate of the DAC capacitors as shown in Figure 19. The ADC uses two T/H circuits to sample pseudodifferentially the input signal on top plate of the capacitors and uses inverters to switch between V_{ref} and GND. The small unit capacitance value is 1 pF.

4. Measurement

The ADC is designed in a standard 0.18 μm CMOS process and in cadence ADEL spectra simulator. Figure 20 shows the digital code output for an input of 1.8 V. Figure 21 is showing the die micrograph. The static performance is characterized through differential nonlinearity (DNL) and integral

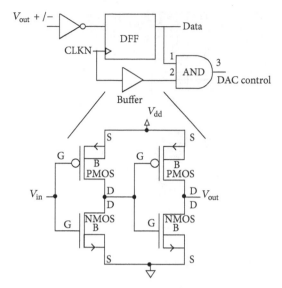

FIGURE 17: DAC control logic.

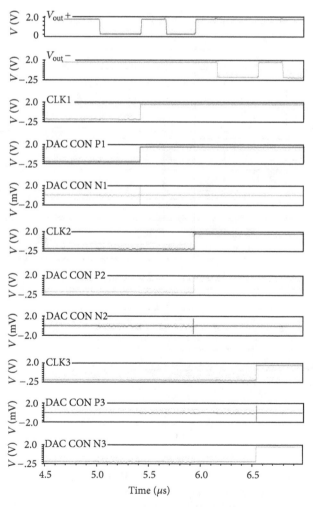

FIGURE 18: DAC control logic output.

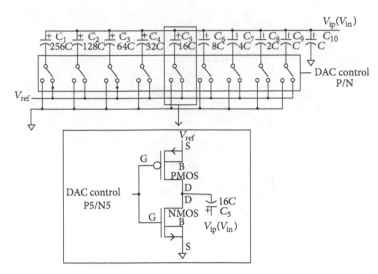

FIGURE 19: Charge redistribution DAC with inverter based switches.

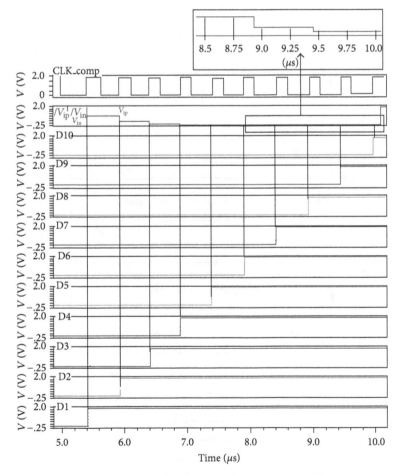

FIGURE 20: Digital code output for an input of 1.8 V.

nonlinearity (INL) measurement. The measured DNL and INL are +0.455/−0.43 LSB and +0.62/−0.58 LSB, respectively, and are shown in Figures 22 and 23. The measured FFT spectrum at input frequency 4.68 KHz and sample frequency of 100 KHz is shown in Figure 24. The measured SNDR is 59.28 dB which equals ENOB of 9.55 bits and the SFDR is 78.74 dB as shown in Figure 24. The total power dissipation of the ADC is 450 nW at 100 KS/s sampling rate and supply voltage of 1.8 V. T/H circuit and the DAC capacitor arrays are consuming 50 nA (approximately); dynamic comparator

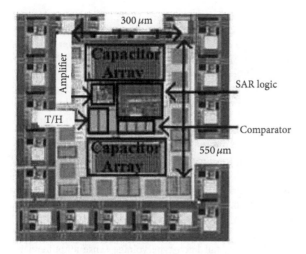

FIGURE 21: Die micrograph.

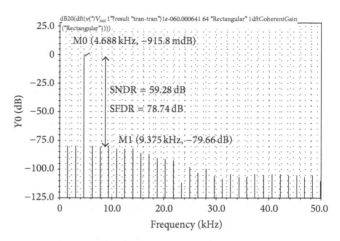

FIGURE 24: Measured FFT of input frequency 4.68 KHz.

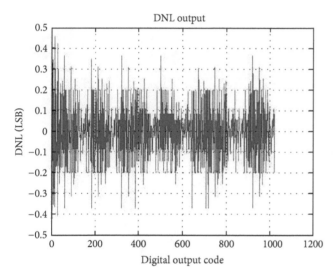

FIGURE 22: DNL plot.

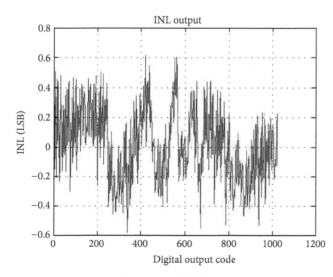

FIGURE 23: INL plot.

is dissipating 100 nA and overall SAR logic circuit consuming 100 nA current. The typical figure of merit (FOM) definition of the ADCs is defined as

$$\text{FOM} = \frac{\text{Power}}{\left(2^{\text{ENOB}} * fs\right)}. \tag{2}$$

The FOM of this work corresponds to 6 fJ/conversion-step.

In 0.18 μm CMOS process the die micrograph occupies an active area of 300 μm × 550 μm. The experimental result provides a 51.2 dB gain for a loss of 14.2 μA of current in the front-end amplifier. The measured power loss including SAR ADC and the front-end amplifier is 21 μwatt. The ADC provides a practical 9.17 ENOB for 56.96 dB SNDR (see Table 2).

5. Conclusion

This paper proposed a low power and high bandwidth front-end amplifier as part of an implantable device for accumulating data from the tissue electrolyte and microelectrode interface. The proposed ADC reduces the power consumption as well as the total capacitance. The ADC delivers 9.55 ENOB with a power consumption of 450 nW at a sampling rate of 100 KS/s. It utilizes an ultralow power design strategy, imposing maximum simplicity in ADC architecture, low capacitive DAC switching scheme. So proposed low power design in this paper suited well the biomedical applications.

Conflict of Interests

The authors declare that there is no conflict of interests regarding the publication of this paper.

References

[1] P.-Y. Robert, B. Gosselin, A. E. Ayoub, and M. Sawan, "An ultra-low-power successive-approximation-based ADC for implantable sensing devices," in *Proceedings of the 49th Midwest Symposium on Circuits and Systems (MWSCAS '06)*, pp. 7–11, August 2007.

[2] X. Zou, X. Xu, J. Tan, L. Yao, and Y. Lian, "A 1-V 1.1-μW sensor interface IC for wearable biomedical devices," in *Proceedings of the IEEE International Symposium on Circuits and Systems (ISCAS '08)*, pp. 2725–2728, May 2008.

[3] H.-H. Ou, Y.-C. Chen, and B.-D. Liu, "A 0.7-V 10-bit 3μW analog-to-digital converter for implantable biomedical applications," in *Proceedings of the IEEE Biomedical Circuits and Systems Conference*, pp. 122–125, December 2006.

[4] O. de Oliveira Dutra and T. C. Pimenta, "Low power low noise bio-amplifier with adjustable gain for digital bio-signals acquisition systems," in *Proceedings of the IEEE 4th Latin American Symposium on Circuits and Systems (LASCAS '13)*, pp. 1–4, March 2013.

[5] K. Iniewski, *VLSI Circuits for Biomedical Applications*, Artech House, 2008.

[6] R. R. Harrison and C. Charles, "A low-power low-noise CMOS amplifier for neural recording applications," *IEEE Journal of Solid-State Circuits*, vol. 38, no. 6, pp. 958–965, 2003.

[7] Y.-K. Chang, C.-S. Wang, and C.-K. Wang, "A 8-bit 500-KS/s low power SAR ADC for bio-medical applications," in *Proceedings of the IEEE Asian Solid-State Circuits Conference (A-SSCC '07)*, pp. 228–231, November 2007.

[8] O. de Oliveira Dutra and T. C. Pimenta, "Low power low noise bio-amplifier with adjustable gain for digital bio-signals acquisition systems," in *Proceedings of the IEEE 4th Latin American Symposium on Circuits and Systems (LASCAS '13)*, pp. 1–4, 2013.

[9] W. Zhao, H. Li, and Y. Zhang, "A low-noise integrated bioamplifier with active DC offset suppression," in *Proceedings of the IEEE Biomedical Circuits and Systems Conference*, pp. 5–8, November 2009.

[10] D.-Q. Zhao, Z.-H. Wu, and B. Li, "A 10-bit low-power differential successive approximation ADC for implantable biomedical application," in *Proceedings of the IEEE International Conference of Electron Devices and Solid-State Circuits (EDSSC '13)*, pp. 1–2, IEEE, Hong Kong, June 2013.

[11] A. Agnes, E. Bonizzoni, P. Malcovati, and F. Maloberti, "An ultra-low power successive approximation A/D converter with time-domain comparator," *Analog Integrated Circuits and Signal Processing*, vol. 64, no. 2, pp. 183–190, 2010.

[12] W. Y. Pang, C. S. Wang, Y. K. Chang, N. K. Chou, and C. K. Wang, "A 10-bit 500-KS/s low power SAR ADC with splitting comparator for bio-medical applications," in *Proceedings of the IEEE Asian Solid-State Circuits Conference (A-SSCC '09)*, pp. 149–152, November 2009.

[13] R. F. Yazicioglu, P. Merken, R. Puers, and C. van Hoof, "A 60 μW 60 nV/$\sqrt{Hz}$ readout front-end for portable biopotential acquisition systems," *IEEE Journal of Solid-State Circuits*, vol. 42, no. 5, pp. 1100–1110, 2007.

[14] R. H. Olsson, M. N. Gulari, and K. D. Wise, "A fully-integrated bandpass amplifier for extracellular neural recording," in *Proceedings of the 1st International IEEE EMBS Conference on Neural Engineering*, pp. 165–168, 2003.

[15] D. Salhi and B. Godara, "A 75dB-gain low-power, low-noise amplifier for low-frequency bio-signal recording," in *Proceedings of the 5th IEEE International Symposium on Electronic Design, Test & Applications (DELTA '10)*, pp. 51–53, January 2010.

[16] C.-C. Liu, S.-J. Chang, G.-Y. Huang, and Y.-Z. Lin, "A 10-bit 50-MS/s SAR ADC with a monotonic capacitor switching procedure," *IEEE Journal of Solid-State Circuits*, vol. 45, no. 4, pp. 731–740, 2010.

[17] H. Balasubramaniam, W. Galjan, W. H. Krautschneider, and H. Neubauer, "12-bit hybrid C2C DAC based SAR ADC with floating voltage shield," in *Proceedings of the 3rd International Conference on Signals, Circuits and Systems (SCS '09)*, pp. 1–5, Medenine, Tunisia, November 2009.

[18] M. Dessouky and A. Kaiser, "Very low-voltage digital-audio $\Delta\Sigma$ modulator with 88-dB dynamic range using local switch bootstrapping," *IEEE Journal of Solid-State Circuits*, vol. 36, no. 3, pp. 349–355, 2001.

[19] A. M. Abo and P. R. Gray, "A 1.5-V, 10-bit, 14.3-MS/s CMOS pipeline analog-to-digital converter," *IEEE Journal of Solid-State Circuits*, vol. 34, no. 5, pp. 599–606, 1999.

[20] R. Teggatz, "Control of body effect in mos transistors by switching source-to-body bias," United States Patent, no. 5786724, 1998.

[21] D. Schinkel, E. Mensink, E. Klumperink, E. Van Tuijl, and B. Nauta, "A double-tail latch-type voltage sense amplifier with 18ps setup+hold time," in *Proceedings of the 54th IEEE International Solid-State Circuits Conference*, pp. 314–301, February 2007.

9

FPGA-Based Synthesis of High-Speed Hybrid Carry Select Adders

V. Kokilavani,[1] K. Preethi,[1] and P. Balasubramanian[2]

[1]*Department of PG Studies in Engineering, S. A. Engineering College (Affiliated to Anna University),*
  Poonamallee-Avadi Road, Veeraraghavapuram, Chennai, Tamil Nadu 600 077, India
[2]*Department of Computer Science and Engineering, S. A. Engineering College (Affiliated to Anna University),*
  Poonamallee-Avadi Road, Veeraraghavapuram, Chennai, Tamil Nadu 600 077, India

Correspondence should be addressed to P. Balasubramanian; spbalan04@gmail.com

Academic Editor: Gianluca Traversi

Carry select adder is a square-root time high-speed adder. In this paper, FPGA-based synthesis of conventional and hybrid carry select adders are described with a focus on high speed. Conventionally, carry select adders are realized using the following: (i) full adders and 2 : 1 multiplexers, (ii) full adders, binary to excess 1 code converters, and 2 : 1 multiplexers, and (iii) sharing of common Boolean logic. On the other hand, hybrid carry select adders involve a combination of carry select and carry lookahead adders with/without the use of binary to excess 1 code converters. In this work, two new hybrid carry select adders are proposed involving the carry select and section-carry based carry lookahead subadders with/without binary to excess 1 converters. Seven different carry select adders were implemented in Verilog HDL and their performances were analyzed under two scenarios, dual-operand addition and multioperand addition, where individual operands are of sizes 32 and 64-bits. In the case of dual-operand additions, the hybrid carry select adder comprising the proposed carry select and section-carry based carry lookahead configurations is the fastest. With respect to multioperand additions, the hybrid carry select adder containing the carry select and conventional carry lookahead or section-carry based carry lookahead structures produce similar optimized performance.

1. Introduction

Carry select adder (CSLA) belongs to the family of high-speed square-root time adders [1, 2] and provides a good compromise between the low area occupancy of ripple carry adders (RCAs) and the high-speed performance of carry lookahead adders (CLAs) [2, 3]. In the existing literature, many flavors of carry select addition have been realized on both ASIC and FPGA platforms with ASIC implementations being predominant. CSLAs usually involve duplication of RCA structures with presumed carry inputs of binary 0 and binary 1 to enable parallel addition and thereby speed up the addition process [3, 4]. To minimize the area metric of CSLAs owing to replication of the RCA structures, an add-one circuit (also called, binary to excess-1 converter, viz. BEC) is introduced [5–7]. Carry select addition can also be performed by utilizing the common Boolean logic (CBL) [8] shared between the sum and carry outputs of a full adder [9].

Nevertheless, due to the serial cascading of full adder modules, the delay metric would not decrease although the area parameter would reduce. Further, optimizations at the device, gate levels [10–15], and realization styles [16, 17] have been carried out to reduce area, improve speed, and minimize the power-delay product of CSLAs on the basis of semicustom and full-custom ASIC-style synthesis. Rather than realizing pure CSLAs, hybrid architectures incorporating carry select and carry lookahead structures have also been proposed [18–21] to improve the design efficiency of CSLAs. Moreover, some FPGA implementations of CSLAs have been attempted [21–23]. Overall, a survey of published literature reveals that CSLAs have been widely implemented using the following topologies and computational elements:

(i) (Conventional) CSLA – full adders and 2 : 1 multiplexers (MUXes)

(ii) CSLA with BEC – Full adders, BECs, and 2 : 1 MUXes

(iii) CSLA based on CBL sharing

(iv) Hybrid CSLA and CLA structures

(v) Hybrid CSLA and CLA including BECs.

In general, CSLAs are composed using a carry select architecture with/without BECs or may consist of a mix of carry select and carry lookahead configurations with/without BECs. CSLAs constructed using pure carry select structures are called "homogeneous CSLAs" and CSLAs realized using a combination of carry select and carry lookahead structures are labeled as "heterogeneous/hybrid CSLAs." The interest behind hybrid CSLAs is supported by the fact that heterogeneous adders tend to better optimize the design metrics compared to homogeneous adders [24]. In a recent work [25], section-carry based CLAs (SCBCLAs) were proposed as an alternative to conventional CLAs; for a 32-bit addition operation, the SCBCLA was found to exhibit reduced propagation delay than the conventional CLA by 15.2%. Motivated by this result, two new hybrid CSLA architectures are proposed in this work, a hybrid CSLA incorporating CSLA and SCBCLA and another hybrid CSLA embedding CSLA, SCBCLA, and BECs. This paper builds upon our prior work [21] by analyzing the performance of different CSLA architectures with respect to diverse input partitions for different addition widths for the case of dual-operand addition and further evaluates the efficacy of the conventional and proposed CSLAs with respect to multioperand additions.

The remaining part of this paper is organized as follows. With 8-bit addition as a running example, Section 2 describes the conventional CSLA topologies with and without BEC logic and also the CSLA based on sharing of CBL. Section 3 presents the architectures of hybrid CSLAs incorporating CLAs and SCBCLAs with/without BEC logic. In Section 4, the performance of different CSLA topologies is evaluated for dual-operand and multioperand additions with operand sizes of 32 and 64-bits. Finally, the conclusions follow in Section 5.

2. Homogeneous CSLA Architectures

The RCA and homogeneous CSLA architectures are shown in Figure 1 for an example case of 8-bit addition. Figure 1(a) depicts an 8-bit RCA, which is formed by a cascade of full adder modules; the full adder [9] is an arithmetic building block that adds an augend and addend bit (say, a and b) along with any carry input (cin) and produces two outputs, namely, sum (Sum) and carry overflow (Cout). Since there is a rippling of carry from one full adder stage to another, the propagation delay of the RCA varies linearly in proportion to the adder width. The CSLA basically partitions the input data into groups and addition within the groups is carried out in parallel; that is, the CSLA is composed of partitioned and duplicated RCAs. It can be seen from Figure 1 that the least significant 4-bit adder stages of RCA and CSLAs are identical. However, the carry produced by the least significant nibble is simply propagated through the more significant nibble in the case of the RCA bit-by-bit, while the carry corresponding to the least significant nibble serves as the selection input for MUXes present in the more significant position in the case of CSLAs.

Figure 1(b) shows the 8-bit conventional CSLA comprising full adders and 2:1 MUXes, henceforth referred to as simply "CSLA." In the case of CSLA shown in Figure 1(b), the full adders present in the most significant nibble position are duplicated with carry inputs (cin) of 0 and 1 assumed; that is, one 4-bit RCA with a carry input ("cin") of 0 and another 4-bit RCA with a carry input ("cin") of 1 are used. Notice that both these RCAs have the same augend and addend inputs. While the least significant 4-bit RCA would be adding the augend inputs (a_3 to a_0) with the addend inputs (b_3 to b_0), the more significant 4-bit RCAs would be simultaneously adding up the augend inputs (a_7 to a_4) with the addend inputs (b_7 to b_4), with presumed carry inputs (cin) of 0 and 1. Due to two addition sets, two sets of sum and carry outputs are produced, one based on 0 as the carry input and another based on 1 as the carry input, which are in turn fed as inputs to the 2:1 MUXes. The number of MUXes used depends on the size of the RCA duplicated. To determine the true sum outputs and the real value of carry overflow pertaining to the most significant nibble position, the carry output (c_4) from the least significant 4-bit RCA is used as the common select input for all the MUXes; thereby the correct result corresponding to either the RCA with 0 as the carry input or the RCA with 1 as the carry input is displayed as output.

Figure 1(c) portrays the 8-bit CSLA containing full adders, 2:1 MUXes, and BEC logic, henceforth identified as "CSLA_BEC". Figure 1(c) also shows the internals of the 5-bit BEC, which is depicted by the circuit shown within the oval. The CSLA_BEC is rather different from the CSLA in that instead of having an RCA with a presumed carry input of 1 in the more significant nibble position, the BEC circuit is introduced. The BEC logic adds binary 1 to the least significant bit of its binary inputs and produces the resultant sum and carry as output. As seen in Figure 1(c), the BEC accepts as inputs the sum and carry outputs of the RCA having a presumed carry input of 0, adds binary 1 to this input, and produces the resulting sum and carry overflow as output. Now the correct result exists between choosing the output of the RCA with a presumed carry input of 0 and the output of the BEC logic. The carry output c_4 of the least significant RCA is used to determine the correct set of the most significant nibble position sum and carry outputs. The logic equations governing the 5-bit BEC are given below. In the equations, $\sim$ signifies logical inversion, $\oplus$ implies logical XOR, and $\bullet$ represents logical conjunction. Consider

$$\text{Sum}_4^1 = \sim \text{Sum}_4^0$$

$$\text{Sum}_5^1 = \text{Sum}_5^0 \oplus \text{Sum}_4^0$$

$$\text{Sum}_6^1 = \text{Sum}_6^0 \oplus \left(\text{Sum}_5^0 \bullet \text{Sum}_4^0 \right)$$

$$\text{Sum}_7^1 = \text{Sum}_7^0 \oplus \left(\text{Sum}_6^0 \bullet \text{Sum}_5^0 \bullet \text{Sum}_4^0 \right)$$

$$c_8^1 = c_8^0 \oplus \left(\text{Sum}_7^0 \bullet \text{Sum}_6^0 \bullet \text{Sum}_5^0 \bullet \text{Sum}_4^0 \right).$$

(1)

The CSLA constructed on the basis of sharing of CBL is depicted through Figure 2, which will be referred to as "CSLA_CBL" henceforth. The CSLA_CBL adder is founded

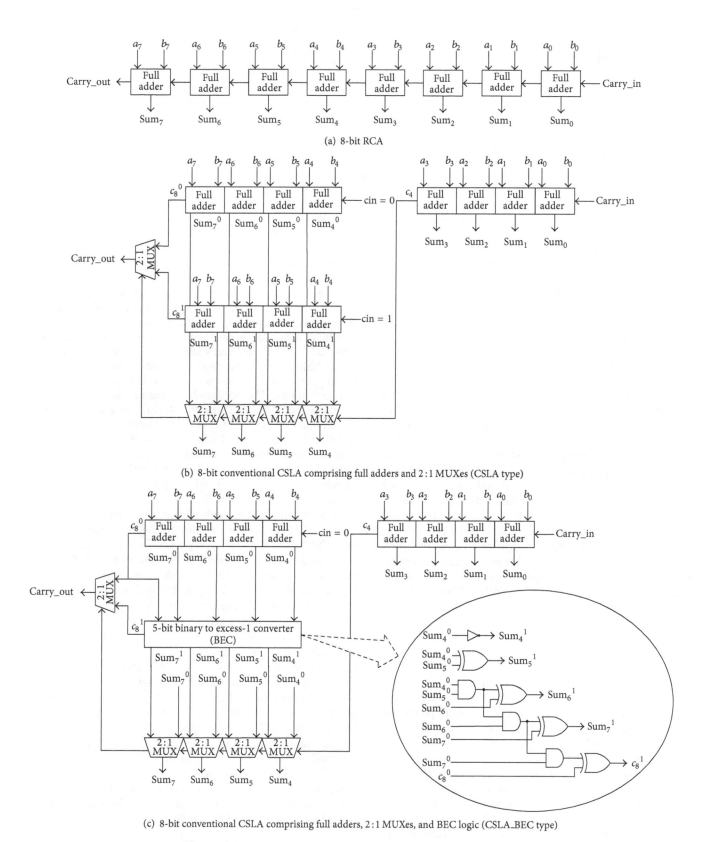

(a) 8-bit RCA

(b) 8-bit conventional CSLA comprising full adders and 2:1 MUXes (CSLA type)

(c) 8-bit conventional CSLA comprising full adders, 2:1 MUXes, and BEC logic (CSLA_BEC type)

FIGURE 1: (a) 8-bit RCA, (b) representative 8-bit homogeneous CSLA, and (c) representative 8-bit homogeneous CSLA with BEC logic.

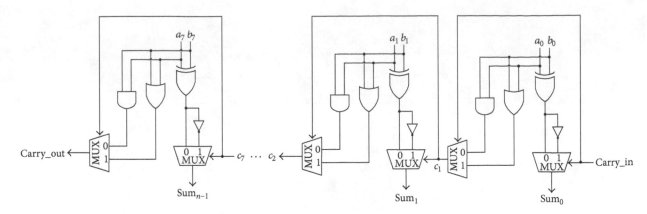

FIGURE 2: 8-bit homogeneous CSLA utilizing shared CBL (CSLA_CBL architecture).

upon utilizing the full adder logic, whose underlying equations are given below with a, b, and cin being the primary inputs, and Sum and Cout being the primary outputs. In (3), "+" implies logical disjunction:

$$\text{Sum} = a \oplus b \oplus \text{cin} \qquad (2)$$

$$\text{Cout} = (a + b) \bullet \text{cin} + (a \bullet b)(\sim \text{cin}). \qquad (3)$$

Referring to (2) and (3), it may be understood that, for a carry input (cin) of 0, (2) and (3) reduce to Sum = $(a \oplus b)$ and Cout = $(a \bullet b)$. With cin = 1, (2) and (3) become Sum =$\sim$ $(a \oplus b)$ and Cout = $(a + b)$. Based on this principle, sum and carry outputs for both possible values of input carries are generated simultaneously and fed as inputs to two 2:1 MUXes. The correct sum and carry outputs are determined by the carry input, serving as the select input for the two MUXes. Though the exorbitant duplicated RCA and RCA with BEC logic structures are eliminated through this approach, leading to savings in terms of area, nevertheless, since the carry propagates from stage-to-stage, the critical data path delay tends to be proportional to the size of the full adders cascade. As a consequence, the delay of the CSLA_CBL adder may be close to that of the RCA which is confirmed by the simulation results given in Section 4.

3. Heterogeneous/Hybrid CSLA Architectures

Apart from synthesizing basic CSLA topologies viz. CSLA, CSLA_BEC, and CSLA_CBL, hybrid CSLA architectures involving CSLA and CLA/SCBCLA were also implemented with the intention of minimizing the maximum propagation path delay. It is well known that a CLA is faster than a RCA, and hence it may be worthwhile to have a CLA as a replacement for the least significant RCA in the CSLA structure. Although the concept of carry lookahead is widely understood, the concept of section-carry based lookahead may not be that well known, and hence to explain the distinction between the two, sample 4-bit lookahead logic realized using these two approaches is portrayed in Figure 3 for an illustration. For details on different section-carry based carry lookahead structures and SCBCLA constructions using them, an avid reader is directed to references [25–27], which

constitute prior works in the realm of synchronous and asynchronous designs.

The section-carry based carry lookahead generator shown enclosed within the circle in Figure 3 produces a single lookahead carry signal corresponding to a "section" or "group" of the adder inputs (hence the term "section-carry"), while the conventional carry lookahead generator encapsulated within the rectangle produces multiple lookahead carry signals corresponding to each pair of augend and addend primary inputs. The section-carry based carry lookahead generator differs from the traditional carry lookahead generator in that bit-wise lookahead carry signals are not required to be computed for the former. The XOR and AND gates used for producing the necessary propagate and generate signals (P_3 to P_0 and G_3 to G_0) are highlighted using dotted lines in Figure 3; these constitute the propagate-generate logic referred to in Figures 4 and 5.

8-bit hybrid CSLAs with/without BEC logic and comprising a CLA in the least significant stage viz. "CSLA-CLA" and "CSLA_BEC-CLA" adder types are shown in Figure 4. On the other hand, 8-bit hybrid CSLAs with/without BEC logic and incorporating a SCBCLA in the least significant stage viz. "CSLA-SCBCLA" and "CSLA_BEC-SCBCLA" adder varieties are portrayed in Figure 5. Both the conventional CLA and SCBCLA constitute three functional blocks: propagate-generate logic, lookahead carry generator, and the sum producing logic. Not only is the carry lookahead generator different for CLA and SCBCLA adders, but the sum producing logic is also different; in case of CLA, the sum producing logic comprises only XOR gates, whereas in the SCBCLA, the sum producing logic consists of full adders and an XOR gate, with the XOR gate providing the sum of the primary inputs a_3, b_3, and c_3. While rippling of carries occurs internally within the carry-propagate adder constituting the SCBCLA and producing the requisite sums, the lookahead carry signal corresponding to an adder section is generated independently (in parallel) and serves as the lookahead carry input for the successive CSLA stage.

4. Results and Discussion

Three homogeneous CSLA architectures viz. CSLA, CSLA_BEC, and CSLA_CBL and four heterogeneous CSLA architectures

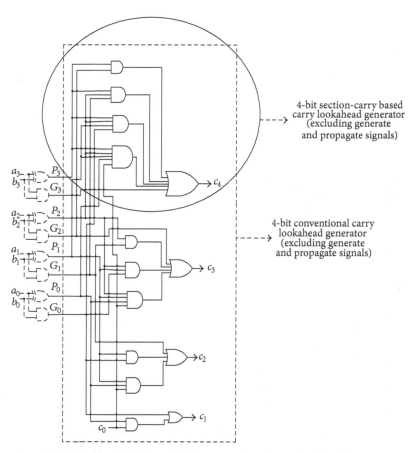

4-bit section-carry based
carry lookahead generator
(excluding generate
and propagate signals)

4-bit conventional carry
lookahead generator
(excluding generate
and propagate signals)

FIGURE 3: 4-bit conventional and section-carry based carry lookahead generators.

viz. CSLA-CLA, CSLA_BEC-CLA, CSLA-SCBCLA, and CSLA_BEC-SCBCLA were described topologically in Verilog HDL similar to previous works [16, 21–23, 25] to perform two kinds of addition operations viz. dual-operand addition and multioperand addition. For dual-operand addition, two binary operands having corresponding sizes of 32-bits and 64-bits were considered. For multioperand addition, addition of four binary operands, each of size 32-bits, and another multioperand addition involving four binary operands with each having size of 64-bits were considered. Moreover, two types of multioperand additions were performed based on (i) carry save adder (CSA) topology, and (ii) bit-partitioned addition scheme. All the adders were synthesized using a 90 nm FPGA (XC3S1600E) [28], with speed optimization specified as the design goal in the Xilinx 9.1i ISE design suite. The critical path delay and area values (in terms of number of basic logic elements viz. BELs) were ascertained after automatic place-and-route. The results of dual-operand additions shall be presented first, followed by the results obtained for multioperand additions.

4.1. Dual-Operand Addition. CSLAs can be implemented on the basis of uniform or nonuniform primary input partitions; accordingly they are labeled as "uniform" or "non-uniform" CSLAs, in a structural sense. "Input partitioning" basically means splitting up of the primary inputs into groups of inputs so as to pave the way for addition to be done in parallel within

the partitions; it should be noted that input partitioning is inherent to all CSLAs except the CSLA_CBL type (shown in Figure 2) which has a regular carry select structure and hence is void of input partitions. Referring to Figure 1(b), it can be seen that 8 pairs of inputs have been split into two uniform or equal-sized groups of 4-input pairs; thus it can be said that the 8-bit CSLA is realized according to a 4-4 input partition.

For synthesis, 3 uniform input partitions (4-4-4-4-4-4-4, 8-8-8-8, and 16-16) and 2 optimum nonuniform input partitions (3-7-6-5-4-3-2-2 [29] and 8-7-6-4-3-2-2 [15]) were considered for realizing the 32-bit CSLAs. Figure 6 visually portrays the variations in propagation delay corresponding to different primary input partitions for the six CSLA types. On the other hand, 4 uniform input partitions viz. 4-4-4-4-4-4-4-4-4-4-4-4-4-4-4-4, 8-8-8-8-8-8-8-8, and 16-16-16-16, 32-32, and a nonuniform input partition viz. 8-10-9-8-7-6-5-4-3-2-2 [29] were considered for realizing the 64-bit CSLAs. Figure 7 depicts the propagation delay variations subject to different primary input partitions for the six CSLA architectures. The trend line highlighted in Figure 6 shows that the uniform 8-8-8-8 input partition consistently paves the way for least propagation delay (varying from 17 ns to 20 ns) with respect to various 32-bit homogeneous and heterogeneous CSLAs. Similarly the trend line indicated in black in Figure 7 conveys that the uniform 16-16-16-16 input partition results in the least data path delay (varying from 27 ns to 29 ns) for the different homogeneous and heterogeneous 64-bit CSLAs.

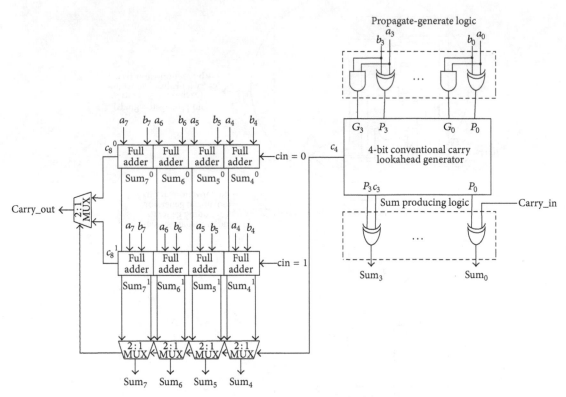

(a) 8-bit hybrid CSLA with a conventional CLA in the least significant stage

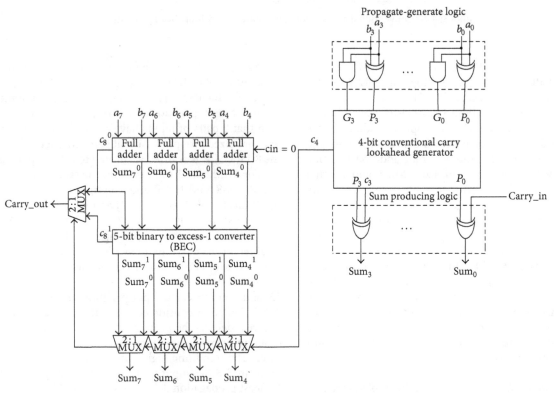

(b) 8-bit hybrid CSLA featuring BEC with a least significant CLA stage

FIGURE 4: Hybrid CSLAs without/with BEC logic comprising a CLA: (a) CSLA-CLA type and (b) CSLA_BEC-CLA type.

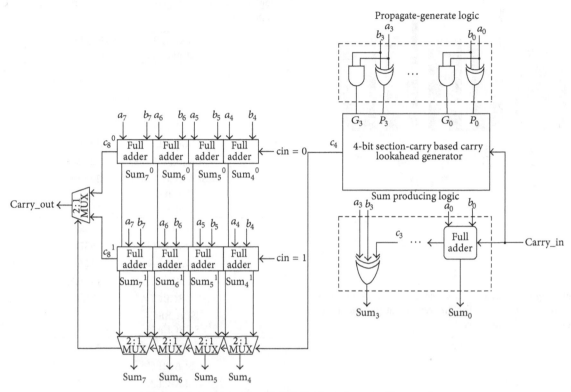

(a) 8-bit hybrid CSLA with 4-bit SCBCLA in the least significant stage

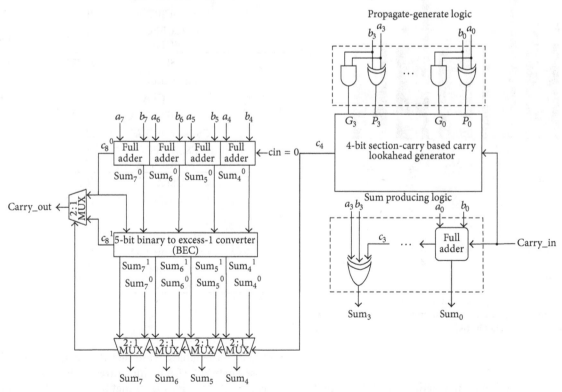

(b) 8-bit hybrid CSLA incorporating BEC with a least significant SCBCLA stage

FIGURE 5: Hybrid CSLAs without/with BEC logic comprising a SCBCLA: (a) CSLA-SCBCLA type and (b) CSLA_BEC-SCBCLA type.

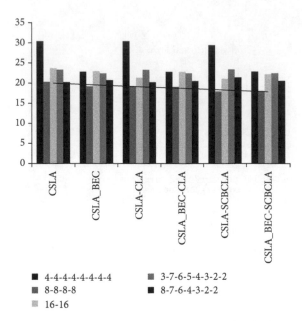

FIGURE 6: Capturing worst-case delay variations of 32-bit homogeneous and heterogeneous CSLAs for different input partitions. X-axis: CSLA type; Y-axis: Delay in ns.

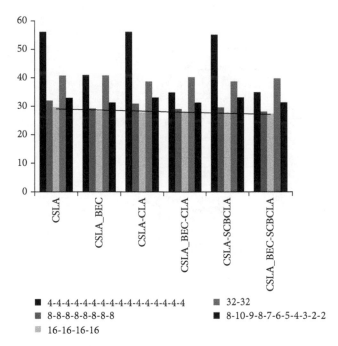

FIGURE 7: Portraying critical path delay variations of 64-bit homogeneous and heterogeneous CSLAs for different input partitions. X-axis: CSLA type; Y-axis: Delay in ns.

TABLE 1: Maximum propagation delay and area (# BELs) of 32-bit homogeneous and heterogeneous CSLAs corresponding to diverse input partitions.

Input partition	Type of CSLA architecture	Critical path delay (ns)	Area (# BELs)
Not applicable	RCA	30.604	**63**
Not applicable	CSLA_CBL	37.604	**63**
4-4-4-4-4-4-4-4	CSLA	30.388	105
	CSLA_BEC	22.820	106
	CSLA-CLA	30.398	106
	CSLA_BEC-CLA	22.781	106
	CSLA-SCBCLA*	29.359	108
	CSLA_BEC-SCBCLA*	22.864	108
8-8-8-8	CSLA	20.280	117
	CSLA_BEC	19.176	104
	CSLA-CLA	19.260	121
	CSLA_BEC-CLA	19.059	104
	CSLA-SCBCLA*	**17.897**	123
	CSLA_BEC-SCBCLA*	18.052	110
16-16	CSLA	23.722	105
	CSLA_BEC	22.986	91
	CSLA-CLA	21.384	114
	CSLA_BEC-CLA	22.835	91
	CSLA-SCBCLA*	21.097	119
	CSLA_BEC-SCBCLA*	22.255	106
3-7-6-5-4-3-2-2	CSLA	23.337	110
	CSLA_BEC	22.411	108
	CSLA-CLA	23.337	110
	CSLA_BEC-CLA	22.411	108
	CSLA-SCBCLA*	23.408	110
	CSLA_BEC-SCBCLA*	22.482	108
8-7-6-4-3-2-2	CSLA	20.218	118
	CSLA_BEC	20.743	111
	CSLA-CLA	20.218	118
	CSLA_BEC-CLA	20.473	111
	CSLA-SCBCLA*	21.403	117
	CSLA_BEC-SCBCLA*	20.544	111

The maximum combinational path delay (also called, "critical path delay") encountered and the total number of BELs consumed by different homogeneous and heterogeneous CSLAs to perform the addition of two 32-bit operands and two 64-bit operands separately is shown in Tables 1 and 2, respectively. The optimum delay and area values are in bold font in the tables. Note that the symbol * signifies the proposed hybrid CSLA architectures in the tables.

From Table 1, it is evident that the CSLA-SCBCLA hybrid adder based on the 8-8-8-8 input partition features the least propagation delay (17.897 ns) amongst all homogeneous and hybrid CSLAs, and hence the 8-8-8-8 input partition is deemed to be optimum. The 32-bit RCA has critical path delay of 30.604 ns, while the 32-bit CSLA_CBL adder is found to have the longest path delay of 37.604 ns. Compared to the maximum delay of the hybrid CSLA-SCBCLA, the hybrid CSLA_BEC-SCBCLA adder which is another proposed hybrid CSLA topology has a comparable speed performance of 18.052 ns. However with respect to area, the RCA and CSLA_CBL structures require less number of BELs than all the CSLAs. Hence it is inferred from Figure 6 and Table 1 that for the addition of two input operands having sizes of 32-bits the hybrid CSLA-SCBCLA adder is preferable over all other homogeneous and heterogeneous CSLAs and the favorable input data partition is 8-8-8-8.

Based on a similar observation, by referring to Figure 7 and Table 2, it can be seen that the 16-16-16-16 input partition is found to be optimum from a delay (i.e., speed) perspective for 64-bit dual-operand addition. The proposed CSLA_BEC-SCBCLA constructed using the 16-16-16-16 input data partition leads to the least latency amongst all other adder topologies; however, the other proposed CSLA viz. CSLA-SCBCLA based on a similar input partition features almost a similar delay metric. In terms of area occupancy though, the 64-bit RCA is optimized. Nevertheless, the RCA encounters considerably more data path delay by 1.6× in comparison with the proposed CSLA_BEC-SCBCLA based on a 16-16-16-16 input partition.

4.2. Multioperand Addition.

The performance of different homogeneous and heterogeneous CSLAs is evaluated based on the case studies of multioperand addition involving 4 binary operands, with respective sizes of 32-bits and 64-bits. Two multioperand addition schemes are considered, one involving the carry save adder (CSA) topology, and another involving the bit-partitioning method.

4.2.1. CSA Based Multioperand Addition.

The structure of an example CSA used to add four n-bit binary numbers is shown in Figure 8. Here, a_{n-1} to a_0, b_{n-1} to b_0, c_{n-1} to c_0, and d_{n-1} to d_0 represent the primary inputs and the sum bits and Sum_{n+1} to Sum_0 represents the primary outputs. The subscript 0 denotes the LSB and the subscript $(n-1)$ denotes the MSB. As shown in Figure 8, there are three adders in three levels to perform the addition of four input operands. In each CSA, the carry output signal of the current bit at a level is not transferred to the next bit adder of the same level as the carry input; instead, the carry output is transferred to the next bit adder in the lower level as the carry input. In the top-level adder, three numbers (a, b, and c) are added simultaneously; that is, the bits corresponding to any number could act as input carries for the full adders of the first level CSA. In the next lower level, an extra number (d) is added. The adder in the bottom level, shown within the ellipse in Figure 8, is a simple RCA which is what portrayed here but it may be any dual-operand adder that can be used to compute the final sum.

Experimentation was performed by having different dual-operand adders viz. RCA and various homogeneous and heterogeneous CSLAs in the final adder stage of the CSA, shown in Figure 8, to analyze their relative performance for two different addition scenarios: (i) addition of four binary operands, each of size 32-bits, and (ii) addition of four binary operands with each having size of 64-bits.

TABLE 2: Maximum propagation delay and area (# BELs) of 64-bit homogeneous and heterogeneous CSLAs corresponding to different input partitions.

Input partition	Type of CSLA architecture	Critical path delay (ns)	Area (# BELs)
Not applicable	RCA	71.555	**127**
Not applicable	CSLA_CBL	70.525	129
4-4-4-4- 4-4-4-4- 4-4-4-4- 4-4-4-4	CSLA	56.091	217
	CSLA_BEC	40.870	209
	CSLA-CLA	56.101	218
	CSLA_BEC-CLA	34.799	215
	CSLA-SCBCLA*	55.062	220
	CSLA_BEC-SCBCLA*	34.882	217
8-8-8-8- 8-8-8-8	CSLA	31.866	251
	CSLA_BEC	29.119	224
	CSLA-CLA	30.846	255
	CSLA_BEC-CLA	29.002	224
	CSLA-SCBCLA*	29.483	257
	CSLA_BEC-SCBCLA*	27.995	230
16-16-16-16	CSLA	29.625	252
	CSLA_BEC	28.259	212
	CSLA-CLA	27.759	261
	CSLA_BEC-CLA	28.029	213
	CSLA-SCBCLA*	27.427	266
	CSLA_BEC-SCBCLA*	**27.322**	227
32-32	CSLA	40.705	217
	CSLA_BEC	40.742	189
	CSLA-CLA	38.591	215
	CSLA_BEC-CLA	40.157	189
	CSLA-SCBCLA*	38.591	247
	CSLA_BEC-SCBCLA*	39.682	219
8-10-9-8-7-6- 5-4-3-2-2	CSLA	32.983	251
	CSLA_BEC	31.204	226
	CSLA-CLA	32.983	251
	CSLA_BEC-CLA	31.204	226
	CSLA-SCBCLA*	33.054	251
	CSLA_BEC-SCBCLA*	31.276	226

The FPGA-based synthesis results viz. delay and area obtained for the addition of four binary operands, each having size of 32-bits, are given in Table 3 with the optimized values in bold font. Since the 8-8-8-8 primary input partition was found to yield the least data path delay, as evident from Figure 6 and Table 1, it was preferred for the various CSLA realizations. It can be seen from Table 3 that the hybrid CSLA_BEC-CLA when used in the final adder stage of the CSA encounters the least propagation delay, with

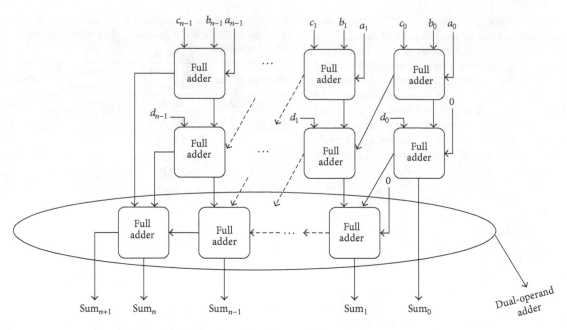

FIGURE 8: CSA topology for addition of four n-bit binary operands.

TABLE 3: Critical path delay and area figures for CSA-based multi-operand addition of four 32-bit operands, with RCA/homogeneous/heterogeneous CSLAs used in the final adder stage.

Input partition	Type of adder architecture	Critical path delay (ns)	Area (# BELs)
Not applicable	RCA	39.842	**190**
Not applicable	CSLA_CBL	39.842	**190**
8-8-8-8	CSLA	27.383	229
	CSLA_BEC	22.455	229
	CSLA-CLA	25.053	229
	CSLA_BEC-CLA	**21.326**	232
	CSLA-SCBCLA*	23.378	227
	CSLA_BEC-SCBCLA*	21.684	233

TABLE 4: Critical path delay and area for CSA-based multioperand addition of four 64-bit operands, with RCA/homogeneous/heterogeneous CSLAs used in the final adder stage.

Input partition	Type of adder architecture	Critical path delay (ns)	Area (# BELs)
Not applicable	RCA	73.792	**382**
Not applicable	CSLA_CBL	71.667	383
16-16-16-16	CSLA	37.034	472
	CSLA_BEC	31.307	462
	CSLA-CLA	33.363	476
	CSLA_BEC-CLA	**30.428**	471
	CSLA-SCBCLA*	32.008	473
	CSLA_BEC-SCBCLA*	30.732	470

the proposed CSLA_BEC-SCBCLA adder closely following it with just a 1.7% delay difference. The conventional CLA, when used in the final adder stage of the CSA as a "homogeneous adder," reports a critical path delay of 34.306 ns. On the contrary, when the conventional CLA is used along with the CSLA inclusive of the BEC as a "heterogeneous adder" (CSLA_BEC-CLA), it enables considerable decrease in maximum data path delay by 37.8% vindicating the observation made in [24] that heterogeneous adders are preferable over homogeneous adders for delay optimization. Although the use of RCA and CSLA_CBL adders in the final adder stage of the CSA helps to minimize the area occupancy compared to their counterparts, they suffer from an exacerbated increase in delay of about 87% over the CSLA_BEC-CLA type.

The synthesis results obtained for the addition of four binary operands, each having sizes of 64-bits, is shown in

Table 4 and the optimized values are in bold font. Since the 16-16-16-16 uniform input partition was found to be delay optimal (refer to Figure 7 and Table 2), it was adopted for implementing all the CSLAs. Again, the CSLA_BEC-CLA variant reports the least propagation delay compared to others as in the previous case, with the proposed CSLA_BEC-SCBCLA reporting almost a similar performance. However due to less logic complexity, the usage of RCA or CSLA_CBL in the final adder stage of the CSA results in the least area occupancy in comparison with the rest, albeit at the expense of a considerable increase in delay by about 1.4x.

4.2.2. Bit-Partitioned Multioperand Addition. In CSAs, rowwise parallel addition is performed where the tree height (i.e., number of adder levels) grows with an increase in

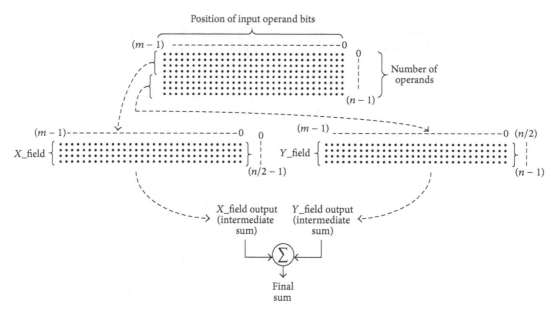

FIGURE 9: Bit-partitioned multioperand addition scheme.

the number of input operands by an approximate linear order. To reduce the logic depth of the adder tree, a bit-partitioning strategy was presented in [30] in the context of self-timed multioperand addition, which involved splitting up of the entire group of data operands into a desired number of subgroups, and the intermediate addition results of the subgroups are finally added to produce the final sum. The bit-partitioning approach basically parallelizes the multioperand addition and is illustrated through Figure 9 for an example scenario where addition of "n" binary operands with each operand having a size of "m" bits is considered whilst assuming "n" to be even. A "dot" represents a bit position in Figure 9.

The entire set of input operands from bit position 0 to bit position $(n-1)$ is divided into two equal-sized groups (for an example) as X_field, which comprises inputs from bit positions 0 to $(n/2-1)$ and the Y_field consisting of inputs from bit positions $(n/2)$ to $(n-1)$. Addition within the individual fields (i.e., X_field and Y_field) is performed simultaneously and the sum bits generated as intermediate outputs from these individual fields (X_field and Y_field) are then added together using a final dual-operand adder to produce the required sum. The bit-partitioning scheme might help to speed-up the addition, especially when several operands have to be added by way of performing parallel column-wise addition of row-wise partitions. For example, considering the addition of 32 data operands, each of size 32 bits, the CSA topology would encounter thirty full adder delays plus the delay associated with the final dual-operand adder. On the other hand, based on the bit-partitioning technique, considering eight partitions with each partition comprising four data operands, the bit-partitioned multioperand adder based upon the CSA topology could encounter a reduced propagation delay of about four full adder delays plus the delay of a dual-operand adder, depending upon the implementation. Also, a high

regularity would be implicit within the overall architecture as the gate-level hardware is being duplicated.

In this work, the bit-partitioning scheme was employed to partition the set of four inputs into two input groups (X_field and Y_field, as shown in Figure 9) and the outputs of X and Y fields were then added to produce the final sum. Several dual-operand adders were used to realize the bit-partitioned addition separately viz. RCA, CSLA_CBL, CSLA, CSLA_BEC, CSLA-CLA, CSLA_BEC-CLA, CSLA-SCBCLA, and CSLA_BEC-SCBCLA. The different bit-partitioned addition structures were individually synthesized using the same FPGA (XC3S1600E). It should be noted that the focus here is only on evaluating the performance of the RCA and different CSLAs as employed for multioperand addition and not to comment upon the efficacy of the bit-partitioning scheme as such (i.e., no comparison with the results of the previous subsection). This is because, as mentioned in the preceding discussions, the bit-partitioning technique is scalable, can be custom-defined, and could potentially benefit in terms of latency reduction primarily for additions involving typically higher dimensions as compared with conventional combinational tree structures.

Table 5 presents the timing and area results obtained for the synthesis of bit-partitioned multi-input addition of 4 binary operands, each of size 32-bits, on the basis of RCA and various homogeneous and heterogeneous CSLAs. Since the 8-8-8-8 uniform input partition was found to be delay-optimum for realizing the 32-bit CSLAs (refer to Figure 6 and Table 1), only this uniform input partition has been considered for implementing the various homogeneous and hybrid CSLAs corresponding to X-field and Y_field of the bit-partitioned multioperand addition. To sum up the outputs of X-field and Y_field, a 33-bit dual-operand adder would be required in which case an extra bit has been added to the most significant position of various CSLA input partitions.

TABLE 5: Critical path delay and area metrics for bit-partitioned multioperand addition of four 32-bit operands, with RCA and various homogeneous/hybrid CSLA architectures used.

Input partition	Type of adder architecture	Critical path delay (ns)	Area (# BELs)
Not applicable	RCA	39.928	**190**
Not applicable	CSLA_CBL	42.241	195
8-8-8-8	CSLA	32.303	458
	CSLA_BEC	29.278	311
	CSLA-CLA	31.727	359
	CSLA_BEC-CLA	28.207	325
	CSLA-SCBCLA*	27.628	365
	CSLA_BEC-SCBCLA*	**27.056**	328

TABLE 6: Critical path delay and area parameters for bit-partitioned multioperand addition of four 64-bit operands, with RCA and various homogeneous/hybrid CSLA architectures used.

Input partition	Type of adder architecture	Critical path delay (ns)	Area (# BELs)
Not applicable	RCA	73.840	**382**
Not applicable	CSLA_CBL	77.946	388
16-16-16-16	CSLA	50.957	748
	CSLA_BEC	46.559	637
	CSLA-CLA	50.426	781
	CSLA_BEC-CLA	45.679	648
	CSLA-SCBCLA*	**45.608**	800
	CSLA_BEC-SCBCLA*	45.665	691

The optimum synthesis metrics obtained for the example multi-input addition are in bold font in Table 5. It can be seen that the proposed CSLA_BEC-SCBCLA paves the way for least computation time (27.056 ns) amongst all. In comparison, the undesirable increases in delay values for other bit-partitioned multioperand adders incorporating RCA, CSLA_CBL, CSLA, CSLA_BEC, CSLA-CLA, CSLA_BEC-CLA, and CSLA-SCBCLA types are found to be 47.6%, 56.1%, 15.9%, 3%, 15.9%, 3%, and 2.1%, respectively. However, the RCA results in the lowest area occupancy (190 BELs) and the CSLA_CBL adder occupies nearly the same area with just 5 more BELs. Nevertheless, the bit-partitioned multioperand adder based upon the RCA pays a 47.6% delay penalty in comparison with that utilizing the CSLA_BEC-SCBCLA.

Table 6 shows the delay and area values obtained for the synthesis of bit-partitioned addition of four input operands of sizes 64 bits, corresponding to different adder architectures, with the CSLAs utilizing the 16-16-16-16 uniform input partition since this partition was found to be delay optimal (refer to Figure 7 and Table 2). With respect to less area, the RCA is found to be the optimum architecture. However,

in terms of less critical path delay, the proposed CSLA-SCBCLA benefits by achieving a good delay reduction of 38.2% compared to the maximum path delay of the RCA based bit-partitioned multioperand adder.

5. Conclusions

CSLA is an important member of the high-speed adder family. In this paper, existing CSLA architectures viz. homogeneous and heterogeneous have been described and two new hybrid CSLA topologies were put forward: (i) carry select-cum-section-carry based carry lookahead adder (CSLA-SCBCLA) and (ii) carry select-cum-section-carry based carry lookahead adder including BEC logic (CSLA_BEC-SCBCLA). The speed performances of the various CSLA structures have been analyzed based on the case studies of 32-bit and 64-bit dual-operand and multioperand additions. Both uniform and nonuniform input data partitions were considered for the various CSLA implementations and FPGA-based synthesis was performed. It has been found for dual-operand additions; the proposed CSLA-SCBCLA/CSLA_BEC-SCBCLA architecture is faster and outperforms all other homogeneous and heterogeneous CSLAs. For bit-partitioned multi-input additions, the proposed CSLA-SCBCLA/CSLA_BEC-SCBCLA architecture promises high speed. Nevertheless, for multioperand addition based on the CSA topology, the conventional CSLA_BEC-CLA and the proposed CSLA_BEC-SCBCLA architectures were found to exhibit an optimized and comparable speed performance. From the inferences derived through this work, it is likely that the proposed hybrid CSLA architectures could achieve enhanced performance over conventional CSLAs for ASIC-based synthesis as well.

Conflict of Interests

The authors declare that there is no conflict of interests regarding the publication of this paper.

Acknowledgment

The authors thank the constructive comments of the reviewers, especially the pointing out of some typos in the initial submitted version by a reviewer, which has helped to improve this paper's presentation.

References

[1] O. J. Bedrij, "Carry-select adder," *IRE Transactions on Electronic Computers*, vol. 11, no. 3, pp. 340–346, 1962.

[2] A. R. Omondi, *Computer Arithmetic Systems: Algorithms, Architecture and Implementation*, Prentice Hall, 1994.

[3] I. Koren, *Computer Arithmetic Algorithms*, A K Peeters/CRC Press, 2nd edition, 2001.

[4] B. Parhami, *Computer Arithmetic: Algorithms and Hardware Designs*, Oxford University Press, New York, NY, USA, 2nd edition, 2010.

[5] T.-Y. Chang and M.-J. Hsiao, "Carry-select adder using single ripple-carry adder," *Electronics Letters*, vol. 34, no. 22, pp. 2101–2103, 1998.

[6] Y. Kim and L.-S. Kim, "64-bit carry-select adder with reduced area," *Electronics Letters*, vol. 37, no. 10, pp. 614–615, 2001.

[7] B. Ramkumar and H. M. Kittur, "Low-power and area-efficient carry select adder," *IEEE Transactions on VLSI Systems*, vol. 20, no. 2, pp. 371–375, 2012.

[8] I.-C. Wey, C.-C. Ho, Y.-S. Lin, and C.-C. Peng, "An area-efficient carry select adder design by sharing the common boolean logic term," in *Proceedings of the International MultiConference of Engineers and Computer Scientists (IMECS '12)*, vol. 2, pp. 1091–1094, March 2012.

[9] P. Balasubramanian and N. E. Mastorakis, "High speed gate level synchronous full adder designs," *WSEAS Transactions on Circuits and Systems*, vol. 8, no. 2, pp. 290–300, 2009.

[10] W. Jeong and K. Roy, "Robust high-performance low-power carry select adder," in *Proceedings of the Asia and South Pacific Design Automation Conference*, pp. 503–506, Kitakyushu, Japan, January 2003.

[11] M. Alioto, G. Palumbo, and M. Poli, "A gate-level strategy to design carry select adders," in *Proceedings of the IEEE International Symposium on Circuits and Systems*, vol. 2, pp. 465–468, IEEE, May 2004.

[12] M. Alioto, G. Palumbo, and M. Poli, "Optimized design of parallel carry-select adders," *Integration, the VLSI Journal*, vol. 44, no. 1, pp. 62–74, 2011.

[13] A. Nève, H. Schettler, T. Ludwig, and D. Flandre, "Power-delay product minimization in high-performance 64-bit carry select adders," *IEEE Transactions on Very Large Scale Integration (VLSI) Systems*, vol. 12, no. 3, pp. 235–244, 2004.

[14] Y. He, C.-H. Chang, and J. Gu, "An area efficient 64-bit square root carry-select adder for low power applications," in *Proceedings of the IEEE International Symposium on Circuits and Systems (ISCAS '05)*, vol. 4, pp. 4082–4085, May 2005.

[15] B. K. Mohanty and S. K. Patel, "Area-delay-power efficient carry select adder," *IEEE Transactions on Circuits and Systems II: Express Briefs*, vol. 61, no. 6, pp. 418–422, 2014.

[16] J. Monteiro, J. L. Güntzel, and L. Agostini, "A1CSA: an energy-efficient fast adder architecture for cell-based VLSI design," in *Proceedings of the 18th IEEE International Conference on Electronics, Circuits and Systems (ICECS '11)*, pp. 442–445, Beirut, Lebanon, December 2011.

[17] Y. Chen, H. Li, K. Roy, and C.-K. Koh, "Cascaded carry-select adder (C²SA): a new structure for low-power CSA design," in *Proceedings of the International Symposium on Low Power Electronics and Design*, pp. 115–118, August 2005.

[18] Y. Wang, C. Pai, and X. Song, "The design of hybrid carry-lookahead/carry-select adders," *IEEE Transactions on Circuits and Systems II: Analog and Digital Signal Processing*, vol. 49, no. 1, pp. 16–24, 2002.

[19] G. A. Ruiz and M. Granda, "An area-efficient static CMOS carry-select adder based on a compact carry look-ahead unit," *Microelectronics Journal*, vol. 35, no. 12, pp. 939–944, 2004.

[20] H. G. Tamar, A. G. Tamar, K. Hadidi, A. Khoei, and P. Hoseini, "High speed area reduced 64-bit static hybrid carry-lookahead/carry-select adder," in *Proceedings of the 18th IEEE International Conference on Electronics, Circuits and Systems (ICECS' 11)*, pp. 460–463, December 2011.

[21] V. Kokilavani, P. Balasubramanian, and H. R. Arabnia, "FPGA realization of hybrid carry select-cum-section-carry based carry lookahead adders," in *Proceedings of the 12th International Conference on Embedded Systems and Applications*, pp. 81–85, 2014.

[22] R. Yousuf and Najeeb-ud-din, "Synthesis of carry select adder in 65nm FPGA," in *Proceedings of the IEEE Region 10 Conference (TENCON '08)*, pp. 1–6, November 2008.

[23] U. Sajesh Kumar and K. K. Mohamed Salih, "Efficient carry select adder design for FPGA implementation," *Procedia Engineering*, vol. 30, pp. 449–456, 2012.

[24] J.-G. Lee, J.-A. Lee, B.-S. Lee, and M. D. Ercegovac, "A design method for heterogeneous adders," in *Embedded Software and Systems*, vol. 4523 of *Lecture Notes in Computer Science*, pp. 121–132, Springer, 2007.

[25] K. Preethi and P. Balasubramanian, "FPGA implementation of synchronous section-carry based carry look-ahead adders," in *Proceedings of the IEEE 2nd International Conference on Devices, Circuits and Systems (ICDCS '14)*, pp. 1–4, IEEE, Combiatore, India, March 2014.

[26] P. Balasubramanian, D. A. Edwards, and W. B. Toms, "Self-timed section-carry based carry lookahead adders and the concept of alias logic," *Journal of Circuits, Systems and Computers*, vol. 22, no. 4, Article ID 1350028, 2013.

[27] P. Balasubramanian, D. A. Edwards, and H. R. Arabnia, "Robust asynchronous carry lookahead adders," in *Proceedings of the 11th International Conference on Computer Design*, pp. 119–124, 2011.

[28] Xilinx, http://www.xilinx.com.

[29] K. K. Parhi, "Low-energy CSMT carry generators and binary adders," *IEEE Transactions on VLSI Systems*, vol. 7, no. 4, pp. 450–462, 1999.

[30] P. Balasubramanian, D. A. Edwards, and W. B. Toms, "Self-timed multi-operand addition," *International Journal of Circuits, Systems and Signal Processing*, vol. 6, no. 1, pp. 1–11, 2012.

10

Design of Low Power and Efficient Carry Select Adder Using 3-T XOR Gate

Gagandeep Singh and Chakshu Goel

ECE Department, Shaheed Bhagat Singh State Technical Campus, Ferozepur, Punjab 152004, India

Correspondence should be addressed to Chakshu Goel; chakshu77@yahoo.com

Academic Editor: Liwen Sang

In digital systems, mostly adder lies in the critical path that affects the overall performance of the system. To perform fast addition operation at low cost, carry select adder (CSLA) is the most suitable among conventional adder structures. In this paper, a 3-T XOR gate is used to design an 8-bit CSLA as XOR gates are the essential blocks in designing higher bit adders. The proposed CSLA has reduced transistor count and has lesser power consumption as well as power-delay product (PDP) as compared to regular CSLA and modified CSLA.

1. Introduction

In today's VLSI circuit designs, there is a significant increase in the power consumption due to the increasing speed and complexity of the circuits. As the demand for portable equipment like laptops and cellular phones is increasing rapidly, great attention has been focused on power efficient circuit designs [1–4]. Adders are the basic building blocks of the complex arithmetic circuits. Adders are widely used in Central Processing Unit (CPU), Arithmetic Logic Unit (ALU), and floating point units, for address generation in case of cache or memory access and in digital signal processing [5–7].

Having adders with fast addition operation and low power along with low area consumption is still a challenging issue. Depending upon the area, delay, and power consumption, the various adders are categorized as ripple carry adder (RCA), carry select adder (CSLA), and carry lookahead adder (CLAA). CSLA provides a compromise between the large area with small delay of CLAA and small area but longer delay of RCA [8]. CSLA uses pair of RCAs for addition, that is, one block of RCA with C_{in} (carry in) = 0 and other block of RCA with C_{in} = 1. Depending on the value of previous carry, the final sum and carry outputs are selected using multiplexer. Due to the pair of RCAs used for each bit addition, the simplest kind of CSLA is not very efficient [9].

Keeping in mind that XOR gates are the building blocks of adders, here in this work, we use a 3T-XOR gate to design an 8-bit CSLA. The main advantage of using 3T-XOR gate is that the power consumption of the circuit decreases due to the large decrease in number of switching transistors (MOSFETs) used in the design of 8-bit CSLA.

This paper is organized as follows. Section 2 presents the earlier works on carry select adder including the detailed structure of regular CSLA as well as modified CSLA. Section 3 explains the proposed CSLA and evaluates the reduction in switching transistors (MOSFETs) count. The implementation details as well as simulation results of proposed CSLA are analyzed in Section 4 and Section 5 concludes the whole work.

2. Earlier Works on Carry Select Adder

In digital adders, the speed of addition is limited due to the time taken by the carry signal to propagate through the adder. The regular carry select adder (R-CSLA) was introduced to mitigate the problem of carry propagation delay by independently generating multiple carries and then selecting the correct sum and carry outputs depending on the value of previous carry [9]. As previously discussed, this type of CSLA (i.e., R-CSLA) was not area efficient due to the use of pair of RCAs (each for C_{in} = 0 and C_{in} = 1) to produce

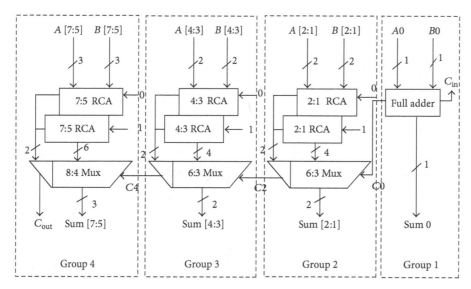

FIGURE 1: Regular 8-bit CSLA.

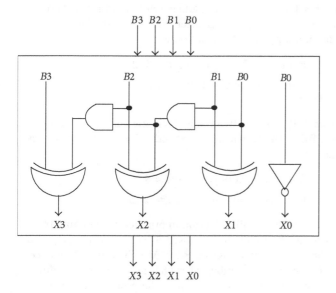

FIGURE 2: Four-bit BEC circuit.

TABLE 1: Truth table of 4-bit BEC.

$I[3:0]$	$X[3:0]$
0000	0001
0001	0010
0010	0011
0011	0100
0100	0101
0101	0110
0110	0111
0111	1000
1000	1001
1001	1010
1010	1011
1011	1100
1100	1101
1101	1110
1110	1111
1111	0000

the final sum and carry output. The 8-bit R-CSLA is shown in Figure 1.

To make low power consumption and an area efficient CSLA, an add-one circuit known as Binary to Excess-1 Converter (BEC) circuit was introduced. This BEC circuit replaced the RCA with C_{in} = 1 used in R-CSLA as lesser numbers of logic gates were used in BEC as compared to n-bit RCA [10–12]. The truth table and circuit diagram of 4-bit BEC are shown in Figure 2 and Table 1, respectively.

The 8-bit modified carry select adder (M-CSLA) using BEC is shown in Figure 3. As shown in the Figure 3, 8-bit M-CSLA was divided into four groups with different bit sizes of RCA and BEC. M-CSLA consists of RCAs (for C_{in} = 0), BEC circuits (for C_{in} = 1), and multiplexers (MUX). One input to the MUX is sum along with carry outputs from RCA and another input to the MUX is sum along with carry outputs

from BEC circuit. The final sum and carry outputs are selected depending upon the value of previous carry which is inputted as the select line to the MUX [10].

3. Proposed Work on CSLA

In this work, we use a modified XOR gate as it forms the basic building block of CSLA. Here, we use a 3-T XOR gate instead of 12-T XOR gate used in previous designs of R-CSLA and M-CSLA which helps in more efficient design of 8-bit CSLA [13]. The circuit diagrams of 3-T XOR gate and proposed 8-bit CSLA are shown in Figures 4 and 5, respectively.

The overall performance of CSLA in terms of power consumption, transistor count, and power-delay product (PDP) can be enhanced by modifying the XOR gate. A single

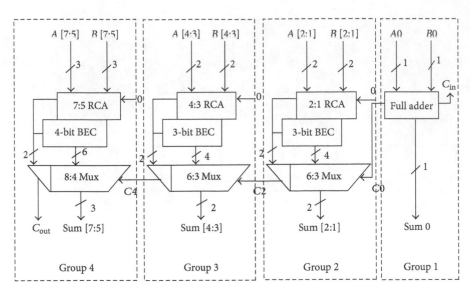

FIGURE 3: Modified 8-bit CSLA.

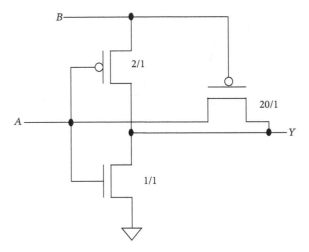

FIGURE 4: Three-T XOR gate.

modified XOR gate (3-T XOR gate) used in this work has 9 lesser transistors as compared to the XOR gate (12-T XOR gate) used in earlier works on CSLA. The proposed 8-bit CSLA is divided into 4 groups as shown in Figure 5.

The total reduction in transistor count for each group is calculated below.

Group 1. It contains one full adder. Each full adder consists of two XOR gates. Therefore total transistor count reduction for group 1 is

number of XOR gates used = 2;

transistor count reduction = 18 (2 ∗ 9).

Group 2. It contains one full adder, one half adder, and one 3-bit Binary to Excess-1 Converter (BEC). The transistor count reduction for group 2 is as follows:

number of XOR gates used = 5 (2 + 1 + 2);

transistor count reduction = 45 (5 ∗ 9).

Group 3. It contains one full adder, one half adder, and one 3-bit Binary to Excess-1 Converter (BEC). The transistor count reduction for group 2 is as follows:

number of XOR gates used = 5 (2 + 1 + 2);

transistor count reduction = 45 (5 ∗ 9).

Group 4. It contains two full adders, one half adder, and one 4-bit BEC. The transistor count reduction for group 4 is as follows:

number of XOR gates used = 8 {(2 ∗ 2) + 1 + 3};

transistor count reduction = 72 (8 ∗ 9).

Therefore the overall reduction in number of switching transistors (MOSFERTs) in proposed 8-bit CSLA as compared to the previously designed 8-bit M-CSLA is 180. Hence, the reduction in number of switching transistors reduces the power consumption as well as the power-delay product (PDP) of 8-bit CSLA.

4. Simulation Results

The proposed 8-bit CSLA has been successfully tested and synthesized in Tanner Tools using 90 nm technology with a supply voltage of 1.0 V. The power consumption and delay time of proposed 8-bit CSLA are calculated for all input conditions and the worst case power consumption as well as delay time is noted down. The power consumption, delay time, and power-delay product (PDP) of proposed 8-bit CSLA are compared with 8-bit R-CSLA and M-CSLA. The results of proposed 8-bit CSLA are also compared with the 8-bit CSA proposed in recent studies [14, 15]. The comparison is shown in Table 2.

It is clear from Table 2 that power consumption of proposed 8-bit CSLA is reduced by 27.7% and 21.7% when compared with R-CSLA and M-CSLA, respectively. The power-delay product (PDP) shows a similar trend as PDP is

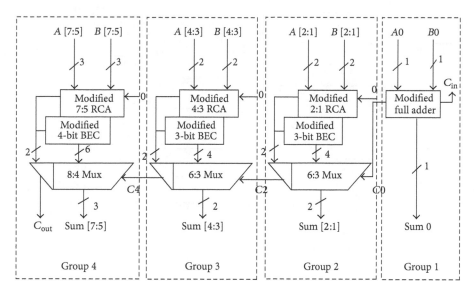

FIGURE 5: Proposed 8-bit CSLA.

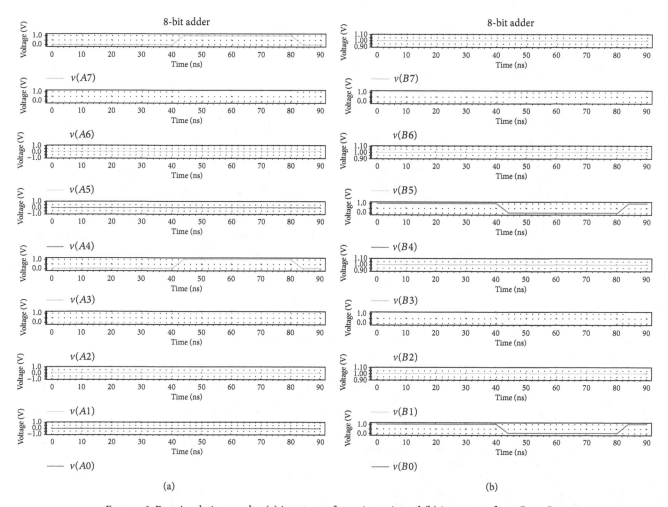

(a)

(b)

FIGURE 6: Postsimulation results: (a) input waveform A_0 to A_7 and (b) input waveform B_0 to B_7.

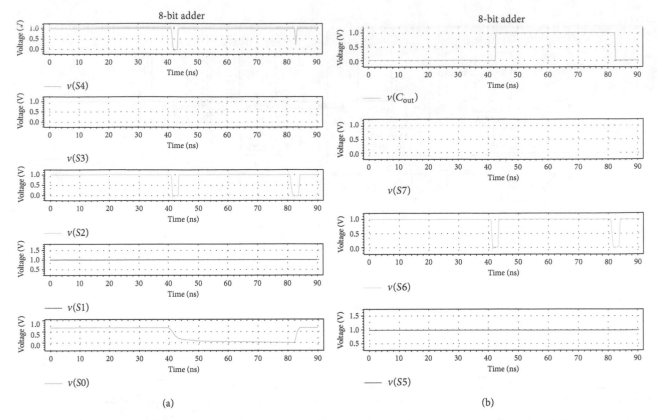

FIGURE 7: Postsimulation results: (a) output waveform S_0 to S_4 and (b) output waveform S_5 to S_7 and C_{out}.

TABLE 2: Comparison of various carry select adders.

Adder	Total power (μW)	Delay (ns)	Power-delay product (10^{-15})
8-bit regular CSLA	203.9	1.719	350.5
8-bit modified CSLA [10]	188.4	1.958	368.5
8-bit CSLA [14]	187.58	1.976	370.8
8-bit CSA using reversible logic [15]	180	17.2	3096
8-bit proposed CSLA	147.4	2.18	321.3

The postsimulation input-output waveforms for the 8-bit proposed CSLA are shown in Figures 6 and 7, respectively. The proposed design is simulated with a 12.5 MHz waveform with rise and fall times of 4 ns.

Figure 8 shows the comparison of various carry select adders in graphical form for the data given in Table 2. We can see from the graph that the proposed CSLA has minimum power-delay product (PDP) as well as the minimum power consumption when compared with regular CSLA, modified CSLA [10], CSLA [14], and reversible logic style based 8-bit CSA [15].

5. Conclusion

A simple approach of enhancing the performance of XOR gate to design an 8-bit CSLA is used in this paper. The proposed CSLA has large decrease in switching transistors (MOSFETs) due to the use of 3-T XOR gate. On comparing this proposed 8-bit CSLA with other existing 8-bit CSLAs like R-CSLA and M-CSLA, there is 27.7% and 21.7% reduction in power, respectively. The power-delay product (PDP) is also reduced by 8.3% and 12.8% when compared with R-CSLA and M-CSLA, respectively. The proposed 8-bit CSLA has the best performance compared with other 8-bit CSLAs present in literature. It would be interesting to design 16-bit CSLA and 32-bit CSLA using 3-T XOR gate.

reduced by 8.3% and 12.8% when compared with R-CSLA and M-CSLA, respectively. Compared with the reversible logic style based 8-bit CSA [15], the proposed design has 18.1% reduction in power consumption and 89.6% reduction in PDP. Compared with the CSLA [14], the proposed CSLA has 21.7% reduction in power consumption and 13.3% reduction in PDP.

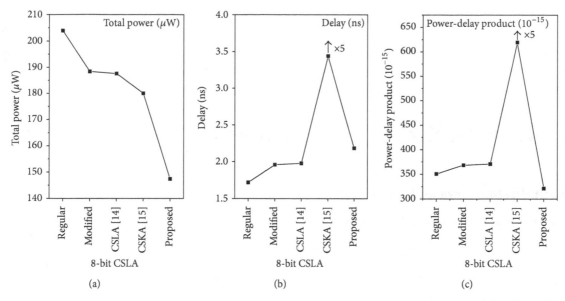

FIGURE 8: Power, delay, and PDP comparison of various CSLAs.

Conflict of Interests

The authors declare that there is no conflict of interests regarding the publication of this paper.

References

[1] K. Navi, M. H. Moaiyeri, R. F. Mirzaee, O. Hashemipour, and B. M. Nezhad, "Two new low-power Full Adders based on majority-not gates," *Microelectronics Journal*, vol. 40, no. 1, pp. 126–130, 2009.

[2] D. Wang, M. F. Yang, W. Cheng, X. G. Guan, Z. M. Zhu, and Y. T. Yang, "Novel low power full adder cells in 180 nm CMOS technology," in *Proceedings of the 4th IEEE Conference on Industrial Electronics and Applications (ICIEA '09)*, pp. 430–433, Xi'an, China, May 2009.

[3] N. Weste and K. Eshraghian, *Principles of CMOS VLSI Design: A Systems Perspective*, Addison-Wesley, Reading, Mass, USA, 1993.

[4] S. Kang and Y. Leblebici, *CMOS Digital Integrated Circuit Analysis and Design*, McGraw-Hill, New York, NY, USA, 3rd edition, 2005.

[5] J. M. Rabaey, A. Chandrakasan, and B. Nikolic, *Digital Integrated Circuits, A Design Perspective*, Prentice Hall, Englewood Cliffs, NJ, USA, 2nd edition, 2002.

[6] J. Uyemura, "CMOS Logic Circuit Design," Kluwer Academic Publishers, New York, NY, USA, 1999.

[7] N. Weste and K. Eshragian, *Principles of CMOS VLSI Design: A Systems Perspective*, Addison-Wesley, Reading, Mass, USA, 1993.

[8] K. Rawat, T. Darwish, and M. Bayoumi, "A low power and reduced area carry select adder," in *Proceedings of the 45th Midwest Symposium on Circuits and Systems*, pp. 1467–1470, August 2002.

[9] O. J. Badrij, "Carry-select Adder," *IRE Transactions on Electronics Computers*, pp. 340–344, 1962.

[10] B. Ramkumar and H. M. Kittur, "Low-power and area-efficient carry select adder," *IEEE Transactions on Very Large Scale Integration (VLSI) Systems*, vol. 20, no. 2, pp. 371–375, 2012.

[11] T.-Y. Chang and M.-J. Hsiao, "Carry-select adder using single ripple-carry adder," *Electronics Letters*, vol. 34, no. 22, pp. 2101–2103, 1998.

[12] Y. Kim and L. S. Kim, "64-bit carry-select adder with reduced area," *Electronics Letters*, vol. 37, no. 10, pp. 614–615, 2001.

[13] S. R. Chowdhury, A. Banerjee, A. Roy, and H. Saha, "A high speed 8 transistor full adder design using novel 3 transistor XOR gates," *International Journal of Electronics, Circuits and Systems*, vol. 2, no. 4, pp. 217–223, 2008.

[14] S. Singh and D. Kumar, "Design of area and power efficient modified carry select adder," *International Journal of Computer Applications*, vol. 33, no. 3, pp. 14–18, 2011.

[15] S. Maity, B. Prasad De, and A. K. Singh, "Design and implementation of low-power high performance carry skip adder," *International Journal of Engineering and Advanced Technology*, vol. 1, no. 4, 2012.

11

FinFETs: From Devices to Architectures

Debajit Bhattacharya and Niraj K. Jha

Department of Electrical Engineering, Princeton University, Princeton, NJ 08544, USA

Correspondence should be addressed to Niraj K. Jha; jha@princeton.edu

Academic Editor: Jaber Abu Qahouq

Since Moore's law driven scaling of planar MOSFETs faces formidable challenges in the nanometer regime, FinFETs and Trigate FETs have emerged as their successors. Owing to the presence of multiple (two/three) gates, FinFETs/Trigate FETs are able to tackle short-channel effects (SCEs) better than conventional planar MOSFETs at deeply scaled technology nodes and thus enable continued transistor scaling. In this paper, we review research on FinFETs from the bottommost device level to the topmost architecture level. We survey different types of FinFETs, various possible FinFET asymmetries and their impact, and novel logic-level and architecture-level tradeoffs offered by FinFETs. We also review analysis and optimization tools that are available for characterizing FinFET devices, circuits, and architectures.

1. Introduction

Relentless scaling of planar MOSFETs over the past four decades has delivered ever-increasing transistor density and performance to integrated circuits (ICs). However, continuing this trend in the nanometer regime is very challenging due to the drastic increase in the subthreshold leakage current (I_{off}) [1–3]. Due to the very narrow channel lengths in deeply scaled MOSFETs, the drain potential begins to influence the electrostatics of the channel and, consequently, the gate loses adequate control over the channel. As a result, the gate is unable to shut off the channel completely in the off-mode of operation, which leads to an increased I_{off} between the drain and the source. The use of thinner gate oxides and high-k dielectric materials helps alleviate this problem by increasing the gate-channel capacitance. However, thinning of gate oxides is fundamentally limited by the deterioration in gate leakage and gate-induced drain leakage (GIDL) [4–6]. Multiple-gate field-effect transistors (MGFETs), which are an alternative to planar MOSFETs, demonstrate better screening of the drain potential from the channel due to the proximity of the additional gate(s) to the channel (i.e., higher gate-channel capacitance) [7–12]. This makes MGFETs superior to planar MOSFETs in short-channel performance metrics, such as subthreshold slope

(S), drain-induced barrier lowering (DIBL), and threshold voltage (V_{th}) roll-off. Improvement in these metrics implies less degradation in the transistor's V_{th} with continued scaling, which in turn implies less degradation in I_{off}.

So far, we have referred to planar MOSFETs built on bulk-Si wafers (or bulk MOSFETs) as planar MOSFETs. Fully-depleted silicon-on-insulator (FDSOI) MOSFETs (planar MOSFETs built atop SOI wafers) avoid the extra leakage paths from the drain to source by getting rid of the extra substrate beneath the channel [13, 14]. Their performance metrics are comparable with those of double-gate FETs (DGFETs), which are MGFETs with two gates. Both offer reduced junction capacitance, higher I_{on}/I_{off} ratio, better S, and improved robustness against random dopant fluctuation (RDF). However, DGFETs have a more relaxed constraint on channel thickness, which makes DGFETs more scalable than FDSOI MOSFETs in the long run [15, 16]. Also, DGFET structures can be built on bulk-Si wafers, as well, which makes DGFETs more attractive to foundries that do not want to switch to an SOI process [17, 18].

Among all MGFETs, FinFETs (a type of DGFET) and Trigate FETs (another popular MGFET with three gates) have emerged as the most desirable alternatives to MOSFETs due to their simple structures and ease of fabrication [19–27]. Two or three gates wrapped around a vertical channel enable easy alignment of gates and compatibility with the standard CMOS fabrication process. In Trigate FETs, an additional selective etching step of the hard mask is involved in order to create the third gate on top of the channel. Although this third gate adds to process complexity, it also leads to some advantages like reduced fringe capacitances and additional transistor width [28–30].

FinFET/Trigate devices have been explored thoroughly in the past decade. A large number of research articles have been published that demonstrate the improved short-channel behavior of these devices over conventional bulk MOSFETs [19–22, 31–33]. Many researchers have presented novel circuit design styles that exploit different kinds of FinFETs [34–48]. Researchers have also explored various symmetric and asymmetric FinFET styles and used them in hybrid FinFET logic gates and memories [49–66]. Newer architectures for caches, networks-on-chip (NoCs), and processors based on such logic gates and memories have also been explored [67–74]. In spite of these advancements in FinFET research, articles that provide a global view of FinFETs from the device level to the topmost architecture level are scarce. Mishra et al. provided such a view at the circuit level [75]. However, FinFETs are not covered at other levels of the design hierarchy. Also, at the circuit level, much progress has been made since the publication of that book chapter. Our article is aimed at a wide range of readers: device engineers, circuit designers, and hardware architects. Our goal is to provide a global view of FinFET concepts spanning the entire IC design hierarchy.

The paper is organized as follows. In Section 2, we review the different types of FinFETs and possible asymmetries that can be designed into their structures. We also discuss the sources of process variations in FinFETs and their impact on FinFET performance. We discuss FinFET process simulation, device simulation, and compact models in Section 3. We describe novel FinFET inverter (INV) and NAND gates, flip-flops, latches, static random-access memory (SRAM), and dynamic random-access memory (DRAM) cells in Section 4. In Section 5, we discuss circuit-level analysis and optimization methodologies and a novel interconnect scheme that leverages FinFETs. We then present a survey of process-voltage-temperature (PVT) variation-aware architecture-level simulation tools in Section 6 and conclude in Section 7.

2. FinFETs

In 1989, Hisamato et al. fabricated a double-gate SOI structure which they called a fully-depleted lean channel transistor (DELTA) [76]. This was the first reported fabrication of a FinFET-like structure. FinFETs have attracted increasing attention over the past decade because of the degrading short-channel behavior of planar MOSFETs [19–24]. Figure 1 demonstrates the superior short-channel performance of FinFETs over planar MOSFETs with the same channel length. Figure 2 shows a conventional planar MOSFET and a FinFET. While the planar MOSFET channel is horizontal, the FinFET channel (also known as the fin) is vertical. Hence, the height

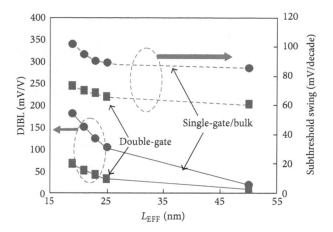

FIGURE 1: DIBL and subthreshold swing (S) versus effective channel length for double-gate (DG) and bulk-silicon nFETs. The DG device is designed with an undoped body and a near-mid-gap gate material [12].

of the channel (H_{FIN}) determines the width (W) of the FinFET. This leads to a special property of FinFETs known as width quantization. This property says that the FinFET width must be a multiple of H_{FIN}, that is, widths can be increased by using multiple fins. Thus, arbitrary FinFET widths are not possible. Although smaller fin heights offer more flexibility, they lead to multiple fins, which in turn leads to more silicon area. On the other hand, taller fins lead to less silicon footprint, but may also result in structural instability. Typically, the fin height is determined by the process engineers and is kept below four times the fin thickness [77, 78].

Although FinFETs implemented on SOI wafers are very popular, FinFETs have also been implemented on conventional bulk wafers extensively [79–81]. Figure 3 shows FinFETs implemented on bulk and SOI wafers. Unlike bulk FinFETs, where all fins share a common Si substrate (also known as the bulk), fins in SOI FinFETs are physically isolated. Some companies prefer the bulk technology because it is easier to migrate to bulk FinFETs from conventional bulk MOSFETs. However, FinFETs on both types of wafers are quite comparable in terms of cost, performance, and yield, and it is premature to pick a winner. From this point on, our discussion will be limited to SOI FinFETs unless otherwise mentioned.

Trigate FETs, referred to interchangeably as FinFETs, in this paper so far, are a variant of FinFETs, with a third gate on top of the fin. Intel introduced Trigate FETs at the 22 nm node in the Ivy-Bridge processor in 2012 [28, 82]. Figure 4 shows a Trigate FET along with a FinFET. The thickness of the dielectric on top of the fin is reduced in Trigate FETs in order to create the third gate. Due to the presence of the third gate, the thickness of the fin also adds to the channel width. Hence, Trigate FETs enjoy a slight width advantage over FinFETs. Trigate FETs also have less gate-source capacitance compared to FinFETs due to additional current conduction at the top surface, but this advantage is diminished by increased parasitic resistance [29].

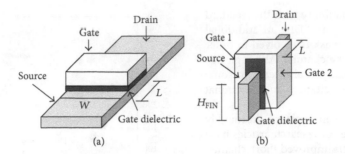

FIGURE 2: Structural comparison between (a) planar MOSFET and (b) FinFET.

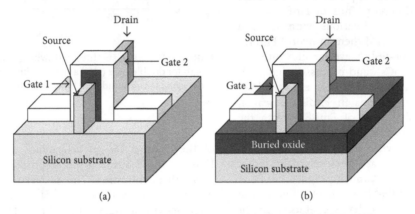

FIGURE 3: Structural comparison between (a) bulk and (b) SOI FinFETs.

Yang and Fossum compared Trigate FETs and FinFETs and argued that FinFETs are superior to Trigate FETs in the long run [83]. They showed that although undoped Trigate FETs may enjoy more relaxed body thickness, they are not competitive with FinFETs in SCE metrics. When trying to achieve comparable SCE metrics, Trigate FETs lose the scaling advantage and suffer from significant layout area disadvantage. However, like the bulk versus SOI debate, it is also premature to declare a clear winner between FinFETs and Trigate FETs. From this point onwards, we will consider FinFETs only unless stated otherwise.

FinFETs can be fabricated with their channel along different directions in a single die. Fabrication of planar MOSFET channels along any crystal plane other than $\langle 100 \rangle$ is difficult due to process variations and interface traps [36, 84]. However, FinFETs can be fabricated along the $\langle 110 \rangle$ plane as well. This results in enhanced hole mobility. $\langle 110 \rangle$-oriented FinFETs can be fabricated by simply rotating the transistor layout by 45° in the plane of a $\langle 100 \rangle$ wafer [85]. Thus, nFinFETs implemented along $\langle 100 \rangle$ and pFinFETs along $\langle 110 \rangle$ lead to faster logic gates since this gives designers an opportunity to combat the inherent mobility difference between electrons and holes. However, this multiorientation scheme has an obvious drawback of increased silicon area [85]. In the following sections, we discuss FinFET classifications and process variations in detail.

2.1. FinFET Classification. There are two main types of FinFETs: shorted-gate (SG) and independent-gate (IG). SG

FinFETs are also known as three-terminal (3T) FinFETs and IG FinFETs as four-terminal (4T) FinFETs. In SG FinFETs, both the front and back gates are physically shorted, whereas in IG FinFETs, the gates are physically isolated (Figure 5). Thus, in SG FinFETs, both gates are jointly used to control the electrostatics of the channel. Hence, SG FinFETs show higher on-current (I_{on}) and also higher off-current (I_{off} or the subthreshold current) compared to those of IG FinFETs. IG FinFETs offer the flexibility of applying different signals or voltages to their two gates. This enables the use of the back-gate bias to modulate the V_{th} of the front gate linearly. However, IG FinFETs incur a high area penalty due to the need for placing two separate gate contacts.

SG FinFETs can be further categorized based on asymmetries in their device parameters. Normally, the workfunctions (Φ) of both the front and back gates of a FinFET are the same. However, the workfunctions can also be made different. This leads to an asymmetric gate-workfunction SG FinFET or ASG FinFET (Figure 6) [86, 87]. ASG FinFETs can be fabricated with selective doping of the two gate-stacks. They have very promising short-channel characteristics and have two orders of magnitude lower I_{off} compared to that of an SG FinFET, with I_{on} only somewhat lower than that of an SG FinFET [49]. Figures 7 and 8 show comparisons of the drain current I_{DS} versus front-gate voltage V_{GFS} curves for SG, IG, and ASG nFinFETs and pFinFETs, respectively, demonstrating the advantages of ASG FinFETs.

Apart from gate-workfunction asymmetry, other asymmetries have also been explored in FinFETs. Goel et al. [57] show that asymmetric drain-spacer-extended (ADSE)

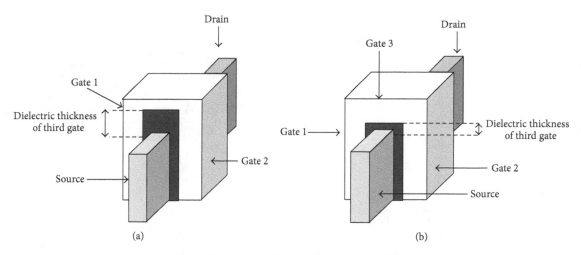

FIGURE 4: Structural comparison between (a) FinFET and (b) Trigate FET.

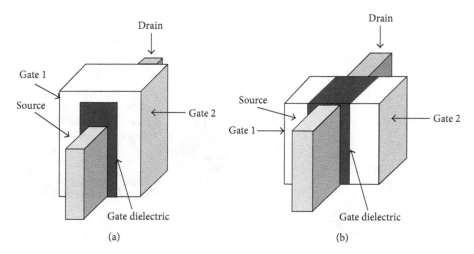

FIGURE 5: Structural comparison between (a) SG and (b) IG FinFET.

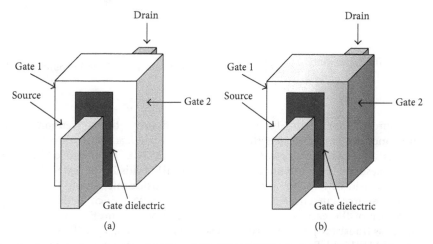

FIGURE 6: Structural comparison between (a) SG and (b) ASG FinFET; shaded gate implies different workfunctions.

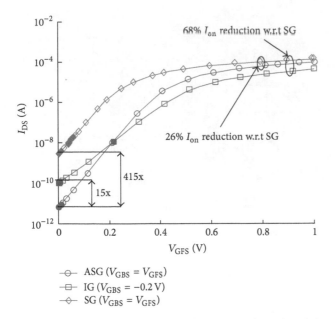

FIGURE 7: Drain current (I_{DS}) versus front-gate voltage (V_{GFS}) for three nFinFETs [49].

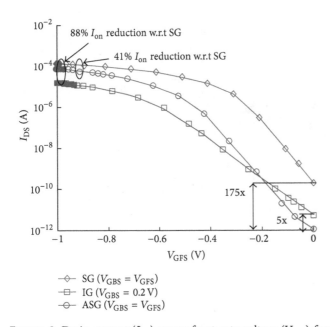

FIGURE 8: Drain current (I_{DS}) versus front-gate voltage (V_{GFS}) for three pFinFETs [49].

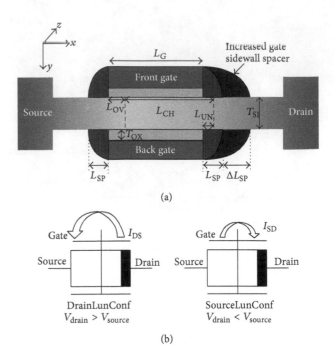

FIGURE 9: Asymmetric drain spacer extension (ADSE) FinFET [57].

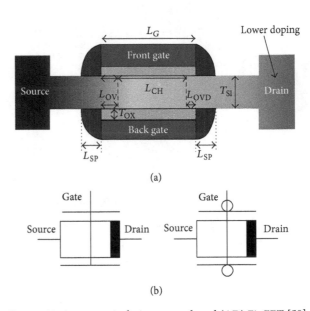

FIGURE 10: Asymmetric drain-source doped (AD) FinFET [58].

FinFETs (Figure 9) can lead to improved short-channel characteristics because of an indirect increase in channel length. However, this improvement comes at the cost of an increased layout area. This asymmetry also destroys the conventional interchangeable source-drain concept in CMOS. An asymmetry is created in the drain-to-source current I_{DS} and source-to-drain current I_{SD} because of the extra underlap. This asymmetry affects FinFET pass transistor performance. Asymmetric drain-source doped (AD) FinFETs (Figure 10), with an order of magnitude difference in the drain and source doping concentrations, have been exploited in [58].

This also destroys the conventional symmetry in I_{DS} and I_{SD}, which again leads to asymmetric FinFET pass transistor performance. SCEs are improved in AD FinFETs because of lower electric fields in the lower-doped drain. FinFETs with asymmetric oxide thickness (ATox) (Figure 11) have also been proposed [88, 89]. Such FinFETs have good subthreshold slopes. Use of IG FinFET (or 4T FinFET) in this context also enables variable V_{th}'s. This asymmetry can be achieved using a ion-bombardment-enhanced etching process. Finally, asymmetric fin-height FinFETs have also been explored [61, 90]. Since the channel width of a FinFET is proportional to

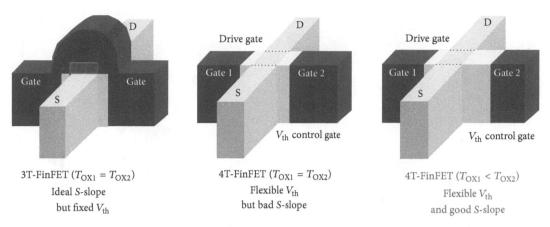

3T-FinFET ($T_{OX1} = T_{OX2}$)
Ideal S-slope
but fixed V_{th}

4T-FinFET ($T_{OX1} = T_{OX2}$)
Flexible V_{th}
but bad S-slope

4T-FinFET ($T_{OX1} < T_{OX2}$)
Flexible V_{th}
and good S-slope

FIGURE 11: Asymmetric oxide thickness (ATox) FinFET [89].

TABLE 1: 22 nm SOI FinFET parameter values.

L_{GF}, L_{GB} (nm)	24
Effective T_{OXF}, T_{OXB} (nm)	1
T_{SI} (nm)	10
H_{FIN} (nm)	40
H_{GF}, H_{GB} (nm)	10
L_{SPF}, L_{SPB} (nm)	12
L_{UN} (nm)	4
N_{BODY} (cm^{-3})	10^{15}
$N_{S/D}$ (cm^{-3})	10^{20}
Φ_{GF}, Φ_{GB} (eV)	4.4(n), 4.8(p)
FP (nm)	50
GP (nm)	92

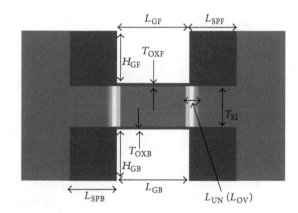

FIGURE 12: A 2D cross-section of a 3D nFinFET with extended source and drain [49].

its fin height, pFinFETs with taller fins can compensate for the inherent mobility mismatch between electrons and holes.

Figure 12 shows a two-dimensional (2D) cross-section of a three-dimensional (3D) FinFET, illustrating various device parameters of interest. Typical values for these parameters are given in Table 1. L_{GF}, L_{GB}, T_{OXF}, T_{OXB}, T_{SI}, H_{FIN}, H_{GF}, H_{GB}, L_{SPF}, L_{SPB}, L_{UN}, N_{BODY}, $N_{S/D}$, Φ_{GF}, Φ_{GB}, FP, and GP refer to the physical front- and back-gate lengths, front- and back-gate effective oxide thicknesses, fin thickness, fin height, front- and back-gate thicknesses, front- and back-gate spacer thicknesses, gate-drain/source underlap, body doping, source/drain doping, front- and back-gate workfunctions, fin pitch, and gate pitch, respectively.

2.2. Process Variations. Reduced feature size and limited photolithographic resolution cause statistical fluctuations in nanoscale device parameters. These fluctuations cause variations in electrical device parameters, such as V_{th}, I_{on}, I_{off}, and so forth, known as process variations. These variations can be inter-die or intra-die, correlated or uncorrelated, depending on the fabrication process. They lead to mismatched device strengths and degrade the yield of the entire die. This is why continued scaling of planar MOSFETs has become so difficult.

In planar MOSFETs, a sufficient number of dopants must be inserted into the channel in order to tackle SCEs. However, this means that RDF may lead to a significant variation in V_{th}. For example, at deeply scaled nodes, the $3(\sigma/\mu)$ variation in V_{th} caused by discrete impurity fluctuation can be greater than 100% [91]. Since FinFETs enable better SCE performance due to the presence of the second gate, they do not need a high channel doping to ensure a high V_{th}. Hence, designers can keep the thin channel (fin) at nearly intrinsic levels (10^{15} cm^{-3}). This reduces the statistical impact of RDF on V_{th}. The desired V_{th} is obtained by engineering the workfunction of the gate material instead. Low channel doping also ensures better mobility of the carriers inside the channel. Thus, FinFETs emerge superior to planar MOSFETs by overcoming a major source of process variation.

FinFETs do suffer from other process variations. Due to their small dimensions and lithographic limitations, FinFETs are subjected to several important physical fluctuations, such as variations in gate length (L_{GF}, L_{GB}), fin-thickness (T_{SI}), gate-oxide thickness (T_{OXF}, T_{OXB}), and gate underlap (L_{UN}) [91–97]. For example, gate oxide is on the etched sidewall of the fin, and may suffer from nonuniformity. The degree of nonuniformity depends on the line-edge roughness (LER) of the fin. LER also causes variations in fin thickness.

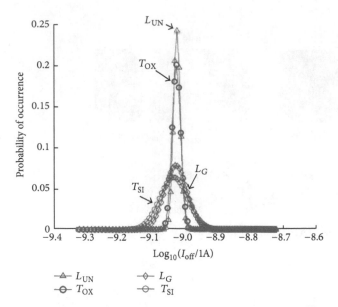

FIGURE 13: Distribution of leakage current (I_{off}) for different process parameters, each varying independently [94].

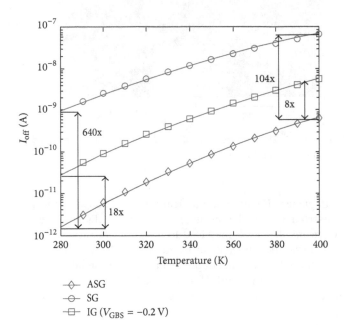

FIGURE 14: I_{off} versus temperature for three nFinFETs [49].

Figure 13 shows the impact of parametric variations on the subthreshold current (I_{off}) of an nFinFET. Xiong and Bokor have studied the sensitivity of electrical parameters to various physical variations in devices designed with a nearly intrinsic channel [91].

Choi et al. have studied temperature variations in FinFET circuits under above-mentioned physical parameters variations [98]. They showed that even under moderate process variations ($3(\sigma/\mu) = 10\%$) in gate length (L_{GF}, L_{GB}) and body thickness (T_{SI}), thermal runaway is possible in more than 15% of ICs when primary input switching activity is 0.4. The effect of temperature variation is more severe in

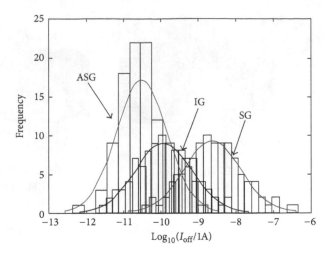

FIGURE 15: Distributions of I_{off} under process variations for three nFinFETs [49].

SOI FinFETs because the oxide layer under the fin has poor thermal conductivity. Hence, heat generated in the fin cannot dissipate easily in SOI FinFETs. Bhoj and Jha have evaluated SG, IG, and ASG FinFETs under temperature variation and found that even though I_{off} degrades for all three FinFETs at a higher temperature, ASG FinFETs still remain the best and retain a 100× advantage over SG FinFETs, as shown in Figure 14 [49]. They also showed the distribution of I_{off} under process variations for the three FinFETs (Figure 15).

3. FinFET Device Characterization

In this section, we discuss various ways of characterizing FinFET devices through simulation. Process simulation followed by device simulation constitutes a technology computer-aided design (TCAD) characterization flow of nanoscale devices, such as FinFETs. Compact models, on the other hand, have been another very popular way of characterizing CMOS devices for decades.

3.1. Process Simulation. Real devices undergo several processing steps. The functionality and performance of the fabricated devices depend on how optimized the process flow is. TCAD process simulation is, therefore, an important step in FinFET device optimization. Process simulation is followed by device simulation. These two simulation steps form an optimization loop in which small changes in the process flow (e.g., time, temperature, doses, etc.) can lead to desirable electrical characteristics of the device. Thus, process simulation helps device engineers explore the parameter space of the process, obviating the need for actual device fabrication. Although 3D process simulation is computationally very expensive, it not only gives good insights into device physics but also provides a cost-effective pre-fabrication process optimization flow.

The Sentaurus process and device simulator from Synopsys is a widely used tool for process simulation [99]. Its 3D process simulation framework is compatible with the mainstream 2D TCAD framework TSUPREM4/MEDICI (also from Synopsys). The 2D framework has been used by designers over the past decade and has been well-calibrated with advanced CMOS libraries. Nawaz et al. have implemented a complete FinFET process flow as a commercially-available process and device simulation environment [100]. As in real devices, all important geometrical features, such as corner roundings and 3D facets, have been implemented in their setup.

Process simulations of large layouts that consist of multiple devices incur extremely high computational costs. A novel layout/process/device-independent TCAD methodology was proposed in [54] in order to overcome the process simulation barrier for accurate 3D TCAD structure synthesis. In it, Bhoj et al. adopt an automated structure synthesis approach that obviates the need for repetitive 3D process simulations for different layouts. In this approach, process-simulated unit devices are placed at the device locations in the layout, eliminating the need for process simulation of the entire layout, thereby reducing computational costs significantly. This structure synthesis approach, followed by transport analysis based capacitance extraction methodology, has been shown to capture accurate parasitic capacitances in FinFET SRAMs and ring oscillators in a practical timeframe [54, 55, 63, 66]. Accurate extraction of parasitic capacitances has led to a comprehensive evaluation of transient metrics of various FinFET SRAM bitcells [55].

3.2. Device Simulation. After process simulation generates a meshed device structure, device simulation is performed on the structure by invoking appropriate transport models. The conventional drift-diffusion transport model is not adequate for capturing SCEs in nanometer MOSFETs and FinFETs. The hydrodynamic model, with quantum corrections (such as density gradient models), has been popular among researchers for FinFET device simulation [101]. Other more accurate models, such as Green's function based solution to Boltzmann's transport equation, impose a drastic computational burden [101]. In order to simulate circuits with multiple devices, Sentaurus device (Synopsys) allows mixed-mode device simulation. Here, individual FinFET devices are connected externally using wires or other circuit elements to form a netlist and coupled transport equations are solved on the entire netlist. This feature enables device engineers to see how the device behaves when used in a circuit.

3.3. Compact Models. Physics based compact models of FinFETs have been a very useful tool for designers. Berkeley short-channel IGFET model (BSIM) and University of Florida double-gate model (UFDG) for SOI multigate MOSFETs and FinFETs were built using TCAD and calibrated using fabricated hardware [102–105]. These models are compatible with commercial circuit simulators, such as simulation program with integrated circuit emphasis (SPICE). Hence, large netlists can be simulated with these models as long as the solution space is within their range. However, device simulation precedes derivation of compact models and is more accurate. Thus, all results presented in this article are based on mixed-mode device simulations.

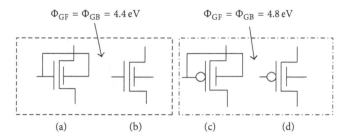

$\Phi_{GF} = \Phi_{GB} = 4.4\,eV$ $\Phi_{GF} = \Phi_{GB} = 4.8\,eV$

(a)   (b)   (c)   (d)

FIGURE 16: Schematic diagrams of (a) SG nFinFET, (b) IG nFinFET, (c) SG pFinFET, and (d) IG pFinFET. Their gate workfunctions are also shown [49].

4. FinFET Standard Cells

After the characterization of individual n/pFinFET devices, we move one level up to characterization of FinFET logic gates, latches, flip-flops, and memory cells, which are the building blocks of any digital integrated circuit [49–51]. IG and ASG FinFETs offer new leakage-delay tradeoffs in FinFET logic gates that can be exploited in low-power or high-performance applications. The schematic diagrams of SG and IG FinFETs are shown in Figure 16. Schematic diagrams of ASG FinFETs are shown in Figure 17. Bhoj and Jha have performed an in-depth analysis and comparison of SG, IG, and ASG FinFET based INV and NAND2 (two-input NAND) gates [49]. These two gates are the most essential building blocks of any logic library because any logic network can be built with just these two gates.

4.1. SG/IG INV. There are four possible configurations of an INV based on how SG and IG FinFETs are combined to implement them. They are called SG, low-power (LP), IGn, and IGp INV. Their schematic diagrams are shown in Figure 18. As suggested by its name, an SG INV has SG n/pFinFETs. It has a highly compact layout. The other three configurations use at least one IG FinFET. The back-gate of an IG pFinFET (nFinFET) is tied to a V_{HIGH} (V_{LOW}) signal. When these signals are reverse-biased, for example, when V_{HIGH} is 0.2 V above V_{DD} and V_{LOW} is 0.2 V below ground, there is a significant reduction in I_{off}. The presence of an IG FinFET also leads to a more complex layout, resulting in 36% area overhead relative to that of an ×2 SG INV (that is double the size of a minimum-sized SG INV). Table 2 compares the normalized area, delay, and leakage of the various INVs. Clearly, SG INV is the best in area and propagation delay (T_p), but incurs much higher leakage current than LP INV. However, LP INV performs poorly in area and propagation delay. IGn INV, however, looks promising based on its intermediate area, delay, and leakage.

4.2. SG/IG NAND2. Similar to INVs, NAND2 gates also have SG (LP) configurations in which all transistors are SG (IG) FinFETs. Since there are more transistors in a NAND2 gate than in an INV, there are more opportunities available for combining SG and IG FinFETs. This leads to various other configurations: MT, IG, IG2, XT, and XT2. Schematic

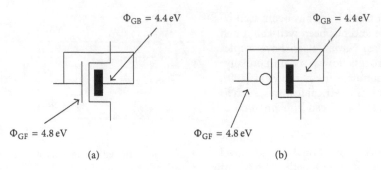

FIGURE 17: Schematic diagrams of ASG: (a) nFinFET and (b) pFinFET. Their gate workfunctions are also shown [49].

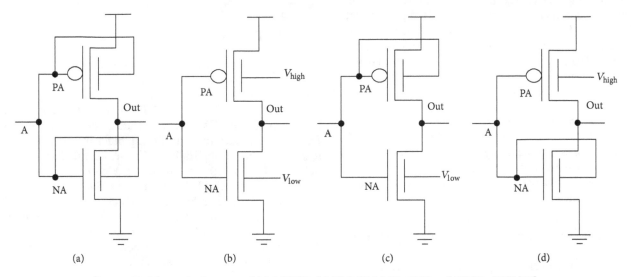

FIGURE 18: Schematic diagrams of (a) SG INV, (b) LP INV, (c) IGn INV, and (d) IGp INV [49].

TABLE 2: Comparison of FinFET INVs [49].

Topology	SG	LP	IGn	IGp
Area	1	1.36	1.36	1.36
Avg. I_{off}	20.92	1	2.75	19.25
T_p	1	3.67	1.67	2.92

TABLE 3: Comparison of FinFET NAND2 gates [49].

Topology	SG	LP	MT	IG	IG2	XT	XT2
Area	1	1.27	1.27	1	1	1.27	1
Avg. I_{off}	18.40	1	7.00	18.40	7.73	18.13	7.73
T_p (Toggle A)	1	4.13	3.80	1.60	2.08	3.20	1.47
T_p (Toggle B)	1	4.50	3.88	1.69	2.02	3.58	1.38
T_p (Toggle AB)	1	3.48	3.09	1	1.55	2.38	1.55

diagrams of SG, LP, and MT NAND2 gates are shown in Figure 19. Schematic diagrams for IG, IG2, XT, and XT2 NAND2 gates are shown in Figure 20. Table 3 shows the normalized area, delay, and leakage of all these NAND2 gates. Again, all comparisons in Table 3 are made relative to ×2 SG NAND2 gate, because it is the largest SG NAND2 gate that can be accommodated in the standard cell height. SG

NAND2 outperforms others in area and propagation delay, but consumes significantly more leakage current than LP NAND2. Out of all the variants, XT2 NAND2 stands out as a reasonable compromise.

4.3. ASG Logic Gates. Bhoj and Jha investigated INV and NAND2 gates with a mix of SG and ASG FinFETs [49]. Schematics/layouts of any SG-FinFET logic gate can be converted to those of an ASG-FinFET logic gate, as shown in Figure 21, without any area overhead. Hence, introduction of ASG FinFETs only impacts leakage and propagation delay. Preserving some of the SG FinFETs in the NAND2S gate (Figure 21(c)) enables leakage-delay tradeoffs, as evident from the leakage-delay spectrum shown in Figure 22 for various logic gates. The pure ASG gates lie in the left half of the spectrum, indicating low leakage, while pure SG gates lie in the bottom half of the spectrum, indicating less delay.

4.4. SG/IG/ASG Latches and Flip-Flops. Brute-force transmission gate (TG) and half-swing (HS) latches and flip-flops (as shown in Figures 23 and 24) implemented with SG, IG, and ASG FinFETs have also been investigated [49, 50]. Tawfik et al. proposed an IG latch by introducing IG FinFETs in the feedback inverter (I3) of the all-SG TG latch in Figure 23(a).

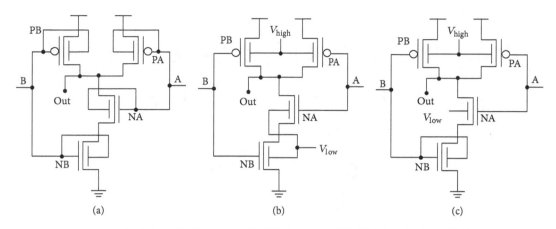

FIGURE 19: Schematic diagrams of NAND2 gates: (a) SG, (b) LP, and (c) MT [49].

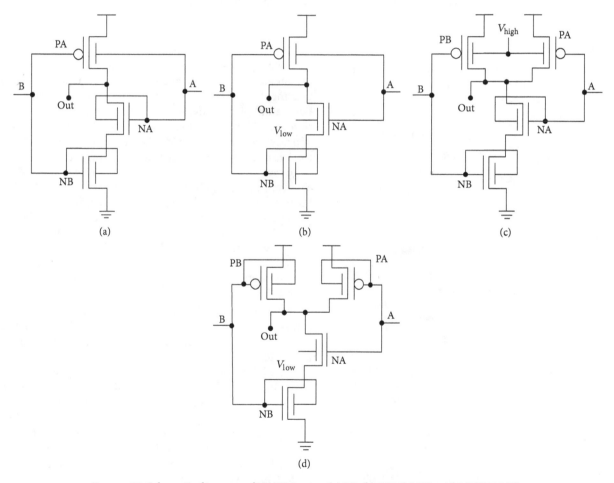

FIGURE 20: Schematic diagrams of NAND2 gates: (a) IG, (b) IG2, (c) XT, and (d) XT2 [49].

With appropriate reverse-biasing of the back gates, the IG FinFETs in I3 are made weaker compared to the drive inverter (I1). As a result, the drive inverter need not be oversized, as conventionally done, ensuring a safe write operation at the same time. At nominal process corners, the IG latch leads to 33% less leakage power and 20% less area compared to the conventional SG latch with almost no degradation in propagation delay and setup time. Similar power and area improvements are obtained for IG flip-flops relative to TG flip-flops (Figure 24(a)). Bhoj and Jha introduced ASG FinFETs in the TG and HS latches and observed similar tradeoffs. Introducing ASG FinFETs in all the latch inverters (I1, I2, and I3) results in a minimum-leakage and maximum-delay configuration. Introducing ASG FinFETs in only I3 leads to a configuration similar to the IG latch. The new configuration reduces leakage power by approximately 50%,

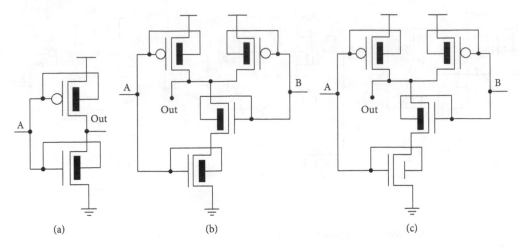

FIGURE 21: Schematic diagrams of ASG FinFET logic gates: (a) INV, (b) NAND2, and (c) NAND2S [49].

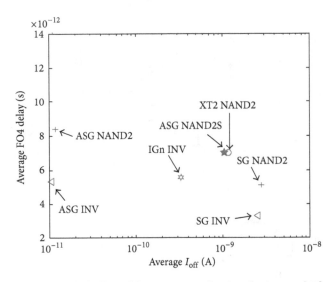

FIGURE 22: The leakage-delay spectrum of various logic gates [49].

but the propagation delay increases by roughly 30%. This configuration also results in area savings as I1 can be sized down, maintaining the desired write stability. Similar results are obtained for ASG flip-flops as well.

As in the case of TG latches and flip-flops, combinations of SG, IG, and ASG FinFETs in inverters (I1 and I2) and nFinFETs (N1 to N4) generate various HS latches (Figure 23(b)) and flip-flops (Figure 24(b)). As expected, the leakage power of the all-ASG configuration is reduced by almost 65%, however, at the expense of doubling of its propagation delay. Using ASG FinFETs in N2/N4 only makes an interesting configuration that results in around 20% improvement in leakage, but only at a negligible cost (less than 5%) in propagation delay. Similar results were obtained for HS flip-flops.

4.5. SRAM. SRAM is a key component of on-chip caches of state-of-the-art microprocessors. In today's multicore processors, typically more than half of the die area is dedicated

to SRAMs [106]. Since SRAMs are built with the smallest transistors possible at a technology node (in order to increase the memory density), statistical fluctuations are extremely detrimental to SRAM performance. Deeply scaled SRAMs, built atop planar MOSFETs, suffer from mismatches in transistor strengths and V_{th} caused by RDF and other sources of process variations. SRAMs also consume most of the chip's total leakage power because of very long idle periods in large memory arrays. Six-transistor (6T) FinFET SRAMs (as shown in Figure 25) have been explored quite thoroughly in the past decade from the point of view of suppressing leakage power and tackling increased variability among bitcells [52–60, 64, 65]. Figure 26 shows the butterfly curves, under process variations, for MOSFET and FinFET based SRAMs. The curves clearly demonstrate that FinFET SRAMs have a superior static noise margin (SNM) because they do not suffer from RDF.

New SRAM bitcell structures have been proposed using a mix of SG, IG, and ASG FinFETs [55, 56, 60, 62]. In [55], FinFET SRAMs have been classified into the following categories: (i) vanilla shorted-gate configurations (VSCs) in which all FinFETs are SG, (ii) independent-gate configurations (IGCs) in which one or more SG FinFETs are replaced with IG FinFETs, and (iii) multiple workfunction shorted-gate configurations (MSCs) in which one or more SG FinFETs are replaced with ASG FinFETs. Table 4 shows the best bitcells from the perspectives of different metrics. RPNM, WTP, I_{READ}, I_{off}, T_R, and T_W refer to the read power noise margin, write-trip power, read current, leakage current, read access time, and write access time of the bitcell, respectively. Out of these, T_R and T_W represent transient metrics whereas the remaining metrics are DC. In Table 4, V (mnp) and A (mnp) refer to VSC and MSC bitcells that have m, n, and p fins in the pull-up (PU), pass-gate (PG), and pull-down (PD) FinFETs, respectively. Pass-gate feedback (PGFB) [59], pull-up write gating (PUWG) [60], split pull-up (SPU) [65], and row-based back-gate bias (RBB) [64] are some popular IGC FinFET SRAM bitcells, as shown in Figure 27. Table 4 also indicates that there is no single SRAM cell that is the best in all the metrics, but it is possible to find a cell that is ahead of

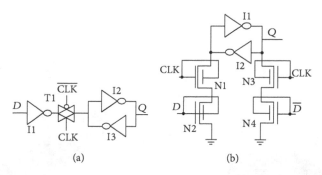

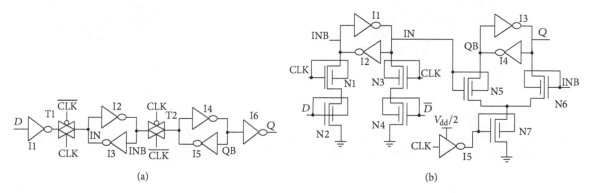

FIGURE 23: Schematic diagrams of FinFET latches: (a) transmission-gate and (b) half-swing [49].

FIGURE 24: Schematic diagrams of FinFET flip-flops: (a) transmission-gate and (b) half-swing [49].

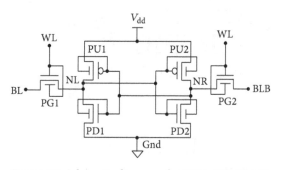

FIGURE 25: Schematic diagram of a 6T FinFET SRAM.

TABLE 4: Comparison of FinFET SRAM cells [55].

Metric	VSC	IGC	MSC
Max. RPNM	V(113)	PGFB-PUWG	A(112)
Min. WTP	V(122)	PGFB-SPU	A(111)
Max. I_{READ}	V(135)	RBB	A(112)
Min. I_{off}	V(111)	RBB	DPG-H
Min. T_R	V(111)	PGFB	A(11)S
Min. T_W	V(111)	PGFB-SPU	A(111)

the others in some of the metrics. A careful look at the absolute values of the metrics reveals that IGC bitcells exhibit superior DC metric values relative to those of VSC bitcells, but their poor transient performance makes them unattractive. On the other hand, MSC bitcells have competitive DC metric values and better transient performance relative to VSC bitcells. Hence, in a nutshell, MSC bitcells may be a good choice for a FinFET SRAM bitcell. Out of all MSC bitcells, A(111) seems to be the most promising one. It is also shown in [55] that the transient behavior of a bitcell is very important to account for. Evaluations based on only DC metrics may lead to incorrect conclusions.

Goel et al. proposed a different FinFET SRAM bitcell using ADSE FinFETs in the access transistors (i.e., the PG FinFETs) [57]. When the extended spacers of the PG FinFETs are placed towards the internal storage nodes (NL and NR) of

the bitcell, it is called contact-underlap-storage (CUS) SRAM. This SRAM exploits the bidirectional current flow in ADSE FinFETs to improve both the read and write margins (by 11% and 6%, resp.). Also, it reduces the leakage current by as much as 57%. However, it suffers from a degraded access time (7%) and cell area (7%).

Moradi et al. proposed a FinFET SRAM bitcell that exploits AD FinFETs [58]. The lowly doped drains of the AD-access transistors are placed towards the storage nodes. This SRAM bitcell is able to resolve the read-write conflicts as the strength of the access transistors varies based on the voltage of the storage nodes. This boosts both read (7.3%) and write (23%) margins. These improvements come at the cost of an increased access time (42%) because the access transistor becomes weak during a read operation. Improvement in subthreshold leakage of this bitcell is also to be noted (2.8×).

Sachid and Hu showed that multiple fin-height FinFETs can be used to design more dense and stable SRAMs [61].

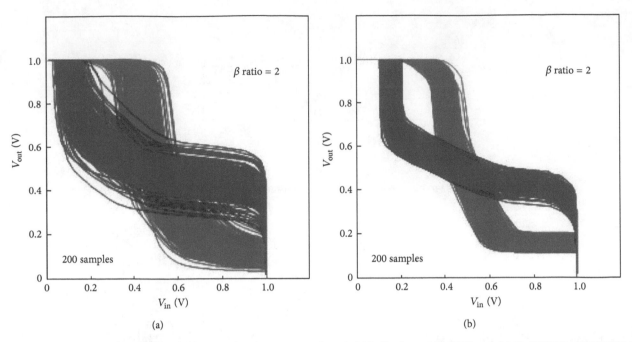

FIGURE 26: Butterfly curves for SRAMs implemented with 20 nm gate-length (a) bulk planar MOSFET and (b) FinFET. The FinFET SRAM exhibits a superior SNM because of smaller V_{th} variation due to the use of an undoped channel [95].

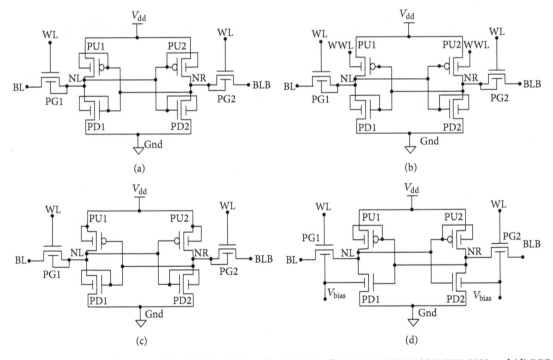

FIGURE 27: Schematic diagrams of FinFET SRAM bitcells: (a) PGFB, (b) PGFB-PUWG, (c) PGFB-SPU, and (d) RBB.

Using multiple fin heights enables better control over the strengths of PU, PG, and PD transistors, leading to a better noise margin, without incurring any area penalty. The drawbacks of this scheme are increased leakage power and process complexity.

4.6. DRAM. One-transistor dynamic random-access memories (1T-DRAMs) have traditionally been used both in off-chip main memory and on-chip caches due to their significant area advantage over SRAMs. With the advent of partially depleted-SOI (PDSOI) technology, a capacitorless 1T-DRAM, also known as floating-body cell (FBC), was proposed. This DRAM leads to a smaller area and a less complicated fabrication process than conventional embedded DRAMs [107–109]. Its functionality is based on the V_{th} shift produced by majority carrier accumulation in the floating

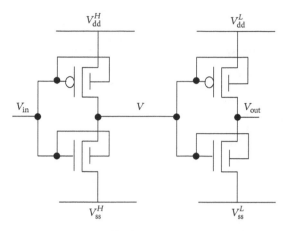

FIGURE 28: Buffer design using TCMS [34].

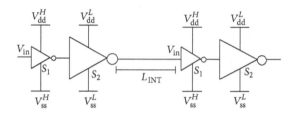

FIGURE 29: Interconnect design using TCMS [34].

body. However, the higher impurity concentration required to suppress SCEs degrades the retention characteristics of planar PDSOI 1T-DRAMs. Double-gate FinFET DRAMs (DG FinDRAM) are able to overcome these scaling issues of 1T-DRAMs [110–112]. The second gate, with the application of an appropriate bias, helps with the accumulation of majority carriers and thereby relaxes the high impurity concentration criterion. FinFET based 1T-DRAMs also exhibit long retention times and large sense margins. Thus, they have emerged as a promising embedded memory alternative.

5. Circuit-Level Analysis

Logic circuit analysis and optimization tools have been implemented using FinFET based standard cell libraries described in the previous section. In this section, we describe them in brief.

5.1. Analysis. FinPrin is a statistical static timing analysis (SSTA) and power analysis tool for FinFET logic circuits that considers PVT variations and is based on accurate statistical models for delay, dynamic power, and leakage power of the logic netlist [113]. It takes a register transfer-level (RTL) or gate-level description of a netlist as an input and estimates leakage/dynamic power and delay distributions (μ and σ for Gaussian distributions) at every node of the netlist, based on the circuit-level parameter values provided in the FinFET design library, such as input and output capacitance, input and output resistance, and leakage current, taking into account the impact of PVT variations. The leakage and

temperature variation models are macromodel based [94], whereas the delay models are based on an SSTA approach [114]. These models also take spatial correlations of the gates into account using a rectangular grid based method [115]. FinPrin's performance has been compared with that of accurate quasi-Monte Carlo (QMC) simulations [116, 117] and was shown to produce very accurate means (μ) and reasonably accurate standard deviations (σ), while enabling a significant computation time speedup (two orders of magnitude).

5.2. Optimization. Optimization of logic circuits is made possible by accurate analysis. Synopsys Design Compiler is commercially used for power/delay optimization of logic circuits, given a standard cell library [99]. In order to exploit the various FinFET design styles, a linear programming based optimization algorithm and tool are proposed in [39]. The algorithm is used to assign gate sizes and FinFET types to the mapped circuit, under a timing constraint, by selecting standard cells from the FinFET design library. Unlike traditional greedy gate-sizing algorithms, this algorithm divides the available slack among gates whose cells may be replaced. It is shown that this approach can achieve 15–30% better power consumption than Synopsys Design Compiler [39].

5.3. Novel Interconnect Structures and Logic Synthesis. Interconnects assume a lot of importance in deeply scaled technology nodes as they govern the delay and power consumption of modern integrated circuits. FinFETs not only provide newer circuit design styles, but also can lead to an efficient interconnect implementation strategy. A mechanism to improve the interconnect efficiency, called threshold voltage control through multiple supply voltages (TCMS), has been proposed in [34]. The TCMS principle is based on the fact that the back-gate bias of a FinFET affects the V_{th} of the front gate. Instead of using the conventional dual-V_{dd} scheme, TCMS uses a slightly higher supply voltage (V_{dd}^H) and a slightly negative supply voltage (V_{ss}^H) along with the nominal supply voltages, V_{dd}^L, and ground (which is referred to as V_{ss}^L for symmetry). TCMS is based on the observation that an overdriven inverter (i.e., whose input is driven by an inverter supplied with V_{dd}^H and V_{ss}^H and whose supply voltage is V_{dd}^L), as shown in Figure 28, has both less leakage and less delay. Less leakage is ensured because of an increase in the V_{th} of the leaking transistor and less delay is ensured because of the higher current drive in the active transistor. The improvement in the drive strength of the active transistor results in improved delay that can be traded off for area and power reduction under a given timing constraint. A chain of such inverter pairs can be formed on the interconnect, as shown in Figure 29, without the need for voltage-level shifters due to the use of higher-V_{th} transistors in the inverter supplied with V_{dd}^H and V_{ss}^H. This scheme enables a significant reduction in subthreshold leakage power in TCMS buffer interconnects. It has been shown that, on an average, TCMS provides overall power savings of 50.4% along with area savings of 9.2% as compared to a state-of-the-art dual-V_{dd} interconnect synthesis scheme [34].

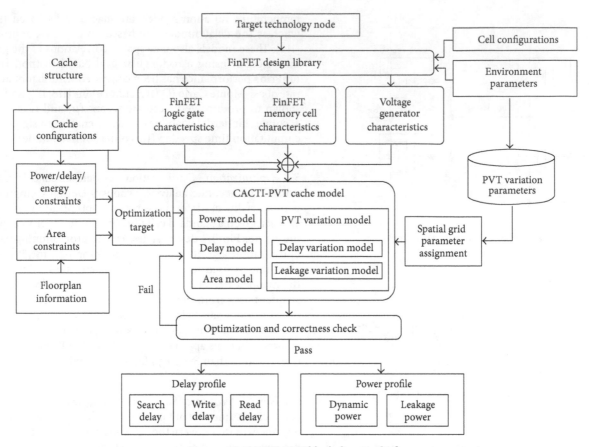

FIGURE 30: CACTI-PVT block diagram [69].

The TCMS principle can also be applied to logic synthesis [35]. In this case, a FinFET logic gate can take advantage of the TCMS principle if its input arrives from a gate supplied with the V_{dd}^H set and its supply voltage belongs to the V_{dd}^L set. Since the opposite scenario leads to a high leakage current, it is avoided. Based on the combinations of supply voltage (V_{dd}^L or V_{dd}^H), input voltage (V_{dd}^L or V_{dd}^H), and threshold voltage (high-V_{th} or low-V_{th}), INV and NAND2 have seven and 25 variants, respectively. As in the case of the interconnects, use of high-V_{th} FinFETs in V_{dd}^H gates that need to be driven by a V_{dd}^L input voltage obviates the need for a voltage-level converter between the V_{dd}^L and V_{dd}^H gates. With the use of a linear programming based optimization algorithm, TCMS leads to an overall power reduction of 3× under relaxed delay constraints.

6. Architecture-Level Analysis

Next, we ascend the design hierarchy to the architecture level. Due to shrinking feature sizes and severe process variations, the delay and power consumption at the chip level are not easy to predict any more [114]. Because of their inherent statistical nature, a yield analysis of an integrated circuit (under a design constraint) has become very important. This analysis estimates the percentage of chips that will meet the given power and delay constraints for the particular chip architecture for a given process. In the following subsections, we discuss PVT-aware simulation tools for various FinFET based architectural components.

6.1. FinFET Based Caches. An integrated PVT variation-aware power-delay simulation framework, called FinCANON [69], has been developed for FinFET based caches and NoCs. It has two components: CACTI-PVT for caches and ORION-PVT for NoCs. CACTI-PVT is an extension of CACT-FinFET [67]. CACTI-PVT can be used to obtain the delay and leakage distributions of FinFET based caches with varying sizes, SRAM cell types, and back-gate biases. The block diagram of CACTI-PVT is shown in Figure 30. It uses a FinFET design library consisting of FinFET logic gates of various sizes and types and different types of FinFET SRAM cells. This library is characterized using accurate device simulation. The process variation models used in CACTI-PVT are calibrated using QMC simulations, along with the rectangular grid-based method to model spatial correlations. Peripheral components implemented with SG FinFETs and SRAM cells implemented with some IG FinFETs or ASG FinFETs provide the best balance between delay and leakage of the FinFET caches.

6.2. FinFET Based NoCs. With increasing number of cores in chip multiprocessors (CMPs), NoCs have emerged as an

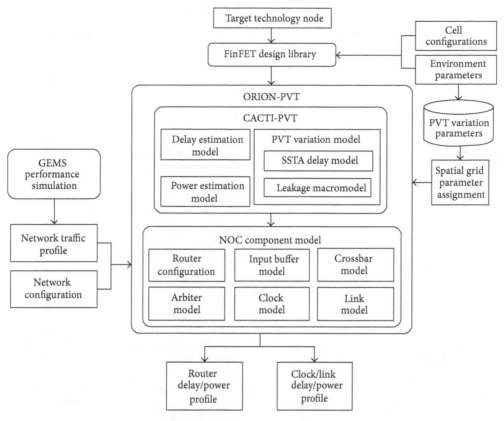

FIGURE 31: ORION-PVT block diagram [69].

effective communication mechanism among the cores. Fin-CANON also includes a performance/power simulation tool, called ORION-PVT, aimed at FinFET NoCs [69]. ORION-PVT, whose block diagram is shown in Figure 31, is an extension of ORION-FinFET [68]. Here, an SSTA technique and a macromodel based methodology are used to model the PVT variations in delay and leakage. It also provides a power breakdown of an on-chip router. Leakage power is found to dominate the total power of the router at higher temperatures.

A FinFET based implementation of a variable-pipeline-stage router (VPSR) is proposed in [70]. VPSR enables dynamic adjustment of the number of pipeline stages in the router based on incoming network traffic. As a result, different flow control digits (flits) may traverse pipeline stages of varying lengths while passing through the router. This leads to enhanced router performance because VPSR adapts its throughput to the network traffic requirement at runtime. VPSR also enables significant savings in leakage power through reverse-biasing (called adaptive back-gate biasing) of the back gates of IG FinFETs in infrequently accessed components of the router.

6.3. FinFET Based Multicore Processors. In the computer architecture domain, the trend has shifted in recent years from uniprocessors to CMPs and multicore systems in order to serve the ever-increasing performance demand. Tools like FinCANON have paved the way for a more powerful tool for characterizing multicore processors. McPAT-PVT is a PVT

variation-aware integrated power-delay simulation tool for FinFET based multicore processors [71]. Figure 32 shows the block diagram of McPAT-PVT. It has two key components: processor model and yield analyzer. The processor model contains power/delay macromodels of various functional units (e.g., arithmetic-logic unit, floating-point unit, memory management unit, etc.) of the processor core. The yield analyzer can predict the yield of a specified processor configuration under PVT variations. Figure 33 zooms into the components of the processor model. The efficacy of this tool has been demonstrated on an alpha-like processor core and multicore simulations based on Princeton Application Repository for Shared-Memory Computer (PARSEC) benchmarks.

7. Conclusion

In this paper, we have explored the impact of FinFETs from the device to architecture level. We learnt about the shortcomings of planar MOSFETs in today's deeply scaled technologies and the advantages of FinFETs as suitable replacements for planar MOSFETs. We looked into FinFET device characteristics and evaluated tradeoffs among SG, IG, and ASG FinFETs, along with other FinFET asymmetries, such as drain-spacer extension, source/drain doping, gate-oxide thickness, and fin height. We learnt about the detrimental impact of PVT variations on FinFET chip performance

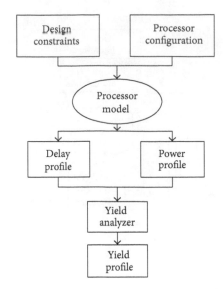

FIGURE 32: McPAT-PVT block diagram [71].

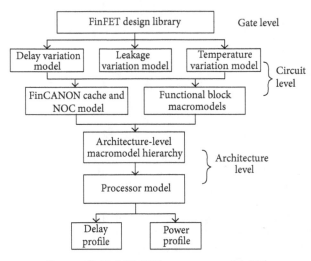

FIGURE 33: McPAT-PVT processor models [71].

and power. We surveyed techniques for characterizing Fin-FET devices and circuits and explored FinFET based logic gates, flip-flops, and memory cells. Finally, we also reviewed PVT variation-aware FinFET circuit- and architecture-level simulation tools. We observed leakage-delay tradeoffs that are possible at each level of the design hierarchy. The availability of a plethora of FinFET styles opens up new design opportunities at each level, which we hope some of the readers will be willing to explore.

Conflict of Interests

The authors declare that there is no conflict of interests regarding the publication of this paper.

Acknowledgment

This work was supported by NSF under Grant nos. CCF-1217076 and CCF-1318603.

References

[1] K. J. Kuhn, "CMOS scaling for the 22nm node and beyond: Device physics and technology," in *Proceedings of the International Symposium on VLSI Technology, Systems and Applications (VLSI-TSA '11)*, pp. 1–2, April 2011.

[2] K. Roy, S. Mukhopadhyay, and H. Mahmoodi-Meimand, "Leakage current mechanisms and leakage reduction techniques in deep-submicrometer CMOS circuits," *Proceedings of the IEEE*, vol. 91, no. 2, pp. 305–327, 2003.

[3] D. J. Frank, R. H. Dennard, E. Nowak, P. M. Solomon, Y. Taur, and H.-S. P. Wong, "Device scaling limits of Si MOSFETs and their application dependencies," *Proceedings of the IEEE*, vol. 89, no. 3, pp. 259–288, 2001.

[4] C. Hu, "Gate oxide scaling limits and projection," in *Proceedings of the IEEE International Electron Devices Meeting*, pp. 319–322, December 1996.

[5] Y.-C. Yeo, T.-J. King, and C. Hu, "MOSFET gate leakage modeling and selection guide for alternative gate dielectrics based on leakage considerations," *IEEE Transactions on Electron Devices*, vol. 50, no. 4, pp. 1027–1035, 2003.

[6] J. Chen, T. Y. Chan, I. C. Chen, P. K. Ko, and C. Hu, "Subbreakdown drain leakage current in MOSFET," *Electron device letters*, vol. 8, no. 11, pp. 515–517, 1987.

[7] "International technology roadmap for semiconductors," 2011, http://www.itrs.net.

[8] T. Skotnicki, J. A. Hutchby, T.-J. King, H.-S. P. Wong, and F. Boeuf, "The end of CMOS scaling: toward the introduction of new materials and structural changes to improve MOSFET performance," *IEEE Circuits and Devices Magazine*, vol. 21, no. 1, pp. 16–26, 2005.

[9] H.-S. P. Wong, D. J. Franks, and P. M. Solomon, "Device design considerations for double-gate, ground-plane, and single-gated ultra-thin SOI MOSFET's at the 25 nm channel length generation," in *Proceedings of the IEEE International Electron Devices Meeting (IEDM '98)*, pp. 407–410, San Francisco, Calif, USA, December 1998.

[10] P. M. Solomon, K. W. Guarini, Y. Zhang et al., "Two gates are better than one," *IEEE Circuits and Devices Magazine*, vol. 19, no. 1, pp. 48–62, 2003.

[11] K. Suzuki, T. Tanaka, Y. Tosaka, H. Horie, and Y. Arimoto, "Scaling theory for double-gate SOI MOSFET's," *IEEE Transactions on Electron Devices*, vol. 40, no. 12, pp. 2326–2329, 1993.

[12] E. J. Nowak, I. Aller, T. Ludwig et al., "Turning silicon on its edge [double gate CMOS/FinFET technology]," *IEEE Circuits and Devices Magazine*, vol. 20, no. 1, pp. 20–31, 2004.

[13] R.-H. Yan, A. Ourmazd, and K. F. Lee, "Scaling the Si MOSFET: from bulk to SOI to bulk," *IEEE Transactions on Electron Devices*, vol. 39, no. 7, pp. 1704–1710, 1992.

[14] Y.-K. Choi, K. Asano, N. Lindert et al., "Ultrathin-body SOI MOSFET for deep-sub-tenth micron era," *IEEE Electron Device Letters*, vol. 21, no. 5, pp. 254–255, 2000.

[15] B. Doris, K. Cheng, A. Khakifirooz, Q. Liu, and M. Vinet, "Device design considerations for next generation CMOS technology: planar FDSOI and FinFET (Invited)," in *Proceedings of the International Symposium on VLSI Technology, Systems, and Applications (VLSI-TSA '13)*, pp. 1–2, April 2013.

[16] C. Hu, "New sub-20 nm transistors: why and how," in *Proceedings of the 48th Design Automation Conference (DAC '11)*, pp. 460–463, June 2011.

[17] J. Markoff, "TSMC taps ARM's V8 on road to 16-nm FinFET," 2012, http://www.eetimes.com/electronicsnews/ 4398727/ TSMC-taps-ARM-V8-in-road to-16-nm-FinFET.

[18] D. McGrath, "Globalfoundries looks to leapfrog fab rival," http://www.eetimes.com/electronicsnews/ 4396720/Globalfou ndries-to-offer-14-nm-process-with-FinFETsin, 2014.

[19] D. Hisamoto, W.-C. Lee, J. Kedzierski et al., "FinFET—a self-aligned double-gate MOSFET scalable to 20 nm," *IEEE Transactions on Electron Devices*, vol. 47, no. 12, pp. 2320–2325, 2000.

[20] B. Yu, L. Chang, S. Ahmed et al., "FinFET scaling to 10 nm gate length," in *Proceedings of the IEEE International Devices Meeting (IEDM '02)*, pp. 251–254, San Francisco, Calif, USA, December 2002.

[21] S. Tang, L. Chang, N. Lindert et al., "FinFET—a quasiplanar double-gate MOSFET," in *Proceedings of the International of Solid-State Circuits Conference*, pp. 118–119, February 2001.

[22] M. Guillorn, J. Chang, A. Bryant et al., "FinFET performance advantage at 22 nm: an AC perspective," in *Proceedings of the Symposium on VLSI Technology Digest of Technical Papers (VLSIT '08)*, pp. 12–13, June 2008.

[23] F.-L. Yang, D.-H. Lee, H.-Y. Chen et al., "5nm-gate nanowire FinFET," in *Proceedings of the Symposium on VLSI Technology—Digest of Technical Papers*, pp. 196–197, June 2004.

[24] X. Huang, W.-C. Lee, C. Kuo et al., "Sub 50-nm FinFET: PMOS," in *Proceedings of the IEEE International Devices Meeting (IEDM '99)*, pp. 67–70, Washington, DC, USA, December 1999.

[25] J.-P. Colinge, *FinFETs and Other Multi-Gate Transistors*, Springer, New York, NY, USA, 2008.

[26] T.-J. King, "FinFETs for nanoscale CMOS digital integrated circuits," in *Proceedings of the IEEE/ACM International Conference on Computer-Aided Design (ICCAD '05)*, pp. 207–210, November 2005.

[27] J. B. Chang, M. Guillorn, P. M. Solomon et al., "Scaling of SOI FinFETs down to fin width of 4 nm for the 10 nm technology node," in *Proceedings of the Symposium on VLSI Technology, Systems and Applications (VLSIT '11)*, pp. 12–13, June 2011.

[28] C. Auth, "22-nm fully-depleted tri-gate CMOS transistors," in *Proceedings of the IEEE Custom Integrated Circuits Conference (CICC '12)*, pp. 1–6, San Jose, Calif, USA, September 2012.

[29] C.-H. Lin, J. Chang, M. Guillorn, A. Bryant, P. Oldiges, and W. Haensch, "Non-planar device architecture for 15 nm node: FinFET or trigate?" in *Proceedings of the IEEE International Silicon on Insulator Conference (SOI '10)*, pp. 1–2, October 2010.

[30] K. Lee, T. An, S. Joo, K.-W. Kwon, and S. Kim, "Modeling of parasitic fringing capacitance in multifin trigate FinFETs," *IEEE Transactions on Electron Devices*, vol. 60, no. 5, pp. 1786–1789, 2013.

[31] J. Gu, J. Keane, S. Sapatnekar, and C. H. Kim, "Statistical leakage estimation of double gate FinFET devices considering the width quantization property," *IEEE Transactions on Very Large Scale Integration Systems*, vol. 16, no. 2, pp. 206–209, 2008.

[32] D. Ha, H. Takeuchi, Y.-K. Choi, and T.-J. King, "Molybdenum gate technology for ultrathin-body MOSFETs and FinFETs," *IEEE Transactions on Electron Devices*, vol. 51, no. 12, pp. 1989–1996, 2004.

[33] T. Sairam, W. Zhao, and Y. Cao, "Optimizing FinFET technology for high-speed and low-power design," in *Proceedings of the 17th Great Lakes Symposium on VLSI (GLSVLSI '07)*, pp. 73–77, March 2007.

[34] A. Muttreja, P. Mishra, and N. K. Jha, "Threshold voltage control through multiple supply voltages for power-efficient FinFET interconnects," in *Proceedings of the 21st International Conference on VLSI Design (VLSI '08)*, pp. 220–227, Hyderabad, India, January 2008.

[35] P. Mishra, A. Muttreja, and N. K. Jha, "Low-power FinFET circuit synthesis using multiple supply and threshold voltages," *ACM Journal on Emerging Technologies in Computing Systems*, vol. 5, no. 2, article 7, 2009.

[36] P. Mishra and N. K. Jha, "Low-power FinFET circuit synthesis using surface orientation optimization," in *Proceedings of the Design, Automation and Test in Europe Conference and Exhibition (DATE '10)*, pp. 311–314, March 2010.

[37] S. Chaudhuri, P. Mishra, and N. K. Jha, "Accurate leakage estimation for FinFET standard cells using the response surface methodology," in *Proceedings of the 25th International Conference on VLSI Design (VLSID '12)*, pp. 238–244, Hyderabad, India, January 2012.

[38] A. Muttreja, N. Agarwal, and N. K. Jha, "CMOS logic design with independent-gate FinFETs," in *Proceedings of the IEEE International Conference on Computer Design (ICCD '07)*, pp. 560–567, October 2007.

[39] M. Agostinelli, M. Alioto, D. Esseni, and L. Selmi, "Leakage-delay tradeoff in FinFET logic circuits: a comparative analysis with bulk technology," *IEEE Transactions on Very Large Scale Integration (VLSI) Systems*, vol. 18, no. 2, pp. 232–245, 2010.

[40] M. Rostami and K. Mohanram, "Dual-Vth independent-gate FinFETs for low power logic circuits," *IEEE Transactions on Computer-Aided Design of Integrated Circuits and Systems*, vol. 30, no. 3, pp. 337–349, 2011.

[41] A. Datta, A. Goel, R. T. Cakici, H. Mahmoodi, D. Lekshmanan, and K. Roy, "Modeling and circuit synthesis for independently controlled double gate FinFET devices," *IEEE Transactions on Computer-Aided Design of Integrated Circuits and Systems*, vol. 26, no. 11, pp. 1957–1966, 2007.

[42] Z. Weimin, J. G. Fossum, L. Mathew, and D. Yang, "Physical insights regarding design and performance of independent-gate FinFETs," *IEEE Transactions on Electron Devices*, vol. 52, no. 10, pp. 2198–2205, 2005.

[43] C.-H. Lin, W. Haensch, P. Oldiges et al., "Modeling of width-quantization-induced variations in logic FinFETs for 22 nm and beyond," in *Proceedings of the Symposium on VLSI Technology (VLSIT '11)*, pp. 16–17, June 2011.

[44] R. A. Thakker, C. Sathe, A. B. Sachid, M. Shojaei Baghini, V. Ramgopal Rao, and M. B. Patil, "A novel table-based approach for design of FinFET circuits," *IEEE Transactions on Computer-Aided Design of Integrated Circuits and Systems*, vol. 28, no. 7, pp. 1061–1070, 2009.

[45] M. Agostinelli, M. Alioto, D. Esseni, and L. Selmi, "Design and evaluation of mixed 3T-4T FinFET stacks for leakage reduction," in *Integrated Circuit and System Design. Power and Timing Modeling, Optimization and Simulation*, L. Svensson and J. Monteiro, Eds., pp. 31–41, Springer, Berlin, Germany, 2009.

[46] J. Ouyang and Y. Xie, "Power optimization for FinFET-based circuits using genetic algorithms," in *Proceedings of the IEEE International SOC Conference*, pp. 211–214, September 2008.

[47] B. Swahn and S. Hassoun, "Gate sizing: FinFETs vs 32 nm bulk MOSFETs," in *Proceedings of the 43rd IEEE Design Automation Conference*, pp. 528–531, San Francisco, Calif, USA, July 2006.

[48] A. N. Bhoj, M. O. Simsir, and N. K. Jha, "Fault models for logic circuits in the multigate era," *IEEE Transactions on Nanotechnology*, vol. 11, no. 1, pp. 182–193, 2012.

[49] A. N. Bhoj and N. K. Jha, "Design of logic gates and flip-flops in high-performance FinFET technology," *IEEE Transactions on Very Large Scale Integration (VLSI) Systems*, vol. 21, no. 11, pp. 1975–1988, 2013.

[50] S. A. Tawfik and V. Kursun, "Characterization of new static independent-gate-biased FinFET latches and flip-flops under process variations," in *Proceedings of the 9th International Symposium on Quality Electronic Design (ISQED '08)*, pp. 311–316, San Jose, Calif, USA, March 2008.

[51] S. A. Tawfik and V. Kursun, "Low-power and compact sequential circuits with independent-gate FinFETs," *IEEE Transactions on Electron Devices*, vol. 55, no. 1, pp. 60–70, 2008.

[52] A. Bansal, S. Mukhopadhyay, and K. Roy, "Device-optimization technique for robust and low-power FinFET SRAM design in NanoScale era," *IEEE Transactions on Electron Devices*, vol. 54, no. 6, pp. 1409–1419, 2007.

[53] A. N. Bhoj and R. V. Joshi, "Transport-analysis-based 3-D TCAD capacitance extraction for sub-32-nm SRAM structures," *IEEE Electron Device Letters*, vol. 33, no. 2, pp. 158–160, 2012.

[54] A. N. Bhoj, R. V. Joshi, and N. K. Jha, "Efficient methodologies for 3-D TCAD modeling of emerging devices and circuits," *IEEE Transactions on Computer-Aided Design of Integrated Circuits and Systems*, vol. 32, no. 1, pp. 47–58, 2013.

[55] A. N. Bhoj and N. K. Jha, "Parasitics-aware design of symmetric and asymmetric gate-workfunction FinFET SRAMs," *IEEE Transactions on Very Large Scale Integration Systems*, vol. 22, no. 3, pp. 548–561, 2014.

[56] K. Endo, S.-I. O'Uchi, T. Matsukawa, Y. Liu, and M. Masahara, "Independent double-gate FinFET SRAM technology," in *Proceedings of the 4th IEEE International Nanoelectronics Conference (INEC '11)*, pp. 1–2, June 2011.

[57] A. Goel, S. K. Gupta, and K. Roy, "Asymmetric drain spacer extension (ADSE) FinFETs for low-power and robust SRAMs," *IEEE Transactions on Electron Devices*, vol. 58, no. 2, pp. 296–308, 2011.

[58] F. Moradi, S. K. Gupta, G. Panagopoulos, D. T. Wisland, H. Mahmoodi, and K. Roy, "Asymmetrically doped FinFETs for low-power robust SRAMs," *IEEE Transactions on Electron Devices*, vol. 58, no. 12, pp. 4241–4249, 2011.

[59] Z. Guo, S. Balasubramanian, R. Zlatanovici, T.-J. King, and B. Nikolić, "FinFET-based SRAM design," in *Proceedings of the International Symposium on Low Power Electronics and Design*, pp. 2–7, August 2005.

[60] A. Carlson, Z. Guo, S. Balasubramanian, R. Zlatanovici, T. J. K. Liu, and B. Nikolic, "SRAM read/write margin enhancements using FinFETs," *IEEE Transactions on Very Large Scale Integration (VLSI) Systems*, vol. 18, no. 6, pp. 887–900, 2010.

[61] A. B. Sachid and C. Hu, "Denser and more stable SRAM using FinFETs with multiple fin heights," *IEEE Transactions on Electron Devices*, vol. 59, no. 8, pp. 2037–2041, 2012.

[62] S. A. Tawfik, Z. Liu, and V. Kursun, "Independent-gate and tied-gate FinFET SRAM circuits: design guidelines for reduced area and enhanced stability," in *Proceedings of the 19th International Conference on Microelectronics (ICM '07)*, pp. 171–174, Cairo, Egypt, December 2007.

[63] A. N. Bhoj, R. V. Joshi, S. Polonsky et al., "Hardware-assisted 3D TCAD for predictive capacitance extraction in 32 nm SOI SRAMs," in *Proceedings of the IEEE International Electron Devices Meeting (IEDM '11)*, pp. 34.7.1–34.7.4, Washington, DC, USA, December 2011.

[64] R. Joshi, K. Kim, and R. Kanj, "FinFET SRAM design," in *Proceedings of the 23rd International Conference on VLSI Design (VLSID '10)*, pp. 440–445, Bangalore, India, January 2010.

[65] R. V. Joshi, K. Kim, R. Q. Williams, E. J. Nowak, and C.-T. Chuang, "A high-performance, low leakage, and stable SRAM row-based back-gate biasing scheme in FinFET technology," in *Proceedings of the 20th International Conference on VLSI Design held jointly with 6th International Conference on Embedded Systems (VLSID '07)*, pp. 665–670, January 2007.

[66] A. N. Bhoj, R. V. Joshi, and N. K. Jha, "3-D-TCAD-based parasitic capacitance extraction for emerging multigate devices and circuits," *IEEE Transactions on Very Large Scale Integration Systems*, vol. 21, no. 11, pp. 2094–2105, 2013.

[67] C. Y. Lee and N. K. Jha, "CACTI-FinFET: an integrated delay and power modeling framework for FinFET-based caches under process variations," in *Proceedings of the 48th ACM/EDAC/IEEE Design Automation Conference (DAC '11)*, pp. 866–871, June 2011.

[68] C.-Y. Lee and N. K. Jha, "FinFET-based power simulator for interconnection networks," *ACM Journal on Emerging Technologies in Computing Systems*, vol. 6, no. 1, article 2, 2008.

[69] C.-Y. Lee and N. K. Jha, "FinCANON: a PVT-aware integrated delay and power modeling framework for FinFET-based caches and on-chip networks," *IEEE Transactions on Very Large Scale Integration Systems*, vol. 22, no. 5, pp. 1150–1163, 2014.

[70] C.-Y. Lee and N. K. Jha, "Variable-pipeline-stage router," *IEEE Transactions on Very Large Scale Integration*, vol. 21, no. 9, pp. 1669–1681, 2013.

[71] A. Tang, Y. Yang, C.-Y. Lee, and N. K. Jha, "McPAT-PVT: delay and power modeling framework for FinFET processor architectures under PVT variations," *IEEE Transactions on Very Large Scale Integration Systems*. In press.

[72] X. Chen and N. K. Jha, "Ultra-low-leakage chip multiprocessor design with hybrid FinFET logic styles," *ACM Journal on Emerging Technologies in Computing Systems*. In press.

[73] A. Tang and N. K. Jha, "Thermal characterization of test techniques for FinFET and 3D integrated circuits," *ACM Journal on Emerging Technologies in Computing Systems*, vol. 9, no. 1, article 6, 2013.

[74] A. Tang and N. K. Jha, "Design space exploration of FinFET cache," *ACM Journal on Emerging Technologies in Computing Systems*, vol. 9, no. 3, pp. 20:1–20:16, 2013.

[75] P. Mishra, A. Muttreja, and N. K. Jha, "FinFET circuit design," in *Nanoelectronic Circuit Design*, N. K. Jha and D. Chen, Eds., pp. 23–54, Springer, New York, NY, USA, 2011.

[76] D. Hisamoto, T. Kaga, Y. Kawamoto, and E. Takeda, "A fully depleted lean-channel transistor (DELTA)—a novel vertical ultra thin SOI MOSFET," in *Proceedings of the International Electron Devices Meeting (IEDM '89)*, pp. 833–836, Washington, DC, USA, December 1989.

[77] M. Alioto, "Comparative evaluation of layout density in 3T, 4T, and MT FinFET standard cells," *IEEE Transactions on Very Large Scale Integration (VLSI) Systems*, vol. 19, no. 5, pp. 751–762, 2011.

[78] N. Collaert, M. Demand, I. Ferain et al., "Tall triple-gate devices with TiN/HfO$_2$ gate stack," in *Proceedings of the Symposium on VLSI Technology*, pp. 108–109, June 2005.

[79] T.-S. Park, H. J. Cho, J. D. Choe et al., "Characteristics of the full CMOS SRAM cell using body-tied TG MOSFETs (Bulk FinFETS)," *IEEE Transactions on Electron Devices*, vol. 53, no. 3, pp. 481–487, 2006.

[80] H. Kawasaki, K. Okano, A. Kaneko et al., "Embedded bulk FinFET SRAM cell technology with planar FET peripheral circuit for hp32 nm node and beyond," in *Proceedings of the Symposium on VLSI Technology (VLSIT '06)*, pp. 70–71, June 2006.

[81] S. Y. Kim and J. H. Lee, "Hot carrier-induced degradation in bulk FinFETs," *IEEE Electron Device Letters*, vol. 26, no. 8, pp. 566–568, 2005.

[82] J. Markoff, "Intel increases transistor speed by building upward," May 2011, http://www.nytimes.com/2011/05/05/science/05chip.html.

[83] J.-W. Yang and J. G. Fossum, "On the feasibility of nanoscale triple-gate CMOS transistors," *IEEE Transactions on Electron Devices*, vol. 52, no. 6, pp. 1159–1164, 2005.

[84] L. Chang, M. Ieong, and M. Yang, "CMOS circuit performance enhancement by surface orientation optimization," *IEEE Transactions on Electron Devices*, vol. 51, no. 10, pp. 1621–1627, 2004.

[85] M. Kang, S. C. Song, S. H. Woo et al., "FinFET SRAM optimization with fin thickness and surface orientation," *IEEE Transactions on Electron Devices*, vol. 57, no. 11, pp. 2785–2793, 2010.

[86] J. Kedzierski, D. M. Fried, E. J. Nowak et al., "High-performance symmetric-gate and CMOS-compatible Vt asymmetric-gate FinFET devices," in *Proceedings of the IEEE International Electron Devices Meeting (IEDM '01)*, pp. 437–440, December 2001.

[87] L. Mathew, M. Sadd, B. E. White, and et al, "FinFET with isolated n+ and p+ gate regions strapped with metal and polysilicon," in *Proceedings of the IEEE International SOI Conference Proceedings*, pp. 109–110, October 2003.

[88] M. Masahara, R. Surdeanu, L. Witters et al., "Demonstration of asymmetric gateoxide thickness four-terminal FinFETs having flexible threshold voltage and good subthreshold slope," *IEEE Electron Device Letters*, vol. 28, no. 3, pp. 217–219, 2007.

[89] M. Masahara, R. Surdeanu, L. Witters et al., "Demonstration of asymmetric gate oxide thickness 4-terminal FinFETs," in *Proceedings of the IEEE International Silicon on Insulator Conference (SOI '06)*, pp. 165–166, October 2006.

[90] Y. Liu, T. Matsukawa, K. Endo et al., "Advanced FinFET CMOS technology: TiN-Gate, fin-height control and asymmetric gate insulator thickness 4T-FinFETs," in *Proceedings of the International Electron Devices Meeting (IEDM '06)*, pp. 1–4, San Francisco, Calif, USA, December 2006.

[91] S. Xiong and J. Bokor, "Sensitivity of double-gate and FinFET-devices to process variations," *IEEE Transactions on Electron Devices*, vol. 50, no. 11, pp. 2255–2261, 2003.

[92] X. Wang, A. R. Brown, B. Cheng, and A. Asenov, "Statistical variability and reliability in nanoscale FinFETs," in *Proceedings of the IEEE International Electron Devices Meeting (IEDM '11)*, pp. 541–544, Washington, DC, USA, December 2011.

[93] E. Baravelli, L. de Marchi, and N. Speciale, "VDD scalability of FinFET SRAMs: robustness of different design options against LER-induced variations," *Solid-State Electronics*, vol. 54, no. 9, pp. 909–918, 2010.

[94] P. Mishra, A. N. Bhoj, and N. K. Jha, "Die-level leakage power analysis of FinFET circuits considering process variations," in *Proceedings of the 11th International Symposium on Quality Electronic Design (ISQED '10)*, pp. 347–355, March 2010.

[95] T. Matsukawa, S. O'uchi, K. Endo et al., "Comprehensive analysis of variability sources of FinFET characteristics," in *Proceedings of the Symposium on VLSI Technology (VLSIT '09)*, pp. 118–119, Honolulu, Hawaii, USA, June 2009.

[96] S. Chaudhuri and N. K. Jha, "3D vs. 2D analysis of FinFET logic gates under process variations," in *Proceedings of the 29th IEEE International Conference on Computer Design (ICCD '11)*, pp. 435–436, Amherst, Mass, USA, November 2011.

[97] S. M. Chaudhuri and N. K. Jha, "3D vs. 2D device simulation of FinFET logic gates under PVT variations," *ACM Journal on Emerging Technologies in Computing Systems*, vol. 10, no. 3, 2014.

[98] J. H. Choi, J. Murthy, and K. Roy, "The effect of process variation on device temperature in FinFET circuits," in *Proceedings of the IEEE/ACM International Conference on Computer-Aided Design (ICCAD '07)*, pp. 747–751, November 2007.

[99] Sentaurus TCAD tool suite, http://www.synopys.com.

[100] M. Nawaz, W. Molzer, P. Haibach et al., "Validation of 30 nm process simulation using 3D TCAD for FinFET devices," *Semiconductor Science and Technology*, vol. 21, no. 8, pp. 1111–1120, 2006.

[101] D. Vasileska and S. M. Goodnick, *Computational Electronics*, Morgan & Claypool, 2006.

[102] N. Paydavosi, S. Venugopalan, Y. S. Chauhan et al., "BSIM—SPICE models enable FinFET and UTB IC designs," *IEEE Access*, vol. 1, pp. 201–215, 2013.

[103] S. Venugopalan, D. D. Lu, Y. Kawakami, P. M. Lee, A. M. Niknejad, and C. Hu, "BSIM-CG: a compact model of cylindrical/surround gate MOSFET for circuit simulations," *Solid-State Electronics*, vol. 67, no. 1, pp. 79–89, 2012.

[104] J. G. Fossum, L. Ge, M.-H. Chiang et al., "A process/physics-based compact model for nonclassical CMOS device and circuit design," *Solid-State Electronics*, vol. 48, no. 6, pp. 919–926, 2004.

[105] J. G. Fossum, M. M. Chowdhury, V. P. Trivedi et al., "Physical insights on design and modeling of nanoscale FinFETs," in *Proceedings of the IEEE International Electron Devices Meeting (IEDM '03)*, pp. 29.1.1–29.1.4, Washington, DC, USA, December 2003.

[106] Y. N. Patt, S. J. Patel, M. Evers, D. H. Friendly, and J. Stark, "One billion transistors, one uniprocessor, one chip," *Computer*, vol. 30, no. 9, pp. 51–57, 1997.

[107] E. Yoshida and T. Tanaka, "A design of a capacitorless 1T-DRAM cell using gate-induced drain leakage (GIDL) current for low-power and high-speed embedded memory," in *Proceedings of the IEEE International Electron Devices Meeting*, pp. 3761–3764, Washington, DC, USA, December 2003.

[108] L. C. Tran, "Challenges of DRAM and flash scaling—potentials in advanced emerging memory devices," in *Proceedings of the 7th International Conference on Solid-State and Integrated Circuits Technology Proceedings (ICSICT '04)*, vol. 1, pp. 668–672, October 2004.

[109] A. N. Bhoj and N. K. Jha, "Gated-diode FinFET DRAMs: device and circuit design-considerations," *ACM Journal on Emerging Technologies in Computing Systems*, vol. 6, no. 4, pp. 12:1–12:32, 2010.

[110] T. Tanaka, E. Yoshida, and T. Miyashita, "Scalability study on a capacitorless 1T-DRAM: from single-gate PD-SOI to double-gate FinDRAM," in *Proceedings of the IEEE International Electron Devices Meeting*, pp. 919–922, December 2004.

[111] M. Bawedin, S. Cristoloveanu, and D. Flandre, "A capacitorless 1T-DRAM on SOI based on dynamic coupling and double-gate operation," *IEEE Electron Device Letters*, vol. 29, no. 7, pp. 795–798, 2008.

[112] E. Yoshida, T. Miyashita, and T. Tanaka, "A study of highly scalable DG-FinDRAM," *IEEE Electron Device Letters*, vol. 26, no. 9, pp. 655–657, 2005.

[113] Y. Yang and N. K. Jha, "FinPrin: analysis and optimization of FinFET logic circuits under PVT variations," in *Proceedings of the 26th International Conference on VLSI Design*, pp. 350–355, January 2013.

[114] H. Chang and S. S. Sapatnekar, "Statistical timing analysis under spatial correlations," *IEEE Transactions on Computer-Aided Design of Integrated Circuits and Systems*, vol. 24, no. 9, pp. 1467–1482, 2005.

[115] A. Agarwal, D. Blaauw, and V. Zolotov, "Statistical timing analysis for intra-die process variations with spatial correlations," in *Proceedings of the International Conference on Computer Aided Design (ICCAD '03)*, pp. 900–907, November 2003.

[116] A. Singhee and R. A. Rutenbar, "From finance to flip flops: a study of fast Quasi-Monte Carlo methods from computational finance applied to statistical circuit analysis," in *Proceedings of the 8th International Symposium on Quality Electronic Design (ISQED '07)*, pp. 685–692, March 2007.

[117] A. Singhee and R. A. Rutenbar, "Why quasi-Monte Carlo is better than Monte Carlo or Latin hypercube sampling for statistical circuit analysis," *IEEE Transactions on Computer-Aided Design of Integrated Circuits and Systems*, vol. 29, no. 11, pp. 1763–1776, 2010.

12

Advances in Microelectronics for Implantable Medical Devices

Andreas Demosthenous

Department of Electronic and Electrical Engineering, University College London, Torrington Place, London WC1E 7JE, UK

Correspondence should be addressed to Andreas Demosthenous; a.demosthenous@ucl.ac.uk

Academic Editor: Sebastian Hoyos

Implantable medical devices provide therapy to treat numerous health conditions as well as monitoring and diagnosis. Over the years, the development of these devices has seen remarkable progress thanks to tremendous advances in microelectronics, electrode technology, packaging and signal processing techniques. Many of today's implantable devices use wireless technology to supply power and provide communication. There are many challenges when creating an implantable device. Issues such as reliable and fast bidirectional data communication, efficient power delivery to the implantable circuits, low noise and low power for the recording part of the system, and delivery of safe stimulation to avoid tissue and electrode damage are some of the challenges faced by the microelectronics circuit designer. This paper provides a review of advances in microelectronics over the last decade or so for implantable medical devices and systems. The focus is on neural recording and stimulation circuits suitable for fabrication in modern silicon process technologies and biotelemetry methods for power and data transfer, with particular emphasis on methods employing radio frequency inductive coupling. The paper concludes by highlighting some of the issues that will drive future research in the field.

1. Introduction

Neuroengineering, the application of engineering techniques to understand, repair, replace, enhance, or otherwise exploit the properties of neural systems, is a topic that is currently generating considerable interest in the research community. The nervous system is a complex network of neurons and glial cells. It comprises the central nervous system (brain and spinal cord) and the peripheral nervous system. Injuries or diseases that affect the nervous system can result in some of the most devastating medical conditions. Conditions, such as stroke, epilepsy, spinal cord injury, and Parkinson's disease, to name but a few, as well as more general symptoms such as pain and depression, have been shown to benefit from implantable medical devices. These devices are used to bypass dysfunctional pathways in the nervous system by applying electronics to replace lost function.

The first implantable medical devices were introduced in the late 1950s with the advent of the heart pacemaker [1, 2] and subsequently the cochlear implant [3, 4]. Both have restored functionality for hundreds of thousands of patients. A pacemaker uses electronics and sensors to continuously monitor the heart's electrical activity and when arrhythmia is detected, electrical stimulus is applied to the heart (via electrodes) to regulate its speed. A cochlear implant uses electronics to detect and encode sound and then stimulate the auditory nerve to enable deaf individuals to hear. Thanks to remarkable advances in microelectronics, electrode technology, packaging, and biomedical signal processing, active implantable medical devices have developed into advanced systems, employing wireless telemetry for transmission of data and sometimes power.

The success of cochlear implants has inspired the development of implantable devices for restoring other basic human sensations. Visual prosthesis translates camera input into electrical stimulation to the visual nervous system to create pixelized vision [5, 6], while vestibular prosthesis connects motion sensors to vestibular nerves to restore balance sensation [7, 8]. Another example is deep brain stimulation (DBS) that has been shown to provide therapeutic benefits for otherwise treatment-resistant neurological disorders such as Parkinson's disease, tremor, and dystonia [9]. In current clinical DBS systems, high frequency stimulation (>100 Hz) produced by a pulse generator (stimulator) is continuously applied via deep brain electrodes to the targeted tissue area in the brain (Figure 1). The characteristics of these pulses (e.g., frequency, pulse duration, and intensity) are programmed into the stimulator (implanted in the chest area) and adjusted via an external programmer. DBS emerged from heart pacemaker technology and hence is also called a brain pacemaker. Current developments are focused on the design of cranial-mounted inductively-powered DBS systems [10] to reduce the length of the leads from the stimulator to the electrodes and novel stimulation techniques using high-density segmented electrodes, which can enable current-steering and electric field shaping capability [11, 12]. In addition, recent papers have reported the development of prototype closed-loop (i.e., sense and stimulate) neuroprosthetic devices for applications such as vestibular prosthesis [13] and epilepsy [14, 15]. In the case of epilepsy, the device uses implantable multielectrode arrays and amplifiers to record electrical signals from neurons in the brain. The recorded data is then processed to extract important events, which, for example, predict the onset of an epileptic seizure, and electrical stimulation is applied to inhibit the attack.

This paper provides a review of advances in microelectronics for implantable medical devices and systems. The focus is limited to neural recording and stimulation circuits suitable for fabrication in modern silicon process technologies and biotelemetry methods for power and data transfer. After this introductory part, Section 2 reviews several techniques and circuit topologies for neural recording, including methods for on-chip data reduction to reduce the bandwidth requirements of the wireless transmission link. Section 3 covers advances in neural stimulation circuits and Section 4 discusses biotelemetry methods including wireless power and data transmission by inductive coupling. Finally, conclusions are drawn in Section 5 including some suggestions for future research.

2. Neural Amplifiers

Neural signals are low frequency and low amplitude signals. For example, the amplitude of the electroneurogram (ENG) recorded with implanted cuff electrodes [16, 17] is typically in the region of $1 \mu V$, with most energy concentrated between 300 Hz and 5 kHz. Even when recording neural activity with penetrating microelectrodes such as the Utah Electrode Array [18, 19] or the NeuroNexus penetrating probes (http://www.neuronexustech.com/), the recorded neural action potentials often have amplitudes of only a few tens of microvolts. Hence, circuits for neural signal amplification must have low noise performance and additionally low power consumption so that battery life is prolonged, especially in implantable systems (e.g., implantable loop recorder for long-term monitoring of the heart's electrical activity [20]). In addition, front-end neural amplifiers are required to reject electrode offsets or common-mode interference. Both clock-based and continuous-time techniques have been used in the design of neural amplifiers.

2.1. Clock-Based Techniques. The noise in CMOS transistors is usually dominated by flicker ($1/f$) noise up to relatively

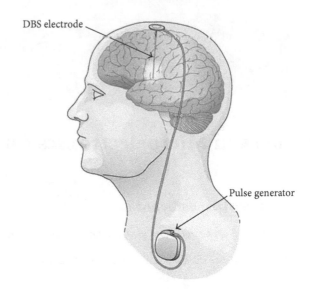

FIGURE 1: A DBS implant typically consists of an implanted pulse generator which generates electrical pulses for stimulation, a set of connection cables, and an electrode rod which delivers the stimulation pulses to the brain target area. Image source: http://cdn.physorg.com/.

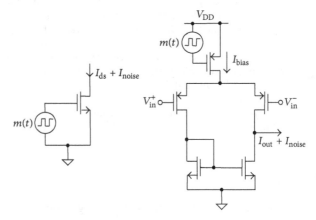

FIGURE 2: Cycling of MOS gate bias (left) and application to an OTA.

high frequencies of the order of several tens of kHz [21]. This is particularly troublesome for the design of low frequency, low noise analog circuits. Typically, p-channel transistors have less $1/f$ noise than n-channel transistors. Various clock-based techniques have been developed to reduce the effects of $1/f$ noise. Noise reduction based on physical effects (switched biasing), chopper modulation and autozeroing, are amongst these techniques.

The *switched biasing* technique (Figure 2) reduces the $1/f$ noise of a MOS transistor by cyclically increasing and decreasing its gate bias so that the device alternates between strong inversion and accumulation [22]. The transistor noise is modulated by the switching signal. The switching operation is represented as a multiplication of the $1/f$ noise current

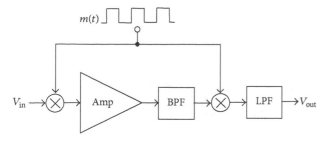

FIGURE 3: Block diagram of a chopper amplifier.

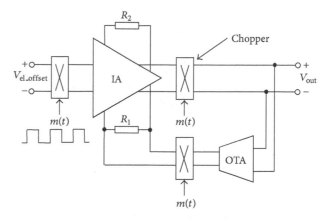

FIGURE 4: Amplifier with $1/f$ noise reduction and electrode offset elimination [28].

(I_{noise}) with the switching signal $m(t)$. For a square-wave signal with 50% duty cycle

$$m(t) = \frac{1}{2} + \frac{2}{\pi}\sin(2\pi f_s t) + \frac{2}{3\pi}\sin(6\pi f_s t) + \frac{2}{5\pi}\sin(10\pi f_s t) + \cdots,$$ (1)

where f_s is the switching frequency. If f_s is set sufficiently high, the baseband noise reduces by half and any modulation effects represented by the sine terms in (1) remain outside the bandwidth of interest and can be removed by filtering. Switched biasing improves noise performance at low frequencies but requires a high speed clock applied to the gate (or bulk) of the transistor with potential problems due to charge feedthrough and additional noise originating from the driver circuit. In addition to reducing the intrinsic $1/f$ noise, the switched biasing technique reduces power consumption [22]. A switched biasing amplifier demonstrating input-referred noise reduction at low frequencies (<100 Hz) is described in [23]. It uses two operational amplifiers (opamps) configured as buffers linked via resistors. The first stage of each opamp is an operational transconductance amplifier (OTA) circuit of the type shown in Figure 2 where switched biasing is applied to the p-channel transistor supplying the tail current I_{bias}. Further noise reduction could be achieved by cycling the voltage bias applied to the bulk terminals of the differential pair input transistors.

The *chopper technique* is also based on signal modulation. The technique enables the design of amplifiers with high common-mode rejection ratio (CMRR) performance. A block diagram of the chopper amplifier is shown in Figure 3 [24]. Before amplification, the amplifier input signal is modulated by a square-wave signal of frequency f_s that is much higher than the baseband frequencies of interest. The chopping signal may be represented as in (1). The upconverted signal is then amplified and bandpass filtered. The modulated signal spectrum is located at frequencies higher than the $1/f$ noise corner. After amplification, the signal is converted back to baseband by multiplication with the same modulation waveform used for upconversion. Lowpass filtering restores the desired signal. The technique reduces both $1/f$ noise and amplifier dc offset voltages, but the noise performance is ultimately limited by the noise floor of the amplifier. In addition, practical nonidealities, including the finite amplifier bandwidth, can lead to signal distortion.

Several integrated neural amplifiers employing the chopper technique have been described for recording from both implanted and surface electrodes [25–30]. The design in [28] employs an ac-coupled chopping technique to reject electrode offsets and achieve low noise performance. The concept of this technique is shown in Figure 4. The system consists of a feedforward stage and a feedback stage. To suppress the $1/f$ noise of the instrumentation amplifier (IA), the feedforward stage employs an input chopper and an output chopper. To eliminate the electrode offset, the feedback stage employs a lowpass OTA stage followed by a chopper stage. The operation of the circuit is as follows. The input differential electrode offset ($V_{\text{el\_offset}}$) is modulated by the input chopper and appears across resistor R_1. By action of the current-feedback the current through R_1 is copied to R_2 and defines the output voltage after demodulation by the output chopper. The lowpass OTA stage filters the dc component of the output and converts it into current. The OTA output current is in turn modulated by a chopper stage. In the steady state, the current supplied by the OTA is $V_{\text{el\_offset}}/R_1$. As a result, no current is supplied by the IA and the current passing through R_2 is zero, so the output (V_{out}) is zero. The neural amplifier in [29] uses chopper modulation and switched-capacitor techniques to reduce the $1/f$ noise and achieve frequency tuning. The circuit is capable of simultaneous recording of extracellular unit spikes (action potentials) and local field potentials (low frequency signals in the 1 Hz to 100 Hz range). Another publication applies chopper stabilization with a distortion cancelling technique to the design of a front-end transimpedance amplifier for current-mode biosensors [30]. It achieves both low noise and low distortion performance at minimum current consumption. The cancellation technique reduces the distortion and, combined with chopping, substantially reduces the $1/f$ noise. For a current consumption of 50 μA, it is shown that the input-referred noise density without cancelling and chopping at 10 Hz is 29 pA/$\sqrt{\text{(Hz)}}$ and reduces to 3 pA/$\sqrt{\text{(Hz)}}$ when both techniques are employed. The total harmonic distortion for a peak current of 1 μA is −55 dB.

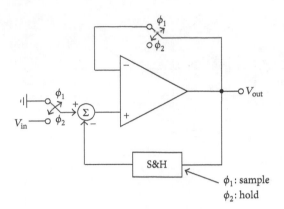

FIGURE 5: Autozero amplifier.

The *autozeroing technique* is shown in Figure 5. During sampling phase (ϕ_1), the amplifier is configured as a unity gain buffer and the input noise is sampled. During the amplification phase (ϕ_2), the noise sample is subtracted from the instantaneous amplifier input noise. As the sampling frequency chosen is higher than the $1/f$ noise frequency the sample is highly correlated to the instantaneous noise and the low frequency noise is cancelled. A detailed analysis of the autozeroing technique is given in [31]. A drawback of the technique is that high frequency white noise is undersampled and folded back into the baseband where it increases the noise floor. A bandpass micropower neural amplifier employing autozeroing and featuring variable-gain capability is presented in [32]. An interesting design of a low voltage, low noise amplifier combining autozeroing and chopping stabilization is described in [33].

2.2. Continuous-Time Techniques. The noise reduction techniques described above require a clock generation circuit and thus suffer from potential problems associated with high frequency interference and clock feedthrough. In addition, high frequency switching circuits can increase the complexity and power consumption of the design. As an alternative, continuous-time techniques have been extensively used in the design of neural amplifiers. The classic circuit is the ac-coupled OTA-based neural amplifier with capacitive feedback shown in Figure 6 [34]. The circuit is built around a single stage OTA in CMOS technology. The ratio of capacitors C_1 and C_2 sets the midband gain of the bandpass response. The input is capacitively coupled through C_1, so any dc offset from the electrode-tissue interface is removed (C_1 should be made much smaller than the electrode impedance to minimize signal attenuation). Transistors M_a–M_d implement MOS pseudoresistors with an extremely large incremental resistance (>10^{12} Ω). This allows the cutoff frequency of the input high-pass filters (i.e., ac-coupled stage) to be set to the millihertz region. The lower cutoff frequency is set by the product of C_2 and the MOS pseudoresistor implemented by M_a and M_b. The upper cutoff frequency is a function of the load capacitance (C_L), the OTA transconductance (G_m), and the midband gain (C_1/C_2). To reduce the effect of $1/f$ noise, the OTA input transistors should be p-channel

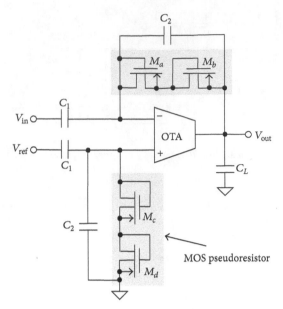

FIGURE 6: OTA-based amplifier with capacitive feedback [34].

devices with large gate areas. Numerous designs of neural amplifier based on the circuit in Figure 6 or with some variations (e.g., in the realization of the pseudoresistors, use of fully-differential topology with one or two stage OTAs, use of current-reuse techniques to double the transconductance, and so forth [35–39]) have been reported in the literature, including commercial amplifier chips by Intan Technologies, LLC (http://www.intantech.com/). Methods for effective optimization of a recording channel in terms of its power consumption, input-referred voltage noise, silicon area, and technology used are discussed in [40]. The design of nanopower OTAs with enhanced linearity is presented in [41].

In the case of recording from a multielectrode array, the total power consumption of the amplifier array (as well as the silicon area) may be reduced by using the partial OTA sharing structure proposed in [42]. In this technique, each of the n amplifiers in the array share the components corresponding to the reference electrode (i.e., pseudoresistors M_c and M_d and capacitors C_1 and C_2 connected to V_{ref} in the amplifier in Figure 6). The silicon area is reduced as a benefit of sharing the bulky capacitor C_1. The improvement factor in terms of silicon area depends on the number of shared amplifiers. In general, for an electrode array size of n, the OTA sharing technique allows an area saving of $(n-1)/(2n) \times 100\%$, a power saving of $(n-1)/(2n) \times 100\%$, and an improvement in noise efficiency factor (NEF) of $[1 - \sqrt{(n+1)/2n}] \times 100\%$, compared to the conventional architecture (i.e., using n neural amplifiers). The NEF is a dimensionless figure of merit (noise-power tradeoff) used to compare different neural amplifier designs. It is defined as [43]

$$\mathrm{NEF} = V_{\mathrm{in,rms}} \sqrt{\frac{2 \cdot I_{\mathrm{tot}}}{\pi \cdot U_T \cdot 4kT \cdot \mathrm{BW}}}, \qquad (2)$$

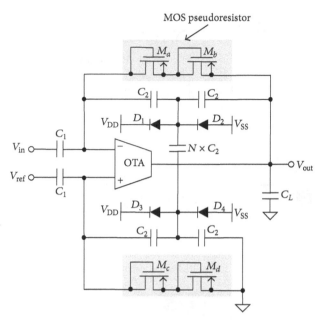

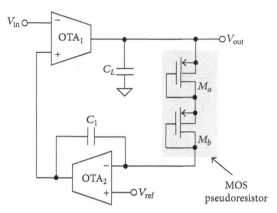

FIGURE 8: Amplifier with active dc rejection [45].

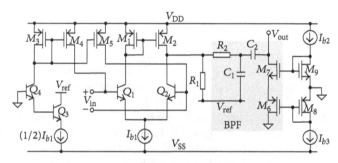

FIGURE 7: Capacitive feedback amplifier with low input capacitance [44].

FIGURE 9: A BiCMOS low noise amplifier [47].

where $V_{in,rms}$ represents the integrated input-referred noise, I_{tot} is the total power consumed by the amplifier, U_T is the thermal voltage, k is Boltzmann's constant, T is the absolute temperature in Kelvin, and BW is the −3 dB bandwidth of the amplifier. To include the supply voltage V_{DD}, the modified metric $NEF^2 \times V_{DD}$ is used [37].

To achieve low noise performance, a large neural amplifier gain is usually required. The conventional ac-coupled neural amplifier (Figure 6) often presents a large input load capacitance (typically 10–20 pF for a midband gain of around 40 dB) to the neural signal source, hence occupying a large silicon area. It suffers from the unavoidable tradeoff between input capacitance and chip area versus the amplifier gain. In the amplifier in Figure 7 [44], this tradeoff is limited by replacing the feedback capacitor with a clamped T-capacitor network. The diodes are used for discharge purposes. Compared to the conventional circuit this amplifier can achieve a given midband gain with less input capacitance, a higher input impedance and smaller silicon area. For a midband gain of 38.1 dB, a neural amplifier employing the topology in Figure 7 used a 1.6 pF input capacitance and a total silicon area of 0.056 mm^2 in 0.35-μm CMOS technology [44].

Electrode offset removal may also be implemented by means of *active feedback* loops. An example neural amplifier topology with active feedback for dc rejection is shown in Figure 8 [45]. It consists of a low noise OTA (OTA$_1$) with an active feedback circuit implemented by a second OTA (OTA$_2$) configured as a Miller integrator. The time-constant of the integrator is set by capacitor C_1 and the MOS pseudoresistor comprising M_a and M_b. The midband gain of the amplifier is the same as the gain of OTA$_1$. The dominant pole of OTA$_1$ sets the amplifier's lowpass cutoff and the common-mode voltage is set by voltage V_{ref}. Another example of a neural amplifier with an active feedback loop to

bypass any dc offset current generated by the electrode-tissue interface is described in [46]. It predominantly makes use of current-mode circuit techniques.

A low noise neural amplifier for implanted cuff electrodes is described in [47]. The circuit schematic is shown in Figure 9. It consists of an input BiCMOS OTA (Q_1, Q_2, M_1, and M_2) terminated in the load resistor R_1, followed by a first-order bandpass filter (for bandwidth restriction). The upper cutoff frequency is set by the combination of resistor R_2 and capacitor C_1, while the lower cutoff frequency is set by capacitor C_2 with the series combination of transistors M_6 and M_7, the latter transistor pair forming a high value (~20 MΩ) active resistor. In addition to eliminating low frequencies below the pass-band of the input neural signal, the high-pass section of the bandpass filter also removes some of the low frequency flicker noise voltage tail and ensures a dc offset-free amplifier output (V_{out}). The dc bias voltages of M_6 and M_7 are provided by the diode-connected transistors M_8 and M_9, respectively, which are in turn biased by the dc current sources I_{b2} and I_{b3}. Circuitry is also included (M_3–M_5 and Q_3-Q_4) to cancel the base currents of Q_1 and Q_2. This neural amplifier achieved a measured input-referred root mean square noise voltage of only 290 nV (noise bandwidth of 1 Hz–10 kHz). A variant of this circuit in CMOS technology using lateral bipolar devices for Q_1 and Q_2 is possible [48]. In general for a given target noise specification the use of lateral pnp bipolar devices (available as parasitic devices in CMOS technology) tends to require a larger silicon area compared to using standard npn bipolar transistors in

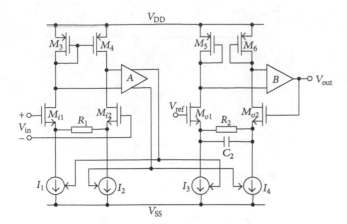

FIGURE 10: A current feedback instrumentation amplifier [51].

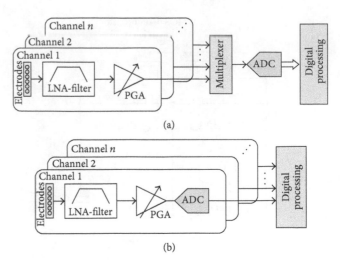

FIGURE 11: Topologies for multichannel neural recording systems.

BiCMOS technology (but BiCMOS technology might incur higher manufacturing costs). A bipolar transistor structure in CMOS technology featuring high matching characteristics is described in [49].

For biomedical front-ends requiring high CMRR performance and accurate gain setting the use of an IA is desirable. An IA may be realized using the classic three-opamp topology. However, the CMRR of the three-opamp IA depends on the matching of the resistors and the need for low output impedance amplifiers can increase power consumption. Another technique for IA design is to employ switched-capacitor circuits [50], but the fold over of noise above the Nyquist frequency can be a major limitation. A popular IA topology for integrated circuits is the *current feedback* technique. In a current feedback IA, the gain is accurately set by the ratio of two resistors and the CMRR does not rely on the matching of resistors. Figure 10 shows the simplified circuit schematic of a current feedback IA [51]. The input transconductor stage uses a simple current mirror load and current sink biasing. The sensing amplifier A serves to exactly balance the drain currents of transistors M_{i1} and M_{i2} by adjusting the complementary currents I_1 and I_2. A direct result of this is that the input differential voltage V_{in} is forced across resistor R_1 and hence M_{i1} and M_{i2} of the input stage essentially act as a unity-gain buffer. Similarly, the high gain amplifier B balances the drain currents of transistors M_{o1} and M_{o2} in the output transconductor stage. Since currents I_3 and I_4 are exact copies of I_1 and I_2, respectively, the output voltage V_{out} appears across resistor R_2. Hence, the dc gain of the IA is given by the ratio R_2/R_1. Placing capacitor C_2 in parallel with R_2 creates a dominant pole, which sets the −3 dB bandwidth of the IA. The CMRR and noise analysis of the circuit are described in [51]. The IA in [52] used the current feedback technique to achieve a CMRR of 99 dB and an input-referred noise of $0.68\,\mu$V rms. It was developed to record neural signals from electrodes on the lumbo-sacral nerve roots which can be used to restore lower-body function to patients with paraplegia after spinal cord injury.

Table 1 compares various integrated neural amplifiers reported in the literature. From the table, it is observed that there is a relationship between current demand and input-referred noise. In general, the lower the noise performance, the higher the current consumption. There is a wide variation in the silicon area used.

2.3. Data Reduction Techniques for Multichannel Neural Recording Front-Ends.

Front-end neural recording interfaces for multichannel (multielectrode) systems are typically based on the two types of architecture in Figure 11 [53]. In the approach in Figure 11(a), the analog front-end circuits which amplify and filter the neural signal acquired from each of the electrodes are grouped in an array of channels. Each of these channels comprises a low noise amplifier (LNA) of the type described in Sections 2.1 and 2.2 and a bandpass filter, followed by a programmable gain amplifier (PGA) to maximize the output swing. The analog outputs from all the channels are then multiplexed in time and converted into digital words by an analog-to-digital converter (ADC). The generated time-multiplexed data frames can be either digitally processed to compress them or sent directly to the output as raw data. In the approach in Figure 11(b), instead of sharing the ADC between the analog outputs of the channels, an ADC is embedded in each channel. Then a common digital processor manages the digitized signals from the channels, classifies (or reduces) the data, and sends it to the output. This solution requires a higher silicon area than the approach in Figure 11(a), but it has some merits in terms of power consumption because of the much lower sampling rate requirement at the digitization stage. In addition, the system in Figure 11(b) has the advantage that is easily scalable by replicating the channels. A very compact circuit implementation for the topology in Figure 11(b) is described in [35], requiring a silicon area of only 0.054 mm^2 per channel in a 130-nm CMOS process technology.

Bandwidth limitation is a key issue for wireless biomedical devices. Wireless transmission of raw data is a major challenge for high channel count recording front-ends. For

TABLE 1: Comparison of integrated neural amplifiers.

Parameter	[34]	[109]	[45]	[110]	[111]	[112]	[29]	[42]	[35]	[37]	[44]	[40]	[39]	[46]	[36]
Year	2003	2004	2007	2009	2009	2010	2010	2011	2012	2012	2013	2013	2013	2013	2013
Fully differential	No	No	No	No	Yes	Yes	Yes	No	Yes	Yes	No	No	No	No	Yes
CMOS process (μm)	0.5	1.5	0.18	0.35	0.13	0.35	0.18	0.18	0.13	0.13	0.35	0.18	0.18	0.18	0.18
Area (mm^2)	0.16	0.107	0.05	N/A	N/A	0.02	N/A	0.063	0.054	0.072	0.065	0.065	N/A	0.76	0.16
Supply voltage (V)	5	3	1.8	1	1	3	1.6	1.8	1.2	1	3	1.8	1/1.8	1	1.2
Supply current (μA)	16	38.27	4.67	1.26	12.5	2.8	43.12	4.4	1.6	12.1	2	6.1	1.2	13	0.36
Gain (dB)	39.5	39.3	49.52	45.7	38.3	33	19.1/37.5	39.4	47.5	40	37.5	48/60	54.8/60.9	44.5/55.9	26
Bandwidth (Hz)	25 m–7.2 k	0–9.1 k	98.4–9.1 k	0.23–7.8 k	23 m–11.5 k	10–5 k	100–8 k	10–7.2 k	167–6.9 k	50 m–10.5 k	1–10 k	0.3–9 k	0.38–5.1 k	1–10 k	80–15 k
Input-referred noise (μV_{rms})	2.2	7.8	5.6	4.43	1.95	6.08	2.36	3.5	3.8	2.2	10.6	5	4	4.4*	8.1
Noise bandwidth (Hz)	N/A	0.1–10 k	1–9.1 k	1–12 k	0.1–25.6 k	10–5 k	1–8 k	10–100 k	1–100 k	0.1–105 k	1–10 k	1–8 k	1–8 k	0.3–10 k	N/A
CMRR (dB)	≥83	N/A	52.68	58	>63	60	79	70.1	83	80	74	48	>60	N/A	>60
PSRR (dB)	≥85	N/A	51.93	40	>63	N/A	62	63.8	70	≥80	55	55	>70	N/A	>80
THD	1%	1.1%	1%	0.53%	1%	N/A	N/A	1%	1%	1%	1%	1.2%	1%	1.03	0.05%
Input range (mV$_{pp}$)	16.7	5	2.4	5.2	1	N/A	N/A	5.7	3.1	1	2.4	1	1.63	**	10
NEF	4	19.4	4.9	2.16	2.48	5.55	6.68	3.35	2.16	2.9	5.78	4.6	1.9	5.45	1.52
NEF^2 × V$_{DD}$	80	1129	43.22	4.53	6.15	92.41	71.39	20.20	5.59	8.41	100.22	38.08	3.61	29.7	2.77

* Referred to the input node of the electrode.
** The amplitude range of the input current is 20 nA$_{pp}$.

example, a neural interface with 100 channels, a 30 kHz sampling frequency per channel, and an 8-bit sample resolution would generate raw data at 24 Mbps. For a 1024-channel system (for cortical neural sensing), the data rate would increase to a massive 250 Mbps. This data rate is beyond the transmission capabilities of existing (wideband) implantable wireless transmitters [54–56]. In the case of extracellularly recorded action potentials (spikes), spike sorting [57] is an efficient process to achieve on-chip data reduction, thereby enabling lower data rate wireless transmission and low power consumption. Neural recording microsystems with data reduction capability typically employ some sort of thresholding to detect and extract the neural information [58, 59]. Within this scheme, detection occurs when the spike's amplitude crosses a specified threshold. The threshold detector can be implemented with analog or digital circuits. For applications where a more detailed classification of multiunit activity into single-unit activity is required, techniques that extract specific biosignal features (e.g., peak-to-peak amplitude, duration, peak-to-zero-crossing time, etc.) are employed. Most spike sorting methods rely on the assumption that each neuron produces a different, distinct shape (as seen by the electrode) that remains constant throughout a recording window. The first step in such techniques is *feature extraction*, in which spikes are transformed into a certain set of features that emphasizes the differences between spikes from different neurons as well as the differences between spikes and noise. Then *dimensionality reduction* takes place, in which feature coefficients that best separate spikes are identified and stored for subsequent processing, while the rest are discarded. Finally, using *clustering* spikes are classified into different groups, corresponding to different neurons, based on the extracted feature coefficients. Implantable spike sorting hardware must be low power and low area. The algorithms implemented in the hardware must be accurate, automatic, real-time, and computationally efficient. A detailed review and comparison of spike sorting algorithms are provided in [57, 60]. An ultra-low power spike sorting digital chip that can perform detection, alignment and feature extraction simultaneously for 64 channels is described in [61]. The chip was implemented in a 90 nm CMOS process and has a power density of $30\,\mu W/mm^2$, which is significantly lower than the power density ($800\,\mu W/mm^2$) known to damage brain cells [62].

Spike sorting can potentially reduce the data rate by several orders of magnitude compared to transmitting raw data. However, the data in the segments without spikes (which contains useful information on the neuronal activities) is lost. To preserve these activities, one solution is to use the discrete wavelet transform to process the data before transmission [63]. This allows the retention of almost all of the data but at the cost of chip area and power consumption. An alternative is the emerging field of compressive sensing [64]. Compressive sensing enables signal reconstruction from a small number of nonadaptively acquired sample measurements corresponding to the information content of the signal rather than to its bandwidth. It has simple compression steps and takes advantage of the signal's sparseness, allowing the signal to be determined from relatively few measurements. Energy-efficient compressive sensing methods for implantable neural recording is a topic of current research [65, 66].

Table 2 compares various multichannel neural recording systems with wireless transmission capability. The design in [55] has the highest number of channels (128) and data rate (90 Mbps). The design in [67] has the lowest power consumption per channel.

3. Neural Stimulators

Electrical stimulus applied to nerves can trigger action potentials. At least two electrodes are required in order to produce current flow. The electrodes are commonly arranged in monopolar or bipolar configuration. In both cases, the active (working) electrode is placed near the nerve to be stimulated. In monopolar stimulation, the indifferent electrode is placed away from the active electrode, whilst in bipolar stimulation the reference electrode is placed near the active electrode.

There are two main modes of stimulation, namely, current-mode and voltage-mode, as shown in Figure 12, although charge-mode stimulation also exists [68]. Current-mode stimulation (Figure 12(a)) is extensively used in implantable stimulators. Active current sources (and sinks) are used to supply the stimulus current to the load (tissue-electrode impedance). For current sources with high output impedance, the stimulus current amplitude is not affected by changes in the load. Examples of integrated current-mode stimulators in CMOS technology for various applications such as nerve root stimulation for spinal cord injury, vestibular prosthesis for balance disorders, and deep brain stimulation for severe movement disorders are described in [11, 69–73]. The current amplitude range is from 1 mA to 16 mA.

In voltage-mode stimulation (Figure 12(b)), the stimulator output is a voltage, and therefore the magnitude of the current delivered to the tissue is dependent on the inter-electrode impedance. Thus, it is difficult to control the exact amount of charge supplied to the load because of impedance variations. In the system described in [74], the stimulator drives the electrodes with a sequence of voltage steps, charging the electrode metal-fluid capacitance. This applies a voltage waveform that is an approximation of the waveform that would appear at the electrode if a current pulse was applied (see Figure 12(a)). Since the charge is delivered to the electrode directly from the capacitors, it avoids the (substantial) power in the current sources of its current-mode counterpart, as well as providing large voltage compliance. However, this method requires large capacitors (which act as voltage sources) that are difficult to implement on-chip. The design in [74] has five $1\,\mu F$ capacitors for a 15-electrode stimulation system, which will increase with more electrodes. It is also difficult to achieve fine resolution compared to current-mode stimulation since voltage-mode is an approximate method of producing a current pulse and increasing the resolution requires more capacitors. Another voltage-mode stimulator is described in [75]. Its architecture features energy recovery enabling power savings of 53%

TABLE 2: Comparison of wireless multichannel neural recording systems.

Parameter	[113]	[59]	[55]	[114]	[115]	[56]	[67]	[54]	[116]	[15]
Year	2007	2009	2009	2010	2010	2012	2012	2013	2013	2013
Number of channels	100	64	128	64	32	96	64	100	64	64
CMOS process (μm)	0.5	0.5	0.35	0.35	0.5	0.13	0.13	0.5	0.13	0.13
Area (mm^2)	27.73	N/A	63.36	8.37	16.27	25	18.4	25.48	12	12
Supply voltage (V)	3.3	1.8	3.3	3	3	1.2	1.2	3	1.2	1.2/3
Total power (mW)	13.5	14.4	6	172	5.85	6.5	0.38	6/90.6	5.03	1.4
Power per channel (μW)	135	225	46.87	269	182.8	67.7	5.93	906	78.5	21.8
Amplifier gain (dB)	40	60	40	65–83	67.8–78	54	47.5–65.5	46	54–60	54–60
Low frequency corner (Hz)	300	10	0.1	1	0.1	280	200	1	10	1
High frequency corner (kHz)	5	9.1	20	10	8	10	6.9	7.8	5	5
Input-referred noise (μV_{rms})	5.1	8	4.9	3.05	4.62	2.2	3.8	2.83	6.5	5.1
ADC resolution (bits)	10	8	9	7.2	8.1	10	7.65	12	7.8	7.6
Sampling frequency (kHz)	15	62.5	640	20	N/A	31.25	27/90	20	57	$\leq$100
Downlink modulation	ASK	FSK	N/A	N/A	N/A	N/A	OOK	N/A	OOK	N/A
Carrier frequency (MHz)	2.64	4/8					40.68		915	
Uplink modulation	FSK	OOK	UWB	FSK	FSK	UWB	LSK	FSK	FSK	UWB
Carrier frequency (GHz)	0.433	0.070–0.2	4	0.4	0.915	N/A	0.004	3.2/3.8	0.915	0–1/3.1–10.6
Data rate (Mbps)	0.33	2	90	1.25	0.709	30	1.92	48	1.5	10
On-chip signal processing	Yes	Yes	Yes	Yes	Yes	No	Yes	No	Yes	Yes
Power source	Inductive	Inductive	Battery	Battery	Inductive	Battery	Inductive	Battery	Battery	Battery
Stimulation capability	No	No	No	No	No	No	No	No	No	Yes

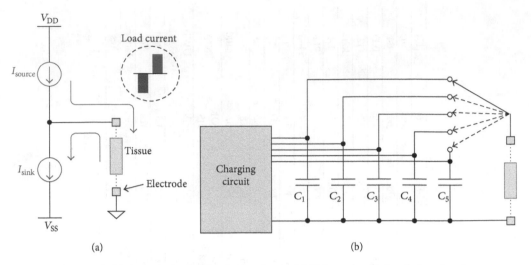

FIGURE 12: (a) Current-mode stimulation circuit. (b) Voltage-mode stimulation circuit.

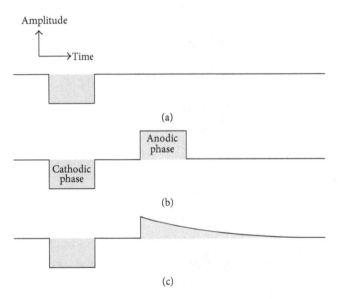

FIGURE 13: Stimulus waveforms. (a) Monophasic. (b) Biphasic with active cathodic and active anodic phases. (c) Biphasic with active cathodic phase and passive anodic phase (exponential decay).

to 66% (depending on the load) compared to traditional current-mode stimulator designs.

Common stimulation waveforms are either monophasic or biphasic (Figure 13). A monophasic stimulus consists of a repeating unidirectional cathodic pulse (this type of stimulus is common in surface electrode stimulation). A biphasic waveform consists of a repeating current pulse that has a cathodic (negative) phase followed by an anodic (positive) phase. The cathodic phase depolarizes nearby axons and triggers the action potential. The succeeding anodic phase reverses the potentially damaging electrochemical processes that can occur at the electrode-tissue interface during the cathodic phase by (ideally) neutralizing the charge accumulated in the cathodic phase, allowing stimulation without tissue damage. The application of charge-balanced waveforms

is very important, especially for implanted electrodes. Usually the stimulus for the cathodic phase is rectangular, supplied by active circuits, while the stimulus for the anodic phase could be either square or exponentially decaying. The rectangular secondary phase is also known as active discharging and the exponentially decaying phase as passive discharging.

3.1. Charge-Balanced Stimulation. Charge imbalance can be caused by many reasons including semiconductor failure, leakage currents due to crosstalk between adjacent stimulating channels (sites), and cable failure. A blocking capacitor in series with each electrode is used for electrical safety against single-fault conditions [76]. The blocking capacitor is also used to achieve (passive) charge balance. Figure 14 shows three current-mode stimulator configurations, each employing a blocking capacitor [69]: (a) dual supplies with both active phases, (b) single supply with both active phases, and (c) single supply with active cathodic phase and passive anodic phase. The programmable current sink I_{stimC} and current source I_{stimA} generate the cathodic and anodic currents, respectively. These currents are driven through the load by the control of switches S_1 and S_2. When only a single supply is available (Figure 14(b)), the anodic and cathodic currents are generated from a single current sink (I_{stim}) by reversing the current paths by switch S_2. Both configurations in Figures 14(a) and 14(b) are (ideally) designed to be charge-balanced to avoid charge accumulation. However, achieving exactly zero net charge after each stimulation cycle is not possible due to mismatch or timing errors and leakage from adjacent stimulating sites. Therefore, it is important to include switch S_3 to periodically remove the residual charge by providing an extra passive discharge phase in which the voltage on the blocking capacitor drives current through the electrodes to fully discharge them. Given the necessity for the third phase in the circuits in Figures 14(a) and 14(b), it is possible to use the passive discharge phase as the main anodic phase as shown in the circuit in Figure 14(c) and the corresponding waveform in Figure 13(c). Note that the use of capacitive

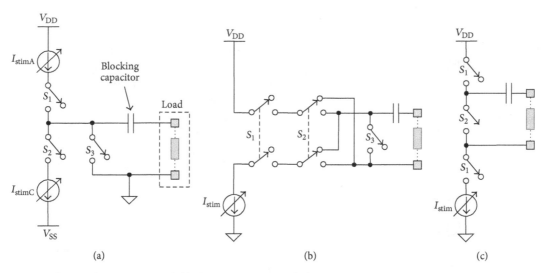

FIGURE 14: Current-mode stimulator circuits with blocking capacitor: (a) dual supplies with active cathodic and active anodic phases; (b) Single supply with active cathodic and active anodic phases; (c) single supply with active cathodic phase and passive anodic phase.

electrodes may also drive the passive discharge phase. However, blocking capacitors may still be deemed necessary to ensure that dc current cannot flow into the electrodes in the event of semiconductor failure due to breakdown voltage or leakage current. Due to the large value required for the blocking capacitors (which can be a few microfarads in the case of stimulators for lower-body applications), they are typically realized as off-chip surface-mount components. For multichannel stimulation implants, a single blocking capacitor per channel may not be sufficient to guarantee safety due to a single-fault failure [77].

For applications where the use of blocking capacitors is not possible due to physical size limitations (e.g., retinal implants with several hundreds of stimulating electrodes), other methods for (active) charge balancing exist (Figure 15).

(1) Dynamic current balancing [78]. In the circuit in Figure 15(a), a current sink is used to generate the cathodic phase and a pMOS transistor (M_1) to generate the anodic phase. Before generating a biphasic current pulse, $S_{cathodic}$ and S_{anodic} are open, and the two sampling switches, S_{samp}, are closed. When the circuit settles, the amplitude of the drain current of M_1 is the same as the current sink (due to the feedback), and the resulting bias voltage (V_{bias}) on the gate of M_1 is sampled and held. Then the two S_{samp} switches open and $S_{cathodic}$ closes to form the cathodic current. Following this, $S_{cathodic}$ opens and S_{anodic} closes. Because the gate voltage of M_1 is held at V_{bias}, an anodic current equal to the amplitude of the (cathodic) sink current passes through S_{anodic} to the load. By optimizing the S&H circuit to reduce errors due to charge effects, a residual dc current error of 6 nA is claimed [78].

(2) Active charge balancer [79]. In the circuit in Figure 15(b), the residual voltage after a biphasic pulse is measured and compared to a safe reference voltage.

If the electrode voltage exceeds the safe window, additional short stimulation current pulses are applied to steer the electrode voltage towards a balanced condition. The charge balancer is typically activated for less than 5% of the operation time. This method has the advantage of providing feedback information on the electrode condition after stimulation.

(3) H-bridge with multiple current sinks [80]. This approach assumes an asymmetric biphasic waveform. By way of example, consider the H-bridge circuit in Figure 15(c). During the high-amplitude cathodic phase, identical current sinks $I_1, I_2, \ldots, I_N$ act in parallel to pass current through the electrodes for a time T. Then during the low-amplitude anodic phase, one of the N current sinks is used to pass current through the electrodes (in the reverse direction) for a time $N \cdot T$. On a single stimulation cycle this would give inaccurate charge balance as in practice the current sinks would not be perfectly matched. However, if the current sink that is active in the anodic phase is sequentially changed after each stimulus waveform, then after N cycles each sink would have been active for the same amount of time during the cathodic and anodic phases, yielding accurate charge balance. The method is claimed to achieve a maximum charge mismatch of 0.45% [80].

(4) Multiphase compensation [71]. The multiphase compensation technique is illustrated by the waveform in Figure 15(d). To generate an asymmetric biphasic pulse, the width of the anodic phase is extended to N times the cathodic width, T. Ideally, the anodic current amplitude should be N times smaller than the cathodic amplitude. The amplitude of the currents is controlled through an ADC. Due to the finite resolution of the ADC, the amplitude of the cathodic

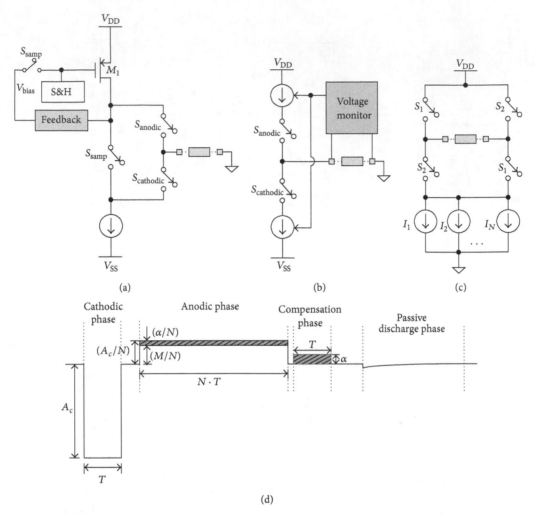

FIGURE 15: Techniques for charge balancing. (a) Dynamic current balancing [78]; (b) active charge balancer [79]; (c) H-bridge with multiple current sinks [80]; (d) multiphase compensation approach [71].

current, A_c, may not be an integer multiple of N, where $A_c = M + \alpha$ and M is the integer multiple of N that is the closest to A_c. Thus, there will be charge balance error between the two phases equal to $(\alpha/N) \cdot N \cdot T = \alpha \cdot T$. In the multiphase compensation technique, an additional shorter anodic pulse of width T and amplitude α is initiated after the anodic phase to compensate for the error. Subsequently, a passive discharge phase can be applied to further reduce any remaining charge imbalance. The method is claimed to achieve a residual dc current error of 4.5 nA [71].

3.2. Other Stimulator Circuits.

A stimulator circuit that is fail-safe with no off-chip blocking capacitors is shown in Figure 16 [72]. The circuit generates an active stimulation phase by high frequency current switching (HFCS), followed by a passive discharge phase. During the active stimulation phase, current I_{stim} (generated by a current generator circuit) is switched alternately through the left and right branches of the charge transfer block (Figure 16(b)). This high frequency switching mechanism allows the size of the blocking capacitors C_1 and

C_2 to be significantly reduced. The circuit operation is as follows. During the low-state of the control signal $\Phi_{\text{stim\_left}}$, switch M_1 is closed and M_3 is open. In this phase, diode D_1 is reverse biased, diode D_2 is forward biased, and current I_{S1} flows and charges up C_1. In the same phase, on the right branch of the charge transfer block, the control signal $\Phi_{\text{stim\_right}}$ is high, and hence switch M_2 is open and D_3, C_2, and M_4 form a closed path which discharges C_2 to one-diode-drop voltage. During the high-state of $\Phi_{\text{stim\_left}}$, $\Phi_{\text{stim\_right}}$ is turned low which causes C_2 to be charged up and C_1 to be discharged. The complementary high frequency currents I_{S1} and I_{S2} generated during the stimulation phase are summed at the anode of the electrode-tissue load. After the stimulation phase, the load is passively discharged via an ac-coupled discharge switch (Figure 16(c)) using depletion-mode transistors M_5–M_7 (connected in parallel for redundancy) which conduct most of the time. The operation of this circuit is as follows. At each negative edge of the control pulse $\Phi_{\text{discharge}}$, negative charge is injected into capacitors C_3 and C_4. On the following positive edge diode D_5 is reverse biased so the charge on

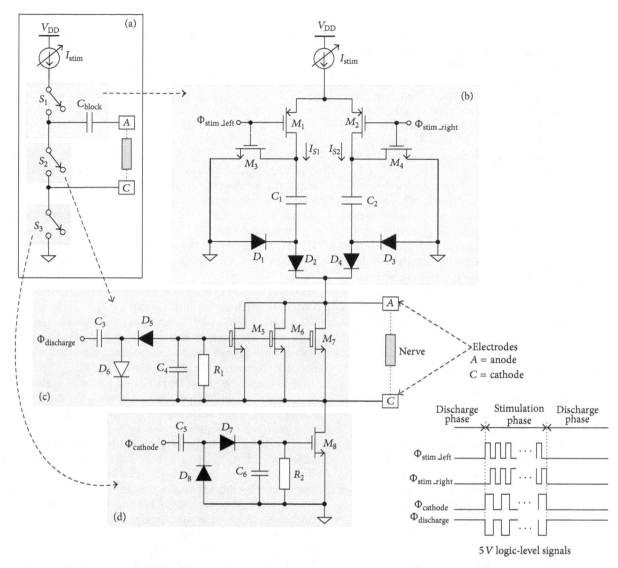

FIGURE 16: Stimulator circuits [72]: (a) configuration with large off-chip blocking capacitor; (b) HFCS charge transfer block; (c) isolated discharge switch (M_5–M_7 are depletion transistors); (d) isolated cathode switch. The combination of (b), (c), and (d) provides a safe stimulator circuit with no off-chip blocking capacitors.

C_4 is retained (apart from the leakage via resistor R_1), and positive charge is injected into C_3 balancing out the injected negative charge on C_3. When a second negative charge arrives it adds to the charge in C_4 and the cycle is repeated. After a couple of cycles enough negative charge is built up on C_4 to switch off M_5–M_7. While the pulses continue, the gate-source voltage of M_5–M_7 remains negative, so they are held off. When the pulses cease the negative charge decays via R_1. An ac-coupled switch is also used for the cathode as shown in Figure 16(d). Its operation is exactly the same as described above, except now it is the positive edge that provides the charge to C_6. Implementation of the HFCS technique requires silicon-on-insulator (SOI) technology which features fully-isolated active and passive devices. The design in [72] used the X-FAB XT06 process technology (http://www.xfab.com/) which has trench isolation. HFCS is suitable for stimulus current amplitudes up to about 1 mA.

The technique can be employed to greatly reduce the size of high-reliability, multichannel stimulator implants sited close to the target nervous tissue (e.g., in the spinal canal or on the brain surface).

There is a demand for high efficiency stimulators in applications which have limited power availability. The basic current-mode stimulator is inherently inefficient and attempts have been made to increase stimulator efficiency. A design is described in [81] which achieves a 2x to 3x reduction in energy consumption compared with the basic current-mode stimulator. It is based on a dc-dc buck voltage converter efficiently providing a variable output for biphasic pulses. The voltage output drive is adjusted by feedback from a sensor detecting the current in the electrode so providing a controlled current independent of the value of the load impedance. The conventional series resistor employed for current sensing wastes energy and is not used. At the output of the dc-dc converter, there is a smoothing capacitor which is in parallel with the electrode load. This capacitor is temporarily disconnected from the converter and the rate of decay of voltage across the capacitor due to the current in the electrode load is detected and used in feedback. This sensing method avoids wasting energy. The system has appreciable sawtooth noise superimposed on the current pulse. Biphasic $400\,\mu A$ current pulses with widths of 1 ms and rise times of about $100\,\mu s$ are shown. Components external to the integrated circuit control unit include an inductor $(39\,\mu H)$ and capacitors (up to $10\,\mu F$).

Another example of minimizing power dissipation while retaining the basic current-mode generator is presented in [82]. It assumes the use of a high frequency (1 MHz) inductive power link. On its secondary coil, zero voltage switching and adjustment of the conduction angle provide a variable voltage supply to the electrode load and current generator. Feedback adjusts the variable voltage supply so that the current generator operates at just above its compliance limit, minimizing the power it uses. The biphasic current is generated using a single sink current generator with switches similar to Figure 15(c). The supply voltage update is near real-time so the quality of the current pulses is high, irrespective of how the load changes during stimulation. Depending on load conditions, 20% to 75% power saving compared to a conventional current-mode stimulator is claimed. In a prototype design, stimulation currents of $20\,\mu A$ to 1 mA with pulse widths of $20\,\mu s$ to $200\,\mu s$ are quoted.

Where cross-coupling between closely spaced, simultaneously operated, stimulating sites must be avoided, floating power supplies are needed. A successful example design, capable of supporting parallel stimulation to electrodes on three semicircular canals for vestibular prosthesis, is described in [83].

4. Delivery of Power and Data to Implants

The requirements imposed on medical devices operating in the body are application specific, but there are a common set of constraints in size, power, and functionality. The interplay between these constraints determines the available processing bandwidth for the electronics, the operating time (in the case of battery operated devices such as pacemakers) and the communication range and bandwidth of the wireless telemetry link. In addition, the location of the device in the body, data rate, frequency, and regulatory standards influence the design complexity and power dissipation of telemetry links.

Long range telemetry links (typically greater than 2 meters) are mainly battery-operated and must conform to strict regulatory standards. They require both high sensitivity receivers and high output power transmitters, both of which result in high power dissipation. In addition, unlike most near-field (short range) inductive links (discussed below) where a stable reference clock can be extracted from the external carrier frequency, long range telemetry links require stable crystal references and frequency synthesizers to generate a local carrier with good frequency stability. These

transceivers are typically operated in dedicated frequency bands such as the U.S. federal communications commission (FCC) approved 402–405 MHz for medical implant communication service (MICS) band (e.g., the ZL70102 Microsemi transceiver). This band has a 300 kHz maximum bandwidth and a maximum output power of −16 dBm [84] and the data rate is limited to about 200 kbps. The industrial, scientific, and medical (ISM) radio bands are also frequently used for medical telemetry transmitters. These include the 902–928 MHz, 2.4–2.4835 GHz, and 5.725–5.875 GHz frequency bands and have transmission ranges up to 10 meters. Lower frequencies require larger antennas, while higher frequencies have higher losses due to tissue absorption. The optimum frequency band for wireless transmitters located in the body is reported to be approximately 900 MHz [85]. Ultra wideband (UWB) is an alternative wireless data transmission method used at very low energy levels for short range, high bandwidth communications. Recently, UWB communication in the 3.1–10.6 GHz band was used to develop low power wireless transmitters in CMOS technology for implantable medical devices [15, 55, 56] for high date rate (10–90 Mbps) transmission from the implant to the external device. Modulation schemes such as on/off keying (OOK) and pulse position modulation (PPM) are used to generate the short pulses.

For short range links (typically up to a few centimetres), low frequency inductive links are used for both power supply and bidirectional data transmission. Examples include cochlear and vestibular implants [3, 86]. These near-field systems can be made small and highly integrated. Modulation schemes such as OOK, amplitude shift keying (ASK), and frequency shift keying (FSK) are often used for data transmission from the external unit to the implanted device [87, 88]. To minimize radio frequency heating due to tissue absorption, these systems are typically operated below 15 MHz with an achievable data rate up to a few Mbps. Transferred power to the implant typically ranges from 10 mW to 125 mW [89–91]. The power transmitted through the tissue should comply with safety standards.

A basic block diagram of a typical architecture for transmitting power/data to an implant via an inductive link is shown in Figure 17. The external transmitter typically consists of a class D or class E power amplifier capable of providing large currents in the tuned primary coil (L_1) from a relatively low voltage. In the implant, the induced voltage that appears across the secondary coil $(L_2$, tuned by a capacitor) is rectified and regulated to provide a power supply for the electronics. The data link from the external transmitter to the implant (the downlink) is often achieved by modulating the envelope of the power carrier to create detectable changes across the secondary coil. The data link from the implant to the external circuit (the uplink) is commonly implemented by load modulation techniques. These techniques utilize the property of the coupled coils in which a change in the load of the secondary circuit is reflected back as changing impedance in the primary, through their mutual inductance M. Examples of modulation techniques for uplink and downlink data transmission are discussed later.

The basic equivalent circuit of the inductive link is shown in Figure 18 [92, 93]. The primary circuit (R_1, L_1, C_1) is

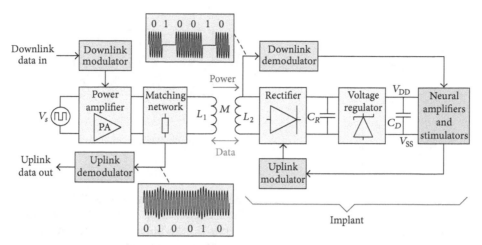

FIGURE 17: Architecture of a wireless inductive telemetry system to transmit power to an implant and data to and from the implant using a single pair of coils.

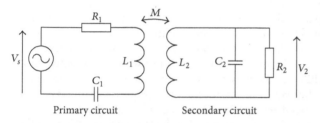

FIGURE 18: Equivalent circuit of the inductive link with a load (R_2).

tuned in series in order to provide a low impedance load to the driving transmitter. Resistor R_1 includes the loss of the inductor L_1 and the output resistance of the voltage source V_s (representing the transmitter driver). At resonance, the voltages related with C_1 and L_1 cancel each other and thus the primary circuit requires small voltage swings at its inputs. Thus, the primary circuit loads the secondary circuit with a small load. The topology of the secondary circuit (L_2, C_2, R_2) is tuned in parallel in order to amplify sufficiently the induced voltage to drive a nonlinear (rectifier) load. Resistor R_2 includes the loss of the inductor L_2 and the load resistance of the implant circuits. Both RLC circuits are tuned to the same resonant frequency. The gain factor of the link at resonance is [93]

$$\left|\frac{V_2}{V_s}\right| = \frac{\sqrt{(R_2/R_1)}}{((k_{\text{crit}}/k) + (k/k_{\text{crit}}))}; \quad k_{\text{crit}} = \sqrt{\frac{C_1 R_1}{C_2 R_2}}, \quad (3)$$

where k is the coupling coefficient and k_{crit} is the critical coupling coefficient. The relative dimensions of the coils and the air gap cause a low coupling coefficient ($k < 0.1$). In addition, the inductances of the primary L_1 and secondary L_2 coils are small and can be generated using coils with a few turns. The typical variations of the gain factor are illustrated in Figure 19. The gain factor is not constant with respect to coupling variations. The link transfers maximum power when the resistance in the primary is equivalent to the reflected secondary resistance, assuming that reactive components are

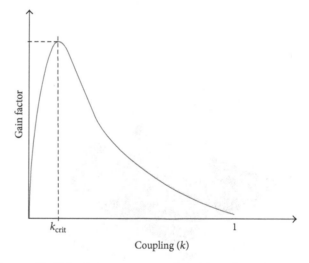

FIGURE 19: Gain factor versus coupling when the primary and secondary circuits are tuned to the same frequency.

being cancelled. It can be shown that efficiency (η) at critical coupling is equal to 50% (Figure 20) [93]. For the purposes of data transfer, the link behaves as a narrow-band bandpass filter. At resonance and assuming that the coils have the same quality factor Q, the bandwidth of the link is $B = \sqrt{2}(f_c/Q)$ where f_c is the carrier frequency [87].

Power transfer (gain factor) optimization and data transfer (bit rate) optimization have contradicting requirements. To illustrate this, the circuit in Figure 18 was simulated in Advanced Design System (Agilent EEsof EDA) with ($R_1 = 1\,\Omega, L_1 = 1\,\mu\text{H}$) and ($R_2 = 1\,\text{k}\Omega, L_2 = 1\,\mu\text{H}$) tuned to $1\,\text{MHz}$ by C_1 and C_2, respectively [94]. The relationships between gain factor (power transfer), carrier frequency, and coupling coefficient are plotted in Figure 21. The optimum gain factor is at the resonant frequency ($1\,\text{MHz}$) and coupling coefficient (k) of around 0.05. The gain factor can be improved by increasing the quality factors. The bandwidth of the inductive link can be increased by lowering the quality factors.

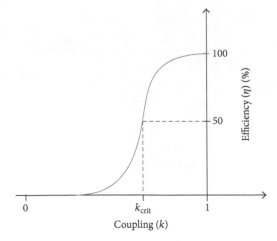

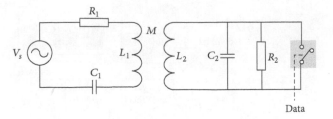

FIGURE 22: Implementation of LSK modulation.

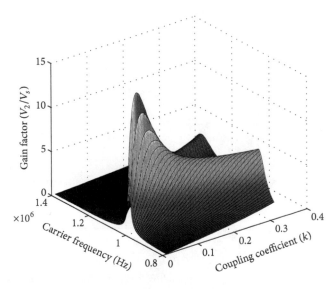

FIGURE 20: Efficiency of the inductive link versus coupling when the primary and secondary circuits are tuned to the same frequency.

FIGURE 21: Inductive link gain factor versus coupling coefficient and carrier frequency simulated in ADS for the equivalent circuit model of Figure 18.

It also increases as the coupling coefficient increases (i.e., the gap between the coils decreases). In practice, the inductive link should be designed to account for the potentially wide variation in the gap between the coils. Coupling compensation techniques (incorporating feedback) regulate the voltage across the secondary coil [95–97]. Power and data links can use separate sets of coils because this allows independent optimization to maximize performance [98]. However, using separate coils has the drawbacks of an increase in the implant footprint and electromagnetic interference. The latter requires the use of complex modulation techniques to minimize its effects and increases system complexity.

The most commonly used technique for uplink data transmission is passive signalling [99] also known as load shift keying (LSK) [100]. This modulation is based on

the reflection of the implant's load to the transmitter via the inductive link. A typical implementation is shown in Figure 22. The binary Data stream shorts the implant coil and the change in impedance is reflected in the transmitter because the implant load is much larger than the on-resistance of the switch transistor. Where only two coils are used (for both power and data) there is a risk of disruption in power delivery if the short is applied for too long to the implant coil. When communication is in idle mode, the link should be optimized for maximum power transfer. The bandwidth of LSK is limited by the coupling factor, the parameters of the coils, and the transient response of the inductive link. Multilevel LSK may be used to increase the data rate [101]. An alternative communication technique for uplink is passive phase shift keying (PPSK) [102]. Unlike LSK, the switch across the secondary coil (Figure 22) closes synchronously with the carrier for half the carrier cycle. The transient response in the primary coil current is detected as a logic "1" signal. An integrated implementation of PPSK in CMOS technology is presented in [103]. The circuit was designed to work at 13.56 MHz with a single set of coils for both data transmission and power delivery. The link can reach a data rate of up to 1/16 of the carrier frequency, that is, 847.5 kbps in this case, making it the fastest data rate achieved by a single wireless (inductively coupled) link used simultaneously for power delivery and communication for implantable devices. Another method for uplink transmission is pulse harmonic modulation (PHM). It achieves a data rate of 20 Mbps for 1 cm coil separation using a carrier frequency of 66.7 MHz [104]. A pattern of very narrow pulses with specific time delays and amplitude is used, which minimizes intersymbol interference across the receiver coil. PHM is one of the fastest data transmission methods currently known via inductively coupled coils. Unfortunately, to implement it, a separate power link is required as this method is carrier-less.

For downlink data transmission, various digital modulation schemes are used. The most common are binary amplitude shift keying (BASK), binary frequency shift keying (BFSK) and binary phase shift keying (BPSK). The simplest in implementation is BASK but it is sensitive to amplitude fluctuation and the bit rate is typically limited to about 10% of the carrier frequency [87, 88]. In BFSK the binary data is represented by constant amplitude, using two different frequencies, where logic "1" is assigned to one frequency and logic "0" to the other. Compared to BASK, BFSK can provide higher data-rate-to-carrier-frequency ratio [105] but it requires a wide passband in the inductive link to allow for the different frequencies, which limits power transfer. The advantage of BPSK over BASK and BFSK is the use of a carrier with fixed amplitude and fixed frequencies [106], enabling efficient and stable power transfer. However, BPSK requires a complicated demodulator, implemented by some kind of phase-locked loop, usually a Costas loop [87]. A comprehensive survey and comparison of modulation techniques is presented in [88].

5. Conclusion and Future Directions

Since the 1950's remarkable efforts have been undertaken in the development of implantable medical devices. Initially most of the successful applications focused on cardiac rhythm management. Today's implantable medical devices provide therapy to treat numerous health conditions. Exciting new applications, for example, in electrical neuromodulation, can be used to treat Parkinson's disease, epilepsy, bladder control, gastrointestinal disorders and numerous psychological disorders such as obsessive-compulsive disorder. Implantable medical devices can now provide a range of pharmacological therapies enabling precise dosage and interval delivery of drugs to more effectively treat patients' conditions while minimizing side effects.

There are many challenges when creating an implantable medical device. These include microelectronic design, electrode technology, packaging, and biomedical signal processing. This paper has focused on the advances in microelectronics over the last decade or so for implantable medical devices and systems. Several examples of neural amplifiers featuring low noise and low power to monitor the small electrical potentials produced from living neurons via electrodes have been discussed. Both clock-based and continuous-time techniques are used in the design of neural amplifiers. Nowadays, implantable neural recording devices include sophisticated signal processing functions on-chip for data compression, such as spike sorting. Analog and mixed analog/digital circuits are used in their implementation. New advanced techniques to further reduce the power and bandwidth requirements of the wireless data transmission link, for example, based on the concept of compressive sensing, will continue to emerge.

Neural stimulation is a key function performed by implantable medical devices for many applications. Some common principles and design techniques for neural stimulators have been presented, including new techniques which avoid the need for off-chip blocking capacitors and methods for achieving charge balancing. There is a need for energy-efficient stimulator circuits to reduce power consumption (two such examples have been discussed) and further developments in this area are anticipated, particularly for applications requiring many stimulation sites (e.g., retinal prosthesis for the blind).

Wireless power and data operation of implantable medical devices are important because they avoid the need for implantable batteries and offer more flexibility to patients. Although major advances have been achieved in the field of wireless communications and wireless powering for implants, further improvements in terms of new techniques that allow

better optimization of the entire system are anticipated. The basic principles of inductively coupled telemetry links commonly used to wirelessly power implants and to provide them with a medium for bidirectional communication have been discussed. Recent developments include the introduction of advanced modulation schemes (e.g., PHM) that allow wideband transmission of data over inductively coupled coils. In addition, transceivers based on conventional wideband wireless radio technology are emerging. These are expected to continue to offer improved performance in terms of an increase in output data rate with lower power consumption requirements, as smaller geometry silicon process technologies are used for the implementation of the implantable circuits. On the subject of providing power to implantable medical devices, it is expected that systems using energy harvested from outside and inside the human body will evolve. An example of such a system that uses the cochlear in the inner ear as a battery source to supply a 2.4 GHz radio transmitter is described in [107]. For such systems ultra-low power circuits are essential.

An important research topic involves closed-loop sense and stimulate systems which will continue to develop, possibly combining both sensing of electrical and chemical responses, for applications such as DBS, epilepsy, and other neurological conditions. Such systems will allow better management of the clinical condition allowing systems to adapt to the varying pathological characteristics. In addition, when new highly miniaturized implantable neural interfaces are developed, a step change in micropackaging techniques will be needed to protect the active area of the integrated circuit from hostile environments experienced in the body. As an example, a recent publication describes a technique for integrated circuit micropackages, dedicated to neural interfaces, based on gold-silicon wafer bonding [108].

The expectation of longer life and a progressively increasing knowledge base will place further dependence on modern healthcare technologies to improve the quality of life of a very large number of patients (both young and old). This will undoubtedly provide a fertile ground for future research.

Conflict of Interests

The author declares that there is no conflict of interests regarding the publication of this paper.

References

[1] C. Ward, S. Henderson, and N. H. Metcalfe, "A short history on pacemakers," *International Journal of Cardiology*, vol. 169, no. 4, pp. 244–248, 2013.

[2] P. E. Vardas, E. N. Simantirakis, and E. M. Kanoupakis, "New developments in cardiac pacemakers," *Circulation*, vol. 127, pp. 2343–2350, 2013.

[3] F. G. Zeng, S. Rebscher, W. V. Harrison, X. Sun, and H. Feng, "Cochlear implants: system design, integration, and evaluation," *IEEE Reviews in Biomedical Engineering*, vol. 1, pp. 115–142, 2008.

[4] B. S. Wilson and M. F. Dorman, "Cochlear implants: a remarkable past and a brilliant future," *Hearing Research*, vol. 242, no.

1-2, pp. 3–21, 2008.

[5] J. M. Ong and L. da Cruz, "The bionic eye: a review," *Clinical & Experimental Ophthalmology*, vol. 40, pp. 6–17, 2012.

[6] T. Guenther, N. H. Lovell, and G. J. Suaning, "Bionic vision: system architectures—a review," *Expert Review of Medical Devices*, vol. 9, no. 1, pp. 33–48, 2012.

[7] C. Wall III, D. M. Merfeld, S. D. Rauch, and F. O. Black, "Vestibular prostheses: the engineering and biomedical issues," *Journal of Vestibular Research: Equilibrium and Orientation*, vol. 12, no. 2-3, pp. 95–113, 2002-2003.

[8] G. Y. Fridman and C. C. Della Santina, "Progress toward development of a multichannel vestibular prosthesis for treatment of bilateral vestibular deficiency," *Anatomical Record*, vol. 295, pp. 2010–2029, 2012.

[9] S. Miocinovic, S. Somayajula, S. Chitnis, and J. L. Vitek, "History, applications, and mechanisms of deep brain stimulation," *JAMA Neurology*, vol. 70, no. 2, pp. 163–171, 2013.

[10] H. M. Lee, H. Park, and M. Ghovanloo, "A power-efficient wireless system with adaptive supply control for deep brain stimulation," *IEEE Journal of Solid-State Circuits*, vol. 48, no. 9, pp. 2203–2216, 2013.

[11] V. Valente, A. Demosthenous, and R. Bayford, "A tripolar current-steering stimulator ASIC for field shaping in deep brain stimulation," *IEEE Transactions on Biomedical Circuits and Systems*, vol. 6, no. 3, pp. 197–207, 2012.

[12] V. Valente, A. Demosthenous, and R. Bayford, "Output stage of a current-steering multipolar and multisite deep brain stimulator," in *Proceedings of the IEEE Biomedical Circuits and Systems Conference (BiOCAS '13)*, pp. 85–88, Rotterdam, The Netherlands, October-November 2013.

[13] J. DiGiovanna, W. Gong, C. Haburcakova et al., "Development of a closed-loop neural prosthesis for vestibular disorders," *Journal of Automatic Control*, vol. 20, pp. 27–32, 2010.

[14] A. Berényi, M. Belluscio, D. Mao, and G. Buzsáki, "Closed-loop control of epilepsy by transcranial electrical stimulation," *Science*, vol. 337, pp. 735–737, 2012.

[15] K. Abdelhalim, H. M. Jafari, L. Kokarovtseva, J. L. Perez Velazquez, and R. Genov, "64-channel UWB wireless neural vector analyzer SOC with a closed-loop phase synchrony-triggered neurostimulator," *IEEE Journal of Solid-State Circuits*, vol. 48, no. 10, pp. 2494–2515, 2013.

[16] G. E. Loeb and R. A. Peck, "Cuff electrodes for chronic stimulation and recording of peripheral nerve activity," *Journal of Neuroscience Methods*, vol. 64, no. 1, pp. 95–103, 1996.

[17] R. B. Stein, D. Charles, L. David, J. Jhamandas, A. Mannard, and T. R. Nichols, "Principles underlying new methods for chronic neural recording," *Canadian Journal of Neurological Sciences*, vol. 2, no. 3, pp. 235–244, 1975.

[18] C. T. Nordhausen, E. M. Maynard, and R. A. Normann, "Single unit recording capabilities of a 100 microelectrode array," *Brain Research*, vol. 726, no. 1-2, pp. 129–140, 1996.

[19] A. S. Dickey, A. Suminski, Y. Amit, and N. G. Hatsopoulos, "Single-unit stability using chronically implanted multielectrode arrays," *Journal of Neurophysiology*, vol. 102, no. 2, pp. 1331–1339, 2009.

[20] S. Mittal, E. Pokushalov, A. Romanov et al., "Long-term ECG monitoring using an implantable loop recorder for the detection of atrial fibrillation after cavotricuspid isthmus ablation in patients with atrial flutter," *Heart Rhythm*, vol. 10, no. 11, pp. 1598–1604, 2013.

[21] Y. Nemirovsky, I. Brouk, and C. G. Jakobson, "1/f noise in CMOS transistors for analog applications," *IEEE Transactions on Electron Devices*, vol. 48, no. 5, pp. 921–927, 2001.

[22] E. A. M. Klumperink, S. L. J. Gierkink, A. P. van der Wel, and B. Nauta, "Reducing MOSFET 1/f noise and power consumption by switched biasing," *IEEE Journal of Solid-State Circuits*, vol. 35, no. 7, pp. 994–1001, 2000.

[23] Y.-J. Min, C.-K. Kwon, H.-K. Kim, C. Kim, and S.-W. Kim, "A CMOS magnetic hall sensor using a switched biasing amplifier," *IEEE Sensors Journal*, vol. 12, no. 5, pp. 1195–1196, 2012.

[24] C. C. Enz, E. A. Vittoz, and F. Krummenacher, "A CMOS chopper amplifier," *IEEE Journal of Solid-State Circuits*, vol. 22, no. 3, pp. 335–342, 1986.

[25] A. Uranga, X. Navarro, and N. Barniol, "Integrated CMOS amplifier for ENG signal recording," *IEEE Transactions on Biomedical Engineering*, vol. 51, no. 12, pp. 2188–2194, 2004.

[26] T. Denison, K. Consoer, W. Santa, A.-T. Avestruz, J. Cooley, and A. Kelly, "A2 μw 100 nV/rtHz chopper-stabilized instrumentation amplifier for chronic measurement of neural field potentials," *IEEE Journal of Solid-State Circuits*, vol. 42, no. 12, pp. 2934–2945, 2007.

[27] Y. Tseng, Y. Ho, S. Kao, and C. N. I. S. Su, "A 0.09 μW low power front-end biopotential amplifier for biosignal recording," *IEEE Transactions on Biomedical Circuits and Systems*, vol. 6, no. 5, pp. 508–516, 2012.

[28] R. F. Yazicioglu, P. Merken, R. Puers, and C. van Hoof, "A 60 μW 60 nV/$\sqrt{Hz}$ readout front-end for portable biopotential acquisition systems," *IEEE Journal of Solid-State Circuits*, vol. 42, no. 5, pp. 1100–1110, 2007.

[29] J. Lee, M. Johnson, and D. Kipke, "A tunable biquad switched-capacitor amplifier-filter for neural recording," *IEEE Transactions on Biomedical Circuits and Systems*, vol. 4, no. 5, pp. 295–300, 2010.

[30] V. Balasubramanian, P. F. Ruedi, Y. Temiz, A. Ferretti, C. Guiducci, and C. C. Enz, "A 0.18 μm biosensor front-end based on 1/f noise, distortion cancelation and chopper stabilization techniques," *IEEE Transaction on Biomedical Circuits and Systems*, vol. 7, no. 5, pp. 660–673, 2013.

[31] C. C. Enz and G. C. Temes, "Circuit techniques for reducing the effects of Op-Amp imperfections: autozeroing, correlated double sampling, and chopper stabilization," *Proceedings of the IEEE*, vol. 84, no. 11, pp. 1584–1614, 1996.

[32] C.-H. Chan, J. Wills, J. LaCoss, J. J. Granacki, and J. Choma Jr., "A novel variable-gain micro-power band-pass auto-zeroing CMOS amplifier," in *Proceedings of the IEEE International Symposium on Circuits and Systems (ISCAS '07)*, pp. 337–340, May 2007.

[33] Y. Masui, T. Yoshida, M. Sasaki, and A. Iwata, "0.6 V supply complementary metal oxide semiconductor amplifier using noise reduction technique of autozeroing and chopper stabilization," *Japanese Journal of Applied Physics*, vol. 46, no. 4B, pp. 2252–2256, 2007.

[34] R. R. Harrison and C. Charles, "A low-power low-noise CMOS amplifier for neural recording applications," *IEEE Journal of Solid-State Circuits*, vol. 38, no. 6, pp. 958–965, 2003.

[35] A. Rodríguez-Pérez, J. Ruiz-Amaya, M. Delgado-Restituto, and A. Rodríguez-Vázquez, "A low-power programmable neural spike detection channel with embedded calibration and data compression," *IEEE Transaction on Biomedical Circuits and Systems*, vol. 6, no. 4, pp. 87–100, 2012.

[36] S. Song, M. J. Rooijakkers, P. Harpe et al., "A 430 nW 64 nV/$\sqrt{Hz}$ current-reuse telescopic amplifier for neural recording applications," in *Proceedings of the IEEE Biomedical Circuits and Systems Conference (BiOCAS '13)*, pp. 322–325, Rotterdam, The Netherlands, October-November 2013.

[37] F. Zhang, J. Holleman, and B. P. Otis, "Design of ultra-low power biopotential amplifiers for biosignal acquisition applications," *IEEE Transactions on Biomedical Circuits and Systems*, vol. 6, no. 4, pp. 344–355, 2012.

[38] S. Song, M. J. Rooijakkers, P. Harpe et al., "A 430nW 64nV/$\sqrt{Hz}$ current-reuse telescopic amplifier for neural recording applications," in *Proceedings of the IEEE Biomedical Circuits and Systems Conference (BiOCAS '13)*, pp. 322–325, Rotterdam, The Netherlands, October-November 2013.

[39] X. Zou, L. Liu, J. H. Cheong et al., "A 100-channel 1-mW implantable neural recording IC," *IEEE Transactions on Circuits and Systems I: Regular Papers*, vol. 60, no. 10, pp. 2584–2596, 2013.

[40] P. Kmon and P. Gryboś, "Energy efficient low-noise multichannel amplifier in submicron CMOS process," *IEEE Transactions on Circuits and Systems I: Regular Papers*, vol. 60, no. 7, pp. 1764–1775, 2013.

[41] J. Gak, M. R. Miguez, and A. Arnaud, "Nanopower OTAs with improved linearity and low input offset using bulk degeneration," *IEEE Transactions on Circuits and Systems I: Regular Papers*, vol. 61, no. 3, pp. 689–698, 2014.

[42] V. Majidzadeh, A. Schmid, and Y. Leblebici, "Energy efficient low-noise neural recording amplifier with enhanced noise efficiency factor," *IEEE Transactions on Biomedical Circuits and Systems*, vol. 5, no. 3, pp. 262–271, 2011.

[43] M. S. J. Steyaert and W. M. C. Sansen, "A micropower low-noise monolithic instrumentation amplifier for medical purposes," *IEEE Journal of Solid-State Circuits*, vol. 22, no. 6, pp. 1163–1168, 1987.

[44] K. A. Ng and Y. P. Xu, "A compact, low input capacitance neural recording amplifier," *IEEE Transaction on Biomedical Circuits and Systems*, vol. 7, no. 5, pp. 610–620, 2013.

[45] B. Gosselin, M. Sawan, and C. A. Chapman, "A low-power integrated bioamplifier with active low-frequency suppression," *IEEE Transactions on Biomedical Circuits and Systems*, vol. 1, no. 3, pp. 184–192, 2007.

[46] C. Y. Wu, W. M. Chen, and L. T. Kuo, "A CMOS power-efficient low-noise current-mode front-end amplifier for neural signal recording," *IEEE Transaction on Biomedical Circuits and Systems*, vol. 7, no. 2, pp. 107–114, 2013.

[47] A. Demosthenous and I. F. Triantis, "An adaptive ENG amplifier for tripolar cuff electrodes," *IEEE Journal of Solid-State Circuits*, vol. 40, no. 2, pp. 412–420, 2005.

[48] R. Rieger, M. Schuettler, D. Pal et al., "Very low-noise ENG amplifier system using CMOS technology," *IEEE Transactions on Neural Systems and Rehabilitation Engineering*, vol. 14, no. 6, pp. 427–437, 2006.

[49] Y. J. Jung, B. S. Park, H. M. Kwon et al., "A novel BJT structure implemented using CMOS processes for high-performance analog circuit applications," *IEEE Transactions on Semiconductor Manufacturing*, vol. 25, no. 4, pp. 549–554, 2012.

[50] M. Degrauwe, E. Vittoz, and I. Verbauwhede, "A micropower CMOS instrumentation amplifier," *IEEE Journal of Solid-State Circuits*, vol. 20, no. 3, pp. 805–807, 1985.

[51] A. Worapishet, A. Demosthenous, and X. Liu, "A CMOS instrumentation amplifier with 90-dB CMRR at 2-MHz using capacitive neutralization: analysis, design considerations, and implementation," *IEEE Transactions on Circuits and Systems I: Regular Papers*, vol. 58, no. 4, pp. 699–710, 2011.

[52] A. Demosthenous, I. Pachnis, D. Jiang, and N. Donaldson, "An integrated amplifier with passive neutralization of myoelectric interference from neural recording tripoles," *IEEE Sensors Journal*, vol. 13, no. 9, pp. 3236–3248, 2013.

[53] A. Rodríguez-Pérez, M. Delgado-Restituto, and F. Medeiro, "A 515 nW, 0–18 dB programmable gain analog-to-digital converter for in-channel neural recording interfaces," *IEEE Transactions on Biomedical Circuits and Systems*, 2013.

[54] Y. Ming, D. A. Borton, J. Aceros, W. R. Patterson, and A. V. Nurmikko, "A 100-channel hermetically sealed implantable device for chronic wireless neurosensing applications," *IEEE Transactions on Biomedical Circuits and Systems*, vol. 7, no. 2, pp. 115–128, 2013.

[55] M. S. Chae, Z. Yang, M. R. Yuce, L. Hoang, and W. Liu, "A 128-channel 6 mW wireless neural recording IC with spike feature extraction and UWB transmitter," *IEEE Transactions on Neural Systems and Rehabilitation Engineering*, vol. 17, no. 4, pp. 312–321, 2009.

[56] H. Gao, R. M. Walker, P. Nuyujukian et al., "HermesE: a 96-channel full data rate direct neural interface in 0.13 μm CMOS," *IEEE Journal of Solid-State Circuits*, vol. 47, no. 4, pp. 1043–1055, 2012.

[57] S. Gibson, J. W. Judy, and D. Marković, "Spike sorting: the first step in decoding the brain: the first step in decoding the brain," *IEEE Signal Processing Magazine*, vol. 29, no. 1, pp. 124–143, 2012.

[58] R. R. Harrison, R. J. Kier, C. A. Chestek et al., "Wireless neural recording with single low-power integrated circuit," *IEEE Transactions on Neural Systems and Rehabilitation Engineering*, vol. 17, no. 4, pp. 322–329, 2009.

[59] A. M. Sodagar, G. E. Perlin, Y. Yao, K. Najafi, and K. D. Wise, "An implantable 64-channel wireless microsystem for single-unit neural recording," *IEEE Journal of Solid-State Circuits*, vol. 44, no. 9, pp. 2591–2604, 2009.

[60] S. Gibson, J. W. Judy, and D. Marković, "Technology-aware algorithm design for neural spike detection, feature extraction, and dimensionality reduction," *IEEE Transactions on Neural Systems and Rehabilitation Engineering*, vol. 18, no. 5, pp. 469–478, 2010.

[61] V. Karkare, S. Gibson, and D. Marković, "A 130-μW, 64-channel neural spike-sorting DSP chip," *IEEE Journal of Solid-State Circuits*, vol. 46, no. 5, pp. 1214–1222, 2011.

[62] T. M. Seese, H. Harasaki, G. M. Saidel, and C. R. Davies, "Characterization of tissue morphology, angiogenesis, and temperature in the adaptive response of muscle tissue to chronic heating," *Laboratory Investigation*, vol. 78, no. 12, pp. 1553–1562, 1998.

[63] K. G. Oweiss, A. Mason, Y. Suhail, A. M. Kamboh, and K. E. Thomson, "A scalable wavelet transform VLSI architecture for real-time signal processing in high-density intra-cortical implants," *IEEE Transactions on Circuits and Systems I: Regular Papers*, vol. 54, no. 6, pp. 1266–1278, 2007.

[64] D. L. Donoho, "Compressed sensing," *IEEE Transactions on Information Theory*, vol. 52, no. 4, pp. 1289–1306, 2006.

[65] Z. Charbiwala, V. Karkare, S. Gibson, D. Marković, and M. B. Srivastava, "Compressive sensing of neural action potentials using a learned union of supports," in *Proceedings of the 8th International Conference on Body Sensor Networks (BSN '11)*, pp. 53–58, Dallas, Tex, USA, May 2011.

[66] Y. Suo, J. Zhang, R. Etienne-Cummings, T. D. Tran, and S. Chin, "Energy-efficient two-stage compressed sensing method for implantable neural recordings," in *Proceedings of the IEEE Biomedical Circuits and Systems Conference (BiOCAS '13)*, pp. 150–153, Rotterdam, The Netherlands, October-November 2013.

[67] A. Rodríguez-Pérez, J. Masuch, J. A. Rodríguez-Rodríguez, M. Delgado-Restituto, and A. Rodríguez-Vázquez, "A 64-channel inductively-powered neural recording sensor array," in *Proceedings of the IEEE Biomedical Circuits and Systems Conference (BiOCAS '12)*, pp. 228–231, Hsinchu, Taiwan, November 2012.

[68] J. Simpson and M. Ghovanloo, "An experimental study of voltage, current, and charge controlled stimulation front-end circuitry," in *Proceedings of the IEEE International Symposium on Circuits and Systems (ISCAS '07)*, pp. 325–328, New Orleans, La, USA, May 2007.

[69] X. Liu, A. Demosthenous, and N. Donaldson, "An integrated implantable stimulator that is fail-safe without off-chip blocking-capacitors," *IEEE Transactions on Biomedical Circuits and Systems*, vol. 2, no. 3, pp. 231–244, 2008.

[70] P. J. Langlois, A. Demosthenous, I. Pachnis, and N. Donaldson, "High-power integrated stimulator output stages with floating discharge over a wide voltage range for nerve stimulation," *IEEE Transactions on Biomedical Circuits and Systems*, vol. 4, no. 1, pp. 39–48, 2010.

[71] D. Jiang, A. Demosthenous, T. Perkins, X. Liu, and N. Donaldson, "A stimulator ASIC featuring versatile management for vestibular prostheses," *IEEE Transactions on Biomedical Circuits and Systems*, vol. 5, no. 2, pp. 147–159, 2011.

[72] X. Liu, A. Demosthenous, and N. Donaldson, "An integrated stimulator with DC-isolation and fine current control for implanted nerve tripoles," *IEEE Journal of Solid-State Circuits*, vol. 46, no. 7, pp. 1701–1714, 2011.

[73] X. Liu, A. Demosthenous, A. Vanhoestenberghe, D. Jiang, and N. Donaldson, "Active Books: the design of an implantable stimulator that minimizes cable count using integrated circuits very close to electrodes," *IEEE Transactions on Biomedical Circuits and Systems*, vol. 6, no. 3, pp. 216–227, 2012.

[74] S. K. Kelly and J. L. Wyatt Jr., "A power-efficient voltage-based neural tissue stimulator with energy recovery," in *Proceedings of the IEEE Solid-State Circuits Conference (ISSCC '04)*, San Francisco, Calif, USA, February 2004.

[75] S. K. Kelly and J. L. Wyatt Jr., "A power-efficient neural tissue stimulator with energy recovery," *IEEE Transactions on Biomedical Circuits and Systems*, vol. 5, no. 1, pp. 20–29, 2011.

[76] X. Liu, A. Demosthenous, and N. Donaldson, "Implantable stimulator failures: causes, outcomes, and solutions," in *Proceedings of the 29th Annual International Conference of the IEEE Engineering in Medicine and Biology Society (EMBC '07)*, pp. 5786–5789, Lyon, France, August 2007.

[77] A. Nonclercq, L. Lonys, A. Vanhoestenberghe, A. Demosthenous, and N. Donaldson, "Safety of multi-channel stimulation implants: a single blocking capacitor per channel is not sufficient after single-fault failure," *Medical & Biological Engineering & Computing*, vol. 50, no. 4, pp. 403–410, 2012.

[78] J.-J. Sit and R. Sarpeshkar, "A low-power blocking-capacitor-free charge-balanced electrode-stimulator chip with lesst than 6 nA DC error for 1-mA: full-scale stimulation," *IEEE Transactions on Biomedical Circuits and Systems*, vol. 1, no. 3, pp. 172–183, 2007.

[79] M. Ortmanns, A. Rocke, M. Gehrke, and H.-J. Tiedtke, "A 232-channel epiretinal stimulator ASIC," *IEEE Journal of Solid-State Circuits*, vol. 42, no. 12, pp. 2946–2959, 2007.

[80] I. Williams and T. G. Constandinou, "An energy-efficient, dynamic voltage scaling neural stimulator for a proprioceptive prosthesis," *IEEE Transactions on Biomedical Circuits and Systems*, vol. 7, no. 2, pp. 129–139, 2013.

[81] S. K. Arfin and R. Sarpeshkar, "An energy-efficient, adiabatic electrode stimulator with inductive energy recycling and feedback current regulation," *IEEE Transactions on Biomedical Circuits and Systems*, vol. 6, no. 1, pp. 1–14, 2012.

[82] U. Çilingiroğlu and S. İpek, "A zero-voltage switching technique for minimizing the current-source power of implanted stimulators," *IEEE Transactions on Biomedical Circuits and Systems*, vol. 7, no. 4, pp. 469–479, 2013.

[83] D. Jiang, A. Demosthenous, T. Perkins, D. Cirmirakis, X. Liu, and N. Donaldson, "An implantable 3-D vestibular stimulator with neural recording," in *Proceedings of the 38th European Solid-State Circuits Conference (ESSCIRC '12)*, pp. 277–280, Bordeaux, France, September 2012.

[84] P. Bradley, "Wireless medical implant technology—recent advances and future developments," in *Proceedings of the 37th European Solid-State Circuits Conference (ESSCIRC '11)*, pp. 54–58, Helsinki, Finland, September 2011.

[85] H. Yu, C.-M. Tang, and R. Bashirullah, "An asymmetric RF tagging IC for ingestible medication compliance capsules," in *Proceedings of the IEEE Radio Frequency Integrated Circuits Symposium (RFIC '09)*, pp. 101–104, Boston, Mass, USA, June 2009.

[86] D. Cirmirakis, D. Jiang, A. Demosthenous, N. Donaldson, and T. Perkins, "A telemetry operated vestibular prosthesis," in *Proceedings of the 19th International Conference on Electronics, Circuits, and Systems (ICECS '12)*, pp. 576–578, Seville, Spain, December 2012.

[87] M. Sawan, Y. Hu, and J. Coulombe, "Wireless smart implants dedicated to multichannel monitoring and microstimulation," *IEEE Circuits and Systems Magazine*, vol. 5, no. 1, pp. 21–39, 2005.

[88] M. A. Hannan, S. M. Abbas, S. A. Samad, and A. Hussain, "Modulation techniques for biomedical implanted devices and their challenges," *Sensors*, vol. 12, no. 1, pp. 297–319, 2012.

[89] K. M. Silay, C. Dehollain, and M. Declercq, "Inductive power link for a wireless cortical implant with biocompatible packaging," in *Proceedings of the 9th IEEE Sensors Conference (SENSORS '10)*, pp. 94–98, Kona, Hawaii, USA, November 2010.

[90] M. A. Adeeb, A. B. Islam, M. R. Haider, F. S. Tulip, M. N. Ericson, and S. K. Islam, "An inductive link-based wireless power transfer system for biomedical applications," *Active and Passive Electronic Components*, vol. 2012, Article ID 879294, 11 pages, 2012.

[91] R. R. Harrison, "Designing efficient inductive power links for implantable devices," in *Proceedings of the IEEE International Symposium on Circuits and Systems (ISCAS '07)*, pp. 2080–2083, New Orleans, La, USA, May 2007.

[92] F. E. Terman, *Electronic and Radio Engineering*, McGraw-Hill, New York, NY, USA, 4th edition, 1955.

[93] N. Donaldson and T. A. Perkins, "Analysis of resonant coupled coils in the design of radio frequency transcutaneous links," *Medical & Biological Engineering & Computing*, vol. 21, no. 5, pp. 612–627, 1983.

[94] D. Cirmirakis, *Novel telemetry system for closed loop vestibular prosthesis [Ph.D. thesis]*, University College London, London, UK, 2013.

[95] D. C. Galbraith, M. Soma, and R. L. White, "A wide-band efficient inductive transdermal power and data link with coupling

insensitive gain," *IEEE Transactions on Biomedical Engineering*, vol. 34, no. 4, pp. 265–275, 1987.

[96] G. Wang, W. Liu, M. Sivaprakasam, and G. A. Kendir, "Design and analysis of an adaptive transcutaneous power telemetry for biomedical implants," *IEEE Transactions on Circuits and Systems I: Regular Papers*, vol. 52, no. 10, pp. 2109–2117, 2005.

[97] P. Si, A. P. Hu, S. Malpas, and D. Budgett, "A frequency control method for regulating wireless power to implantable devices," *IEEE Transactions on Biomedical Circuits and Systems*, vol. 2, no. 1, pp. 22–29, 2008.

[98] G. Simard, M. Sawan, and D. Massicotte, "High-speed OQPSK and efficient power transfer through inductive link for biomedical implants," *IEEE Transactions on Biomedical Circuits and Systems*, vol. 4, no. 3, pp. 192–200, 2010.

[99] N. Donaldson, "Passive signalling via inductive coupling," *Medical & Biological Engineering & Computing*, vol. 24, no. 2, pp. 223–224, 1986.

[100] Z. Tang, B. Smith, J. H. Schild, and P. H. Peckham, "Data transmission from an implantable biotelemeter by load-shift keying using circuit configuration modulator," *IEEE Transactions on Biomedical Engineering*, vol. 42, no. 5, pp. 524–528, 1995.

[101] W. Xu, Z. Luo, and S. Sonkusale, "Fully digital BPSK demodulator and multilevel LSK back telemetry for biomedical implant transceivers," *IEEE Transactions on Circuits and Systems II: Express Briefs*, vol. 56, no. 9, pp. 714–718, 2009.

[102] L. Zhou and N. Donaldson, "A fast passive data transmission method for ENG telemetry," *Neuromodulation*, vol. 6, no. 2, pp. 116–121, 2003.

[103] D. Cirmirakis, D. Jiang, A. Demosthenous, N. Donaldson, and T. Perkins, "A fast passive phase shift modulator for inductively coupled implanted medical devices," in *Proceedings of the 38th European Solid-State Circuits Conference (ESSCIRC '12)*, pp. 301–304, Bordeaux, France, September 2012.

[104] M. Kiani and M. Ghovanloo, "A 20-Mb/s pulse harmonic modulation transceiver for wideband near-field data transmission," *IEEE Transactions on Circuits and Systems II: Express Briefs*, vol. 60, no. 7, pp. 382–386, 2013.

[105] M. Ghovanloo and K. Najafi, "A wideband frequency-shift keying wireless link for inductively powered biomedical implants," *IEEE Transactions on Circuits and Systems I: Regular Papers*, vol. 51, no. 12, pp. 2374–2383, 2004.

[106] Z. Luo and S. Sonkusale, "A novel BPSK demodulator for biological implants," *IEEE Transactions on Circuits and Systems I: Regular Papers*, vol. 55, no. 6, pp. 1478–1484, 2008.

[107] P. P. Mercier, A. C. Lysaght, S. Bandyopadhyay, A. P. Chandrakasan, and K. M. Stankovic, "Energy extraction from the biologic battery in the inner ear," *Nature Biotechnology*, vol. 30, no. 12, pp. 1240–1243, 2012.

[108] N. Saeidi, M. Schuettler, A. Demosthenous, and N. Donaldson, "Technology for integrated circuit micropackages for neural interfaces, based on gold-silicon wafer bonding," *Journal of Micromechanics and Microengineering*, vol. 23, Article ID 075021, 12 pages, 2013.

[109] P. Mohseni and K. Najafi, "A fully integrated neural recording amplifier with DC input stabilization," *IEEE Transactions on Biomedical Engineering*, vol. 51, no. 5, pp. 832–837, 2004.

[110] W. S. Liew, X. D. Zou, L. B. Yao, and Y. Lian, "A 1-V 60μW 16-channel interface chip for implantable neural recording," in *Proceedings of the 31st IEEE Custom Integrated Circuits Conference (CICC '09)*, pp. 507–510, San Jose, Calif, USA, September 2009.

[111] S. Rai, J. Holleman, J. N. Pandey, F. Zhang, and B. Otis, "A 500μW neural tag with 2μV$_{rms}$ AFE and frequency-multiplying MICS/ISM FSK transmitter," in *Proceedings of the IEEE International Solid-State Circuits Conference (ISSCC '09)*, vol. 1, pp. 212–213, San Francisco, Calif, USA, February 2009.

[112] F. Shahrokhi, K. Abdelhalim, D. Serletis, P. L. Carlen, and R. Genov, "The 128-channel fully differential digital integrated neural recording and stimulation interface," *IEEE Transactions on Biomedical Circuits and Systems*, vol. 4, no. 3, pp. 149–161, 2010.

[113] R. R. Harrison, P. T. Watkins, R. J. Kier et al., "A low-power integrated circuit for a wireless 100-electrode neural recording system," *IEEE Journal of Solid-State Circuits*, vol. 42, no. 1, pp. 123–133, 2007.

[114] A. Bonfanti, M. Ceravolo, G. Zambra et al., "A multi-channel low-power IC for neural spike recording with data compression and narrowband 400-MHz MC-FSK wireless transmission," in *Proceedings of the 36th European Solid-State Circuits Conference (ESSCIRC '10)*, pp. 330–333, Seville, Spain, September 2010.

[115] S. B. Lee, H.-M. Lee, M. Kiani, U.-M. Jow, and M. Ghovanloo, "An inductively powered scalable 32-channel wireless neural recording system-on-a-chip for neuroscience applications," *IEEE Transactions on Biomedical Circuits and Systems*, vol. 4, no. 6, pp. 360–371, 2010.

[116] K. Abdelhalim, L. Kokarovtseva, J. L. P. Velazquez, and R. Genov, "915-MHz FSK/OOK wireless neural recording SoC with 64 mixed-signal FIR filters," *IEEE Journal of Solid-State Circuits*, vol. 48, no. 10, pp. 2478–2493, 2013.

13

The European Legislation Applicable to Medium-Range Inductive Wireless Power Transmission Systems

Frédéric Broydé,[1] Evelyne Clavelier,[1] and Lucie Broydé[2]

[1] *Tekcem, 78580 Maule, France*
[2] *ESIGELEC, Saint-Étienne-du-Rouvray, 76800, France*

Correspondence should be addressed to Frédéric Broydé; fredbroyde@excem.fr

Academic Editor: Frederick Mailly

Medium-range inductive wireless power transmission systems allow a sufficient power transfer without requiring close proximity between a primary coil and a secondary coil. We briefly investigate the range of a typical system and its radiated emission, from the perspectives of electromagnetic compatibility (EMC) and human exposure requirements. We then discuss the applicable legislation in the European Union, the main question being the applicability of the R&TTE or radio equipment directives. Our conclusion is that this applicability depends on multiple parameters, among which is the presence of a self-tuning capability or of a transmitter control based on telemetry.

1. Introduction

Varying electric fields and magnetic fields can be used for wireless power transmission (WPT) from the antenna of a power-transmitting unit to the antenna of a power-receiving unit. WPT using electrically small coils as antennas is usually referred to as *inductive WPT*. Here, the antenna used in the power-transmitting unit is a primary coil and the receiving antenna is a secondary coil, as shown in Figure 1. The primary coil and the secondary coil form a transformer having a primary and a secondary which are mechanically separable. When the two are placed in proper orientation and proximity, the coupling becomes sufficient to allow an adequate power transmission.

In a short-range inductive WPT system, power transmission takes place only when the power-transmitting unit and the power-receiving unit have a well-defined position with respect to each other and are in mechanical contact with each other. Thus, the primary and secondary coils are typically a few millimeters apart and a good efficiency can be obtained. Short-range WPT systems are for instance used in chargers for the rechargeable batteries of hand-held battery-operated items, such as electric tooth brushes, radio transceivers, and cellular telephones. This approach allows the hand-held item

to be completely sealed and uses no electrical connection between the mains and the hand-held item.

This paper is about medium-range inductive WPT systems, which do not require a close proximity between the primary and secondary coils during power transmission. For instance, an experimental medium-range inductive WPT system (MRIWPTS) was used to power-feed a light bulb over a distance of 2 m [1–3]. The added convenience of positioning-free operation entails an increased emission level, which can exceed electromagnetic compatibility (EMC) and human exposure requirements (in this paper, *human exposure* refers to human exposure to electromagnetic fields).

In the European Union (EU), the legislation applicable to an electrical apparatus is set forth in the applicable directive(s), the national legislations transposing the provisions of these directives, and other national legislations. A manufacturer who wants to place a MRIWPTS on the market in the UE must first determine the applicable directive(s), which determine the set of applicable harmonized standards, and the applicable conformity assessment procedures.

The main subject of this paper is a discussion of the directive applicable to a MRIWPTS, the main question being the applicability of the radio equipment and telecommunications terminal equipment (R&TTE) directive [4] or of

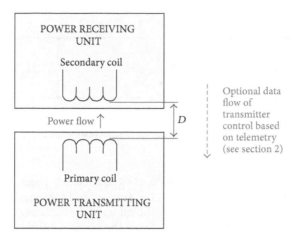

FIGURE 1: An inductive WPT system.

the radio equipment directive which will replace it [5]. If the R&TTE or radio equipment directive is applicable, it covers electromagnetic compatibility (EMC), as well as safety and protection of health issues, including human exposure. In the opposite case, EMC is usually covered by the EMC directive [6, 7], while safety and protection of health, including human exposure, are usually covered by the low voltage directive [8, 9]. Our discussion will not take into account the fact that the MRIWPTS may be a part of a product having other functions influencing the applicability of the R&TTE directive.

WPT is very useful to power-feed implantable medical devices. It is therefore worth mentioning that a device covered by the medical devices directive [10] or by the active implantable medical devices directive [11] is not subject to the EMC directive and low voltage directive, but it may be subject to the R&TTE or radio equipment directives.

The paper is organized as follows. Section 2 presents the typical characteristics of inductive WPT systems and provides a structural definition of MRIWPTSs. Section 3 presents a computation of the power gain, which is a measure of the efficiency of a WPT system, as a function of the distance between the coils. Section 4 addresses the radiated emission of medium-range WPT systems. Section 5 explains that one should consider that radio waves are used in WPT. Sections 6 and 7 discuss the applicability of the R&TTE and radio equipment directives.

2. Overview of Inductive WPT Systems

Let us use D to denote the shortest distance between the primary coil and the secondary coil (including their magnetic circuits, if present). Let us use the wording *resonant coil* to designate a coil used in a series or parallel resonant circuit comprising a capacitor, or a coil used at its self-resonance frequency. An inductive WPT system may use the following:

(i) nonresonant coils in the power-transmitting unit and in the power-receiving unit, in which case a reasonable efficiency requires that a magnetic circuit made of a magnetic material is used to obtain a strong coupling between the coils (i.e., a coefficient of coupling close to 1), D being a gap much smaller than the section of the magnetic circuit;

(ii) one resonant coil and one nonresonant coil, in which case a reasonable efficiency does not require a magnetic circuit made of a magnetic material, and D may be increased compared to two nonresonant coils;

(iii) resonant coils in the power-transmitting unit and in the power-receiving unit, in which case a reasonable efficiency does not require a magnetic circuit made of a magnetic material, and D may be further increased compared to a device using only one resonant coil, up to a few times the largest dimension of one of the coils.

In current inductive WPT systems, the primary coil is excited by a high-frequency current generated by an electronic circuit, typically a switched-mode or resonant inverter operating at a frequency above 20 kHz. Such a design typically implements resonant coils in the power-transmitting unit and/or in the power-receiving unit. For instance,

(i) many transponders for radio frequency identification (RFID) use inductive WPT to supply enough power to sustain the operation of the transponder, typically 10 μW to 1 mW [12];

(ii) implantable medical device typically uses inductive WPT with resonant coils in the frequency range 1 MHz to 20 MHz to deliver 5 mW to 250 mW to the implant [13–16];

(iii) a Wireless Power Consortium has issued a specification for inductive WPT systems operating in the 110 kHz to 205 kHz frequency range and capable of delivering up to 5 W at a distance of about 5 mm between the primary and secondary coils [17];

(iv) WPT systems intended to be used as battery chargers for electric vehicles are being designed, for instance, a 30 kW system operating at a distance of 45 mm [18].

Inductive WPT systems using resonant coils in the power-transmitting unit and in the power-receiving unit provide the best efficiency for a given size of the coils and a given distance between the coils, but they require an accurate tuning. Additionally, power is wasted if the primary coil is excited when the power-receiving unit is not present or does not need to receive power. In order to avoid this situation and to compensate the effect of component tolerances, temperature changes, component drift, and the effect of nearby conducting or magnetic items, it might be desirable to use

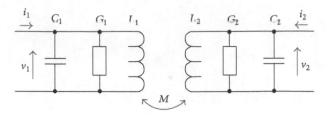

FIGURE 2: Model for the case of parallel resonance.

a WPT system using a *self-tuning capability* of the power-transmitting unit or a more elaborate *transmitter control based on telemetry* involving communication from the power-receiving unit to the power-transmitting unit. According to the self-tuning capability, the power-transmitting unit is capable of

(i) sensing the presence of a nearby secondary coil connected to a circuit comprising only a capacitor and a resistance, and varying or activating/deactivating the power in the primary coil accordingly; and/or

(ii) dynamically seeking resonance and optimizing power transfer with a nearby secondary coil connected to a circuit comprising only a capacitor and a resistance.

Of course, the self-tuning capability must also be able to operate in the presence of the intended power-receiving unit, which typically comprises a rectifier and other nonlinear circuits. We see that the self-tuning capability helps to obtain a good efficiency.

The above-mentioned specification of the Wireless Power Consortium uses *transmitter control based on telemetry*, involving a digital communication using load modulation, referred to as "backscatter modulation" in [17]. This specification also uses a form of *self-tuning capability*, referred to as "analog ping" in [17].

An arbitrary minimum value of D could be used to distinguish a MRIWPTS from a short-range inductive WPT system. However, this approach does not lead to general consequences from the design standpoint. In this paper, a MRIWPTS is defined by a structural attribute: the primary coil does not comprise a magnetic circuit made of a magnetic material.

3. Power Transfer of Inductive WPT Systems

We will now assess the typical efficiency which might be expected from a MRIWPTS using resonant coils in the power-transmitting unit and in the power-receiving unit. Since the largest dimension of each coil is small compared to wavelength, we can use a lumped-element circuit model to derive the currents through the coils. We will assume a simple equivalent linear circuit for the power-transmitting unit and the power-receiving unit. This is not a severe limitation since we only want to assess the behavior at the nominal frequency.

We can use the model shown in Figure 2 to investigate a MRIWPTS using a parallel resonant circuit comprising the primary coil, and another parallel resonant circuit comprising the secondary coil. In Figure 2, L_1 is the inductance of the primary coil, L_2 is the inductance of the secondary coil,

and G_1 and G_2 are conductances representing the losses in the primary and secondary coils, respectively. C_1 and C_2 are the values of the capacitors used to obtain parallel resonance. Since the circuit of the secondary coil is floating, we have not introduced the capacitance between the windings in the model.

We introduce the coefficient of coupling k, the quality factors Q_1 and Q_2, and the radian frequencies ω_{01} and ω_{02}, given by

$$k = \frac{M}{\sqrt{L_1 L_2}}, \tag{1}$$

$$Q_1 = \frac{1}{G_1}\sqrt{\frac{C_1}{L_1}}, \qquad Q_2 = \frac{1}{G_2}\sqrt{\frac{C_2}{L_2}}, \tag{2}$$

$$\omega_{01} = \frac{1}{\sqrt{C_1 L_1}}, \qquad \omega_{02} = \frac{1}{\sqrt{C_2 L_2}}. \tag{3}$$

If we consider the problem for which i_2 is determined by a load of conductance G_{2L}, in such a way that $i_2 = -G_{2L}v_2$, we define the power gain as

$$a_P = \frac{\mathrm{Re}\left(v_2 \bar{i}_2\right)}{\mathrm{Re}\left(v_1 \bar{i}_1\right)}, \tag{4}$$

where the bar denotes the complex conjugate. A simple analysis shows that if the circuits are tuned to the same frequency ω_0, that is, $\omega_0 = \omega_{01} = \omega_{02}$, and that the coupling is weak, that is, $k \ll 1$, an acceptable efficiency is only obtained for $Q_1 \gg 1$ and $Q_2 \gg 1$. Using these assumptions, we find that [19], at $\omega = \omega_0$,

$$a_P \approx \frac{K^2 G_2 G_{2L}}{\left(G_{2L} + G_2\left[K^2 + 1\right]\right)\left(G_{2L} + G_2\right)} \tag{5}$$

in which, following Terman [20, Sections 3 and 5] and Hochmair [13], we have used the dimensionless coupling parameter K defined by

$$K = k\sqrt{Q_1 Q_2}. \tag{6}$$

In order to maximize the approximate power gain given by (5) for G_2 and K fixed, we must use $G_{2L} = G_{2L\,\mathrm{MAX}}$ with

$$G_{2L\,\mathrm{MAX}} = G_2\sqrt{1 + K^2}. \tag{7}$$

Figure 3 shows the coefficient of coupling k and the exact power gain a_P as a function of the distance D between the primary and secondary coils in a coaxial configuration, for $Q_1 = Q_2 = 100$ and parallel resonance at the frequency $f_0 = 27.120$ MHz. This computation considers 2 identical circular coils of mean radius $r = 0.1$ m each having a single turn and no magnetic core. We assume that, at each D, the power-receiving unit absorbs a sinusoidal current and adjusts G_{2L} so that it takes on the value given by (7).

We can use the model shown in Figure 4 to investigate a MRIWPTS using a series resonant circuit comprising the primary coil L_1 and a series resonant circuit comprising the secondary coil L_2. Again, we have not introduced the capacitance between the windings in the model. This WPT system requires a separate analysis because, contrary to appearances, the circuits shown in Figures 2 and 4 are not dually related. In Figure 4, R_1 and R_2 are resistances representing the losses in the primary and secondary coils, respectively. C_1 and C_2 are the values of the capacitors used to obtain series resonance.

We introduce the quality factors Q_1 and Q_2 given by

$$Q_1 = \frac{1}{R_1}\sqrt{\frac{L_1}{C_1}}, \qquad Q_2 = \frac{1}{R_2}\sqrt{\frac{L_2}{C_2}}, \qquad (8)$$

and we will use again ω_{01} and ω_{02} given by (3). If we consider the problem for which i_2 is determined by a load of resistance R_{2L}, in such a way that $v_2 = -R_{2L}i_2$, we may use the power gain given by (4). In practice, as above, we can assume that the circuits are tuned to the same frequency $\omega_0 = \omega_{01} = \omega_{02}$, and we find that [19], at $\omega = \omega_0$,

$$a_P = \frac{K^2 R_2 R_{2L}}{(R_{2L} + R_2[K^2 + 1])(R_{2L} + R_2)}. \qquad (9)$$

In order to maximize the power gain for R_2 and K fixed, we must use $R_{2L} = R_{2L\,\mathrm{MAX}}$ with

$$R_{2L\,\mathrm{MAX}} = R_2\sqrt{1 + K^2}. \qquad (10)$$

We can compute the power gain a_P as a function of the distance D between the primary and secondary coils, for the geometry and the coils defined above in the explanations for Figure 3, using series resonance at the frequency $f_0 = 27.120$ MHz. Assuming that, at each D, the power-receiving unit absorbs a sinusoidal current and adjusts R_{2L} so that it takes on the value given by (10), the exact value of the power gain a_P differs by less than 0.5% from the results obtained for parallel resonance. Thus, Figure 3 also applies to series resonance.

The results presented above are very similar to the ones presented in [19] for a design having a similar geometry but operating at $f_0 = 149$ kHz. To obtain a reasonable efficiency, the ratio of the shortest distance between the coils to the largest dimension of the largest coil will always be less than 4.0 in an ideal experiment.

There are two other simple configurations which may be used to obtain resonant coils in the power-transmitting unit and in the power-receiving unit: parallel resonance in the power-transmitting unit with series resonance in the power-receiving unit and vice versa [18]. The latter seems to be the most popular scheme. However, more complex structures can also be used, for instance, involving two coupled coils in the power-transmitting unit and in the power-receiving unit [2], or a dual resonant circuit [16, 17].

4. Emission of a MRIWPTS

We now want to determine the radiated emission of the MRIWPTS, to investigate EMC and human exposure. This emission is mainly characterized by the magnetic field intensity H, which will be computed in free space. We assume that the excitation of the resonant circuit comprising the primary coil is adjusted in such a way that a power of 1 W is delivered to the load of the power-receiving unit, in addition to the earlier assumption that the power-receiving unit absorbs the maximum power from the resonant circuit comprising the secondary coil.

If we measure H at a distance d of the primary coil, d being much larger than D and the largest dimension of the primary and secondary coils, the free space emission is close to the emission of a magnetic dipole, for which the total radiated power $\mathcal{W}$ is given by Stratton [21, page 438]:

$$\mathcal{W} = \frac{\hbar^4 \eta_0}{6\pi}|\mathcal{M}|^2, \qquad (11)$$

where $\mathcal{M}$ is the r.m.s. magnetic moment of the dipole, $\eta_0 \approx 376.7\,\Omega$ is the intrinsic impedance of free space, and $\hbar$ is the wave number ω/c_0 where c_0 is the velocity of light in free space. For the configuration used above, comprising two coaxial circular coils, we have

$$|\mathcal{M}| = \pi r^2 (i_{L1} + i_{L2}), \qquad (12)$$

where i_{L1} and i_{L2} are the r.m.s. currents flowing in the primary coil and in the secondary coil, respectively.

Figure 5 shows $\mathcal{W}$ computed as a function of D, in the case of the configurations shown in Figures 2 and 4. In the case of parallel resonance shown in Figure 2, i_{L1} is the sum of the currents flowing through L_1 and G_1, and i_{L2} is the sum of the currents flowing through L_2 and G_2. In the case of series resonance shown in Figure 4, we have $i_{L1} = i_1$ and $i_{L2} = i_2$. Once $\mathcal{W}$ is computed, the maximum r.m.s. magnetic field strength at the distance d, denoted by H_{MAX}, is given by Broydé and Clavelier [22]:

$$H_{\mathrm{MAX}} = \frac{1}{d}\sqrt{\frac{3\mathcal{W}}{8\pi\eta_0}\frac{\ell}{(\hbar d)^2}}, \qquad (13)$$

where

$$\ell = \begin{cases} 2\sqrt{(\hbar d)^2 + 1} & \text{if } \hbar d \le (\hbar d)_C \\ \sqrt{(\hbar d)^4 - (\hbar d)^2 + 1} & \text{if } \hbar d > (\hbar d)_C, \end{cases} \qquad (14)$$

$$(\hbar d)_C \equiv \sqrt{\frac{5 + \sqrt{37}}{2}} \approx 2.354. \qquad (15)$$

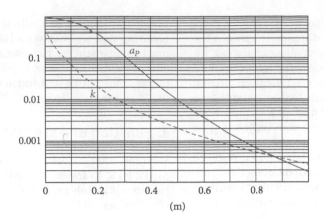

FIGURE 3: Coefficient of coupling k and power gain a_p versus D.

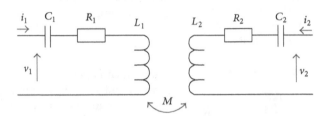

FIGURE 4: Model for the case of series resonance.

The field strength given by (13)–(15) is exact for a magnetic dipole at any frequency, in which case d is the distance between the dipole and the point where the field is measured. It applies to our problem only if the distance d between the center of the primary coil and the measurement point is much larger than D and the largest dimension of the primary and secondary coils. At a distance d which is not much larger than D and the largest dimension of the primary and secondary coils, we need to compute H at each point in space. Figure 6 shows a map of H, in the case of the configuration shown in Figure 2, for coaxial coils and $D =$ 0.3 m. Here, we have $D/2r = 1.5$ and $a_p = 0.163$. To obtain this map, we have used the exact analytical formula for the field intensity produced by each circular loop in the near-field region [21, p. 263, Problem 4]. The reference level relating to the limitation of exposure of the general public defined in [23] at the frequency $f_0 = 27.120$ MHz, namely, 73 mA/m or 97.27 dB(μA/m), is 0 dB in Figure 6. Figure 7 shows H, expressed in dB(μA/m), for $D = 0.3$ m, as a function of the coordinate z along the axis, the center of the primary lying at $z = 0$ m and the center of the secondary lying at $z = 0.3$ m. In Figure 7, the solid curve is the result of the exact near-field computation used to obtain Figure 6, and the dashed curve is obtained using the exact magnetic dipole formulas (13)–(15). We note that, in Figures 6 and 7, the field strength is approximately as high near the secondary coil as it is near the primary coil.

According to Figure 5, the radiated emission of the MRIWPTS computed here is much higher than the one obtained in [19] at $f_0 = 149$ kHz. For instance, for $D = 1$ m,

the total radiated power is about 40 dBm in Figure 5 (at f_0 = 27.120 MHz), as opposed to only −30 dBm in [19] (at f_0 = 149 kHz). With regard to human exposure, the field levels are much lower in Figure 6 than in the corresponding map of [19], but the reference levels defined in [23] are also different: 73 mA/m at $f_0 = 27.120$ MHz versus 5 A/m at $f_0 = 149$ kHz. The bottom line is that the volume where the reference level relating to the limitation of exposure of the general public is exceeded, which corresponds to the interior of the 0 dB curve in Figure 6, is larger in the case of the 27.120 MHz design studied in this paper than it is for the 149 kHz design.

Figure 8 shows a WPT experiment corresponding to the configuration considered in Figures 6 and 7: two single-turn coils of radius 0.1 m separated by a distance $D = 0.3$ m are used at 27.120 MHz. The secondary coil is suspended by a thin string and connected to a 110 V neon lamp in parallel with an air dielectric adjustable capacitor. In this simple experiment, the transmitted power is used to power-feed the lamp, and the power received by the lamp is not measured because the lamp is highly nonlinear. The primary coil is connected to the antenna port of an antenna tuner which receives about 6 W at its transmitter port. The hand of one of the authors is placed between the coils, no effect of the field being perceived. The lamp was then replaced with a resistor of 13.35 kΩ (measured value) and the power received by the transmitter port of the antenna tuner was set to about 48 W, to obtain a power received by the resistor equal to 1 W. An H-field probe was inserted between the coils, at $z = 0.15$ m. This caused a reduction of the power received by the resistor, which was not compensated by any new adjustment. The field

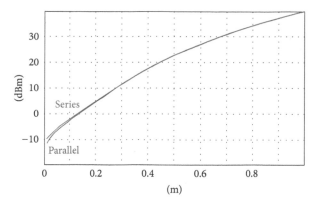

FIGURE 5: Total radiated power $\mathcal{W}$ as a function of the distance D between the coils, for 1 W delivered to the load.

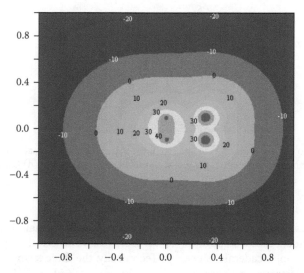

FIGURE 6: Map of H in a plane containing the common axis of the coils, for $D = 0.3$ m. The origin is the center of the primary coil. Horizontal and vertical axes: cartesian coordinates in meters. Colors in 10 dB steps, 0 dB corresponding to about 97.27 dB(μA/m) or 0.073 A/m.

between the coils was measured to be about 126 dB(μA/m). The agreement with Figure 7 is acceptable if we consider that the actual values of Q_1 and Q_2 are not known and that the H-field probe disturbs the field, as evidenced by the received power variation.

If we assume that a MRIWPTS is subject to the R&TTE directive, we may have to use the standards applicable to inductive loop systems in the frequency range 9 kHz to 30 MHz, under this directive [24, 25]. In this case, for the MRIWPTS at $f_0 = 27.120$ MHz, the radiated emission limit is 42 dB(μA/m), measured at 10 m. If we assume $D = 0.3$ m and a power of 1 W delivered to the load of the power-receiving unit, as in Figure 6 to Figure 8, the total radiated power is about 11 dBm in Figure 5, for which the field value given by the exact magnetic dipole formulas (13)–(15) at 10 m is about 46 dB(μA/m). The emission limit is exceeded by about 4 dB so that the design must be revised.

If we assume that a MRIWPTS is subject to the EMC directive, we may, for instance, be allowed to use the EN 55011 standards [26]. In this case, for the MRIWPTS at $f_0 = 27.120$ MHz, no radiated emission limit is applicable because

an ISM frequency is used. Here, we only have to take into account human exposure requirements.

In this section we have had a closer look at two key parameters of a MRIWPTS, the total radiated power and the magnetic field distribution close to the coils, which are dependent on the distance D. We have seen that the applicable technical requirements regarding radiated emission and human exposure play an essential role in an actual design. This is why it is important to correctly determine the applicable legislation.

5. Inductive WPT and Electromagnetic Waves

The term *radio wave* is defined in Article 2 of the R&TTE directive, as a special case of *electromagnetic wave*. Unfortunately, electromagnetic wave is neither defined in the directive, nor unambiguously defined in the literature. The generally accepted definition in textbooks on electromagnetic theory is that electromagnetic waves are the vectors **E** and **H** solutions of the vector wave equation [21, Section 7.1]. Any

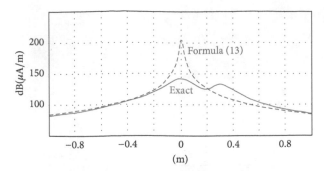

FIGURE 7: H along the common axis of the coils, for $D = 0.3$ m. The origin is the center of the primary coil.

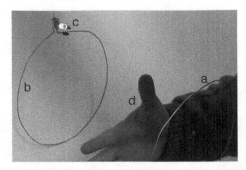

FIGURE 8: A WPT experiment according to the configuration considered in the paper, for $D = 0.3$ m, at 27.120 MHz: (a) is a part of the primary coil, (b) is the secondary coil, (c) is a neon lamp connected in parallel with the secondary coil, and (d) is an experimenter's hand. The lamp is being lit by wireless power.

such electromagnetic wave propagates. This concept is very general: all field distributions are included, except static ones.

This question of vocabulary is not academic since, depending on the meaning given to "electromagnetic waves," different European directives may be applicable, which entail different requirements and different costs (engineering, conformity assessment, and manufacturing costs). This may be why the wording *nonradiative*, which suggests that electromagnetic waves are not generated, has been associated by some authors to their wireless power transfer concepts [3, 27].

The particular fields produced by a harmonic electric or magnetic dipole can be written as the sum of three terms: the "quasistatic" term in $1/r^3$, the "induction" term in $1/r^2$, and the "radiation" term in $1/r$ [21, Sections 8.5, and 8.6], where r denotes the distance between the dipole and the measurement point. For instance, in the case of a magnetic dipole of particular interest in this paper, if we assume and suppress a time-harmonic dependence $e^{j\omega t}$, the magnetic field applied by the magnetic dipole at a location defined by the vector $\mathbf{r}$ is given by

$$\mathbf{H} = \frac{e^{-j\hbar r}}{4\pi} \left\{ \left(\frac{1}{r^3} + \frac{j\hbar}{r^2} \right) [3\,(\mathbf{u} \cdot \mathbf{M})\,\mathbf{u} - \mathbf{M}] - \frac{\hbar^2}{r} [\mathbf{u} \times (\mathbf{u} \times \mathbf{M})] \right\}, \quad (16)$$

where $\mathbf{u}$ is a unit vector in the direction of $\mathbf{r}$, such that $\mathbf{r} = r\mathbf{u}$, where $\hbar$ is the wave number and where $\mathbf{M}$ is the magnetic moment vector. At a distance from the electric or

magnetic dipole less than $\lambda/2\pi$, where λ is the wavelength, $\hbar r$ is less than 1 so that the "quasi-static" term predominates. Technically, the three terms propagate, because they contain the propagation factor $e^{-j\hbar r}$ which indicates the propagation of a spherical wave at the phase velocity c_0.

More generally, if we consider an arbitrary time-harmonic source of variable electric and magnetic fields, the total field at any point is a volume integral of several terms, each term propagating everywhere [21, Section 8.14] [28, Section 7.9]. For instance, the magnetic field applied by the source is given by

$$\mathbf{H} = \frac{1}{4\pi} \iiint_{\text{Source}} \left(\frac{1}{r^2} + \frac{j\hbar}{r} \right) e^{-j\hbar r} \mathbf{u} \times \mathbf{j}_T \, dV, \quad (17)$$

where $\mathbf{j}_T$ denotes the current density from the free charges and the polarization charges. Here, the contribution of a volume element dV has a term in $1/r^2$ and a term in $1/r$, and both propagate at the velocity of light in free space. Thus, propagation takes place everywhere, in the far-field region and in the near-field region of the source. Thus, we should not be misled by the word "static": in the region where the field is quasistatic, the field can be approximately computed as a static field, but the field nevertheless propagates.

Definitions being arbitrary, different points of view are of course possible; hence, the wording "nor unambiguously defined" is used at the beginning of this section. We note that the perspective of elementary explanations on electromagnetic waves, which mostly focus on plane waves and what happens in the far-field region of an antenna where the radiation term dominates, suggests that electromagnetic wave and propagation only exist far from the sources. As shown above, this is misleading. This perspective does not lead to a general definition of "electromagnetic wave."

Engineers do not usually consider capacitors, inductors, or transformers as devices in which electromagnetic waves occur. This is because these circuit elements are usually introduced in circuit theory without reference to the more general framework of electromagnetic theory. This is also because electromagnetic propagation is not needed to explain their basic properties. However, we must also face the fact that if we consider *electromagnetic wave* in every sense of the word, the coupling between the primary coil and the secondary coil of a MRIWPTS is caused by electromagnetic waves.

6. Applicability of the R&TTE Directive

The R&TTE directive, which will remain applicable until 12 June 2016 [5, Art. 50], is applicable to a *radio equipment*, defined as a product or component capable of communication by means of the emission and/or the reception of radio waves, where *radio waves* mean electromagnetic waves of frequencies from 9 kHz to 3000 GHz, propagated in space without artificial guide.

The fact that, in the meaning of the R&TTE directive, inductive coupling uses electromagnetic waves cannot be questioned, since harmonized standards exist for the application of the R&TTE directive to inductive loop systems [24, 25]. This confirms the analysis of Section 5.

Let us now have a closer look at the concept of electromagnetic waves propagated in space without artificial guide. In this requirement, "propagated in space without artificial guide" obviously refers to the capability of communication covering a sufficient distance in any direction, because this capability implies that interactions with other items capable of emission and/or reception of electromagnetic waves cannot be controlled by the user of the product or other interested parties, as opposed to transmission through an electrical interconnection. In other words, if the range is too short, the capability of communicating with electromagnetic waves should be ignored. This is one of the consequences of the wording "in space without artificial guide": there must be enough space (i.e., empty gap).

No such minimum distance appears in the R&TTE directive. However, the current "Guide to the R&TTE Directive 1999/5/EC" excludes cochlear implants of the "current technology" from the scope of the R&TTE directive [29, Section 1.3.5], on the basis of a report prepared by the authors, which has been made public by the European Commission [30]. According to this report, a product comprising a transmitter and a receiver could possibly not be regarded as being in the scope of the R&TTE directive, when the following conditions are simultaneously fulfilled:

(a) the receiver is intended to operate, at a given time, with a single transmitter of the intended type of transmitter, which we will refer to as "the intended transmitter";

(b) the receiver has a maximum range below 10 cm for the intended type of transmitter so that, in a given configuration, the intended transmitter is the only transmitter of the intended type capable of communication with the receiver, as if the transmitted signal was sent through an electrical interconnection;

(c) the receiver is so insensitive that it cannot, in normal electromagnetic environments, receive electromagnetic waves coming from other sources than the intended transmitter.

Any practical MRIWPTS is likely to operate at frequencies higher than 9 kHz for power transmission. At this stage, we may conclude that such a MRIWPTS uses radio waves unless it meets the conditions (a), (b), and (c) defined above.

To determine in which case a MRIWPTS is radio equipment subject to the R&TTE directive, we must now discuss whether a MRIWPTS is capable of communication. At first glance, it seems that the electromagnetic waves used for power transmission do not convey any information and that no communication is therefore taking place. However, we must also consider the possible implementation of a self-tuning capability of the power-transmitting unit or a more elaborate transmitter control based on telemetry, as explained in Section 2.

An accepted interpretation of the R&TTE directive concerning radars states that radars are *capable of communication*. The rationale for this interpretation says that it results from the assumption that "*communication* is considered merely the act of transmitting or receiving signals." Presumably, such transmitted or received signal is implicitly modulated so as to convey information. This interpretation is in line with the fact that the International Telecommunication Union (ITU) covers radiolocation (radars). We must therefore consider that a MRIWPTS using transmitter control based on telemetry is capable of communication so that it falls within the scope of the R&TTE directive if it operates at frequencies higher than 9 kHz for power transmission and if it does not meet the conditions (a), (b), and (c).

The self-tuning capability does not involve modulation conveying information so that we may consider that the self-tuning capability does not imply that a MRIWPTS is capable of communication. Of course, a WPT without self-tuning capability or transmitter control based on telemetry involves no modulation conveying information. Thus, we may conclude that a MRIWPTS which does not use transmitter control based on telemetry need not be subject to the R&TTE directive and that, if it is not capable of radio communication for another purpose than the one discussed here, it does not fall within the scope of the R&TTE directive.

7. Applicability of the Radio Equipment Directive

The radio equipment directive, which will replace the R&TTE directive on 13 June 2016 [5, Art. 50], uses different definitions. It will be applicable to *radio equipment*, defined as an electrical or electronic product, which alone or completed with an accessory such as antenna intentionally emits and/or receives radio waves for the purpose of radio communication and/or radiodetermination, where *radio communication* means communication by means of radio waves, *radiodetermination* means the determination of the position, velocity, and/or other characteristics of an object or the obtaining of information relating to those parameters, by means of the propagation properties of radio waves, and *radio waves* mean electromagnetic waves of frequencies lower than 3 000 GHz, propagated in space without artificial guide.

In the radio equipment directive, the minimum frequency of 9 kHz has disappeared from the definition of radio waves, so that a MRIWPTS uses radio waves unless it meets the conditions (a) (b) and (c) defined in Section 4.

At this point, we do not need to discuss the communication capabilities of the MRIWPTS, because the self-tuning capability and transmitter control based on telemetry are clearly intended for radio determination. Thus a MRIWPTS using the self-tuning capability or transmitter control based on telemetry falls within the scope of the radio equipment directive if it does not meet the conditions (a), (b), and (c).

We may also conclude that if a MRIWPTS uses neither a self-tuning capability nor transmitter control based on telemetry, it need not be subject to the radio equipment directive, and that, if it is not capable of radio communication for another purpose than the one discussed here, it does not fall within the scope of the radio equipment directive.

8. Conclusion

For a MRIWPTS providing a reasonable efficiency, the ratio of the shortest distance between the coils to the largest dimension of the largest coil will always be less than 4,0 in an ideal experiment and less than 2,0 in a real-world application. For a MRIWPTS operating at the 27.120 MHz frequency, a transmitted power of only 1 W, and a distance of 0.3 m between the coils, we have shown that a radiated emission limit may or may not be exceeded, depending on the applicable legislation. We have also shown that, for this MRIWPTS, the reference levels relating to the limitation of exposure of the general public may be exceeded near the coils.

We have presented an analysis of the applicability of the R&TTE and radio equipment directives. The result of this analysis can be summarized into the following simplified statements:

   (i) a MRIWPTS falls within the scope of the R&TTE directive only if it uses transmitter control based on telemetry with a range exceeding 10 cm or another form of radio communication subject to the R&TTE directive;

   (ii) a MRIWPTS will be subject to the radio equipment directive only if it uses the self-tuning capability or transmitter control based on telemetry, with a range exceeding 10 cm, or another form of radio communication subject to the radio equipment directive.

Conflict of Interests

The authors declare that there is no conflict of interests regarding the publishing of this paper.

References

[1] International Patent Application Number PCT/US2006/026480 of 5 July 2006 (WO 2007/008646), "Wireless non-radiative energy transfer," Inventors: Joannopoulos et al.

[2] A. Kurs, A. Karalis, R. Moffatt, D. J. Joannopoulos, P. Fisher, and M. Sojacic, "Wireless power transfer via strongly coupled magnetic resonnances," *Science*, vol. 317, pp. 83–86, 2007.

[3] A. Karalis, J. D. Joannopoulos, and M. Soljačić, "Efficient wireless non-radiative mid-range energy transfer," *Annals of Physics*, vol. 323, no. 1, pp. 34–48, 2008.

[4] Directive 1999/5/EC of the European Parliament and of the Council of 9 March 1999 on radio equipment and telecommunications terminal equipment and the mutual recognition of their conformity.

[5] Directive 2014/53/EU of the European Parliament and of the Council of 16 April 2014 on the harmonisation of the laws of the Member States relating to the making available on the market of radio equipment and repealing Directive 1999/5/EC.

[6] Directive 2004/108/EC of the European Parliament and of the Council of 15 December 2004 on the approximation of the laws of the Member States relating to electromagnetic compatibility and repealing Directive 89/336/EEC.

[7] Directive 2014/30/EU of the European Parliament and of the Council of 26 February 2014 on the harmonisation of the laws of the Member States relating to electromagnetic compatibility (recast).

[8] Directive 2006/95/EC of the European Parliament and of the Council of 12 December 2006 on the harmonization of the laws of Member States relating to electrical equipment designed for use within certain voltage limits.

[9] Directive 2014/35/EU of the European Parliament and of the Council of 26 February 2014 on the harmonisation of the laws of the Member States relating to the making available on the market of electrical equipment designed for use within certain voltage limits.

[10] Council Directive 93/42/EEC of 14 June 1993 concerning medical devices.

[11] Council Directive 90/385/EEC of 20 June 1990 on the approximation of the laws of the Member States relating to active implantable medical devices.

[12] R. Want, "An introduction to RFID technology," *IEEE Pervasive Computing*, vol. 5, no. 1, pp. 25–33, 2006.

[13] E. S. Hochmair, "System optimization for improved accuracy in transcutaneous signal and power transmission," *IEEE Transactions on Biomedical Engineering*, vol. 31, no. 2, pp. 177–186, 1984.

[14] G. A. Kendir, W. Liu, G. Wang et al., "An optimal design methodology for inductive power link with Class-E amplifier," *IEEE Transactions on Circuits and Systems I: Regular Papers*, vol. 52, no. 5, pp. 857–866, 2005.

[15] M. Sawan, Y. Hu, and J. Coulombe, "Wireless smart implants dedicated to multichannel monitoring and microstimulation," *IEEE Circuits and Systems Magazine*, vol. 5, no. 1, pp. 21–39, 2005.

[16] R. F. Xue, K. W. Cheng, and M. Je, "High-efficiency wireless power transfer for biomedical implants by optimal resonant

load transformation," *IEEE Transactions on Circuits and Systems I: Regular Papers*, vol. 60, no. 4, pp. 867–874, 2013.

[17] *System Description—Wireless Power Transfer—Volume I: Low Power—Part 1: Interface Definition*, Version 1.0.3, Wireless Power Consortium, 2011.

[18] C. Wang, O. H. Stielau, and G. A. Covic, "Design considerations for a contactless electric vehicle battery charger," *IEEE Transactions on Industrial Electronics*, vol. 52, no. 5, pp. 1308–1314, 2005.

[19] F. Broydé and E. Clavelier, "The emission level of medium-range inductive wireless power transmission systems," in *Proceedings of the CEM International Symposium on Electromagnetic Compatibility, Session 1B*, Rouen, France, April 2012.

[20] F. E. Terman, *Radio Engineers' Handbook*, McGraw-Hill, New York, NY, USA, 1943.

[21] J. A. Stratton, *Electromagnetic Theory*, McGraw-Hill, 1941.

[22] F. Broydé and E. Clavelier, "Maximum electric and magnetic field strengths at a given distance from some ideal antennas," *IEEE Antennas & Propagation Magazine*, vol. 50, no. 6, p. 38, 2008, and "Correction", IEEE Antennas & Propagation Magazine, vol. 50, no. 6, pp. 38, 2008.

[23] Council Recommendation 1999/519/EC of 12 July 1999 on the limitation of exposure of the genral public to electromagnetic fields (0 Hz to 300 GHz).

[24] EN 300 330-1 V1.7.1 (2010-02), Electromagnetic compatibility and Radio spectrum Matters (ERM); Short Range Devices (SRD); Radio equipment in the frequency range 9 kHz to 25 MHz and inductive loop systems in the frequency range 9 kHz to 30 MHz; Part 1: Technical characteristics and test methods.

[25] EN 300 330-2 V1.5.1 (2010-02), Electromagnetic compatibility and Radio spectrum Matters (ERM); Short Range Devices (SRD); Radio equipment in the frequency range 9 kHz to 25 MHz and inductive loop systems in the frequency range 9 kHz to 30 MHz; Part 2: Harmonized EN under article 3.2 of the R&TTE Directive.

[26] EN 55011:2009 + Amendment A1:2010. Industrial, scientific and medical (ISM) radio-frequency equipment—Radio disturbance characteristics—Limits and methods of measurement.

[27] E. M. Thomas, J. D. Heebl, and C. Pfeiffer, "A power link study of wireless non-radiative power transfer systems using resonant shielded loops," *IEEE Transactions on Circuits and Systems I: Regular Papers*, vol. 59, no. 9, pp. 2125–2136, 2012.

[28] J. van Bladel, *Electromagnetic Fields*, Hemisphere Publishing Corporation, New York, NY, USA, 1985.

[29] "Guide to the R&TTE Directive 1999/5/EC, Version of 20 April 2009," http://ec.europa.eu/enterprise/sectors/rtte/documents/index_en.htm.

[30] "Technical assistance relating to the application of the EMC Directive and of the R&TTE Directive—Specific Technical report no. 1—Emission of cochlear implants—Part B, 3rd Edition, May 13, 2005," Excem document 05032501D, http://ec.europa.eu/enterprise/sectors/rtte/documents/guidance/index_en.htm.

14

Design and Build of an Electrical Machines' High Speed Measurement System at Low Cost

Constantinos C. Kontogiannis and Athanasios N. Safacas

Department of Electrical and Computer Engineering, Patras University, 26443 Patras, Greece

Correspondence should be addressed to Constantinos C. Kontogiannis; ckontogiannis@edynamics.gr

Academic Editor: Liwen Sang

The principal objective of this paper is to demonstrate the capability of high speed measurement and acquisition equipment design and build in the laboratory at a very low cost. The presented architecture employees highly integrated market components eliminating thus the complexity of the hardware and software stack. The key element of the proposed system is a Hi-Speed USB to Serial/FIFO development module that is provided with full software and driver support for most popular operating systems. This module takes over every single task needed to get the data from the A/D to the user software gluelessly and transparently, solving this way the most difficult problem in data acquisition systems which is the fast and reliable communication with a host computer. Other ideas tested and included in this document offer Hall Effect measuring solutions using some excellent features and very low cost ICs widely available on the market today.

1. Introduction

Today, every researcher, engineer, student, or specialist professional that works with electrical machines needs to study and understand deeper the machine operation transient phenomena and how those affect the machine itself as well as the neighbouring electrical equipment. During the past two decades using data acquisition cards in combination with commercial and academic software for the machine signals sampling conversion and storage has proven to be a very efficient means for the study of the transient and steady state operation of electrical machines. Almost in every university, enterprise R&D department, or other institutions, a laboratory that studies electrical machines is equipped with data acquisition cards and software for storing, converting, and presenting the sampled signals captured during the electrical machine operation [1–3]. In order to capture machine signals, a fast electronic sampling card is necessary to interface the selected currents and voltages of the real world to a host computer via its simultaneously sampling analog to digital converters. A fast communication port such as Gigabit Ethernet or Hi-Speed Universal Serial Bus with rates up to 480 Mbit/sec is the most convenient highway to transfer real time captured data to the host computer for further processing.

Fast data transfer is crucial when concurrent sampling of several electrical signals with higher harmonic content, sometimes up to 20th order, is the case. Modern microprocessors and fast computer peripherals make it now possible to study voltage and current harmonic content for more signals at the same time and in greater harmonics order depth.

In the general case commercial products are used for the signals sampling and transfer to a Host PC, while implementation for both fast [1, 2] and slow [3] sampling systems can be found in the literature. None of them though combines both fast data rates up to 400 Mbit/sec and easy implementation at a remarkably low cost that would allow a laboratory to reproduce unlimited number of sampling systems for its engineers or students.

Electrical machines and power engineers now are quite familiar with embedded systems programming for signal measurements, processing, and control, but fast data communications with a Host PC via a fast interface such as Universal Serial Bus, involve driver level programming on the host side. This is usually a problem for non-software engineers, especially when speed optimization should be taken to the limit in order to make a full exploit of the peripheral transfer bandwidth.

In order to overcome this difficulty, engineers are obligated to purchase commercial products consisting of both the data acquisition cards and the necessary dedicated driver software that is executed on the Host PC. This driver transfers the captured data from the sampling card to the Host PC transparently with no need for any user action in this procedure. This sounds great since the job is done without any demands for further user interference. The problem is that these hardware and software sampling systems are not consumer products and therefore their cost is driven to significantly high values. Consequently, although it may be easy for a commercial company to buy some sets for its employees, it is not equally feasible for an academic laboratory to provide adequate sets of cards and software licenses for every student or researcher experimenting with electrical machine measurements.

This work proposes an alternative platform consisting of both software [4] and hardware [5] modules that can be easily built in the laboratory, able to transfer multiple signals of electrical machine measurements to the host PC via a Hi-Speed Universal Serial Bus data rate. The proposed Data Acquisition (DAQ) system involves license and royalty free components that can be found on the market at a very low budget. The manufacturer of the USB transfer module used [3] provides the USB driver and several software sample applications for most popular operating systems and high level programming languages, respectively.

2. Sampling System Architecture Overview

The Architecture of the sampling card can be divided in five discrete cascaded hardware and software layers.

The 1st layer (electrical interface) is attached to the current and voltage signals of the electrical machine converting the real measured amounts into voltages adjusted to be suitable for the CMOS and TTL level A/D converters of the next layer.

The 2nd layer (sampling and digitizing layer) is connected to the 1st layer and reads the analog voltages that correspond to the measured currents and voltages. The voltage signals are sampled and digitized by the MCU analog to digital converters and then they are pushed upstream on the 8-bit parallel FIFO of the next layer.

The 3rd layer (transfer layer) reads the bytes arriving to its 8-bit FIFO and transfers them via USB with a speed up to 400 Mbit/sec to the next layer that is a piece of software (driver) running on the Host PC.

The 4th layer (driver layer) is the software that receives the bulk—so far meaningless data—in a receive buffer on the Host PC and keeps them available for the top 5th layer to consume.

The 5th layer (application layer) pumps the 4th layer's buffers and reconstructs the measurements data into a human friendly format for further presentation and processing (Figure 1).

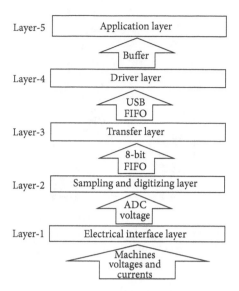

FIGURE 1: Five layers stack overview of the built electrical machines measurement hardware and software systems.

2.1. Layer 1: Electrical Interface. During transient and steady state study of an electrical machine, the main physical amounts of interest are the various currents and voltages on the machine terminals and on some types of machines the excitation field voltages and currents as well.

Terminal currents in the general case are typically measured after they are galvanically isolated and stepped down by LEM hall effect sensors that are quite expensive and occupy a great deal of space on the printed circuit board. In this work another approach is presented in current measurement by means of using a special Hall Effect sensor IC [6] integrated with a single ceramic package suitable for easy PCB mounting. The IC ACS712 is manufactured by Allegro Microsystems Inc. and is a fully integrated Hall Effect based linear current sensor with 2.1 kV rms voltage isolation and a very low resistance current contactor. The ACS712 device accuracy is optimized through the close proximity of the magnetic signal to the Hall transducer. A precise proportional voltage is provided by the low offset, chopper stabilized BiCMOS Hall IC, which is programmed for accuracy after packaging. The internal resistance of the conductive path is 1.2 mΩ typical, providing low power loss. Below are stated the main characteristics of the ACS712 AC/DC current measurement IC (Figure 2).

(i) Up to 30 A AC/DC direct measurement.

(ii) Low-noise analog signal path.

(iii) Device bandwidth is set via FILTER pin.

(iv) 50 kHz bandwidth.

(v) Total out err. 1.5% TA = 25°C, 4% at −40°C to 85°C.

(vi) Small footprint, low-profile SOIC8 package.

(vii) 1.2 mΩ internal conductor resistance.

(viii) 2.1 kV RMS minimum isolation voltage.

(ix) 5.0 V, single supply operation.

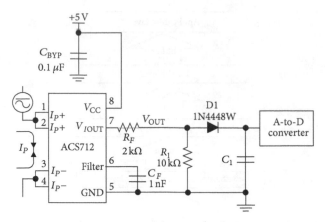

FIGURE 2: ACS712 outputs an analog signal, VOUT varying linearly with the AC or DC primary sensed current.

(x) 66 to 185 mV/A output sensitivity.

(xi) Output voltage proportional to AC or DC currents.

(xii) Factory-trimmed for accuracy.

(xiii) Extremely stable output offset voltage.

(xiv) Nearly zero magnetic hysteresis.

(xv) Ratiometric output from supply voltage.

The wide bandwidth of the ACS712 chip allows for measurements of 50 ksamples/sec [6] driven to the next layer A/D converter for sampling and digitizing.

If currents lower than 30 A AC/DC are considered for measurement then there is no need for any external part or current transformer to be used as the machine terminal current is driven directly through the ACS712 pins 1-2 and 3-4. The output is a positive voltage signal that contains an offset for zero in order to represent all current variations in the positive variation range of the output voltage.

The Quiescent output voltage is the output of the sensor when the primary current is zero. For a unipolar supply voltage, it nominally remains at $V_{CC}/2$. Thus, $V_{CC} = 5$ V translates into $V_{IOUT}(Q) = 2.5$ V. Variation in $V_{IOUT}(Q)$ can be attributed to the resolution of the Allegro linear IC quiescent voltage trim and thermal drift (Figure 3).

For the voltage measurements a fast optocoupling circuit [7] was designed to offer an interface with basic isolation for the board and the mains voltages. The optocoupler used is the Avago Technologies Inc. HCNR200 (Figure 4).

The HCNR200/201 high-linearity analog optocoupler consists of a high-performance AlGaAs LED that illuminates two closely matched photodiodes. The input photodiode can be used to monitor and, therefore, stabilize the light output of the LED. As a result, the nonlinearity and drift characteristics of the LED can be virtually eliminated. The output photodiode produces a photocurrent that is linearly related to the light output of the LED. The close matching of the photo-diodes and advanced design of the package ensure the high linearity and stable gain characteristics of the optocoupler.

The HCNR200/201 can be used to isolate analog signals in a wide variety of applications that require good stability, linearity, bandwidth, and low cost (Figure 5).

(i) Low nonlinearity: 0.01%.

(ii) Low gain temperature coefficient: −65 ppm/°C.

(iii) Wide bandwidth (DC to >1 MHz).

(iv) Worldwide approval (5 kV rms/1 min, 1414 Vpeak).

2.2. Layer 2: Sampling and Digitizing. After the current and voltage signals are collected and conditioned into appropriate level signals (0–5 VAC), the second layer is responsible for converting the analog signals to digital and to output the data bytes on an 8-bit bus to the transfer layer FIFO.

For the Sampling and Digitizing Layer the designer can use any microcontroller or DSP platform that is already used in other laboratory applications and experiments, provided there are adequate ADC channels with a resolution of at least 10 bits and that the microprocessor is fast enough to perform a sampling rate on the 8 analog channels that exceeds eight times the 50 ksamples/sec of each ACS712 channel which sets the limits for all system sampling speed.

As known, sampling is the process of converting a continuous-time signal $x(t)$ into a discrete time sequence $x[n]$. The $x[n]$ is obtained by extracting $x(t)$ every T seconds where T is the sampling period (Figure 6).

The relationship between $x(t)$ and $x[n]$ is given by (Figure 7)

$$X[n] = x(t)|_{t=nT} = x(nT), \quad n = \ldots, -1, 0, 1, 2, \ldots. \quad (1)$$

The maximum available sampling bandwidth to avoid aliasing is limited from the ACS712 maximum response frequency that is 50 kHz [6]. This corresponds to a sampling microcontroller minimum rate of 100 kHz. In practice we always perform some oversampling to avoid any case of aliasing. In the presented test system built the sampling and digitizing layer is implemented by an STM32F103C ARM-Cortex M3 microcontroller with clock speed up to 72 MHz.

The sampling rate of the CPU for each one of the eight 12 bit-ADC channels is set to 400 kHz. Since the output FIFO is 8-bit wide, every sample has to occupy 2 consequent bytes on the output pins. This means that the number of Kbyte/sec on the sampling layer output has to be 800 Kbyte/sec or 6.4 Mbit/sec, and the transfer of the Host has to be synchronized on 6.4 Mbit/sec for each channel. The total channel bandwidth on the next layer will have to be at least 51.2 Mbit/sec on the USB bus, since every byte has eight bit (Figure 8).

As layer 2 for Sampling and digitizing a development board from Olimex with an STM32F103C M3 Cortex onboard was used to collect the analog signals and output them on the 8-pin bus of the layer 3 FIFO. The layer 3 FTDI UM232H header board was soldered on the prototyping area of the STM23-106P board (Figure 9).

2.3. Layer 3: Transfer. The fast transfer of the acquired data to the host computer at speeds that exceed 50 Mbit/sec is

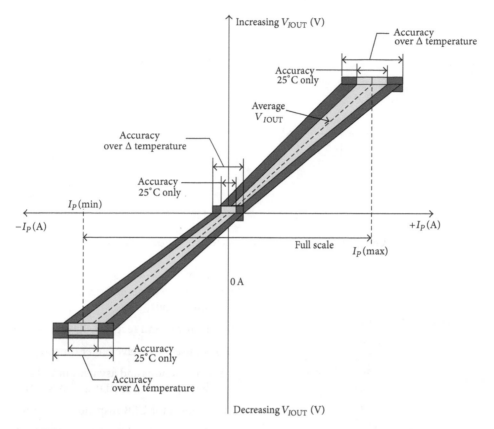

FIGURE 3: ACS712 output voltage versus sensed current. Accuracy at zero amperes and at full scale current [6].

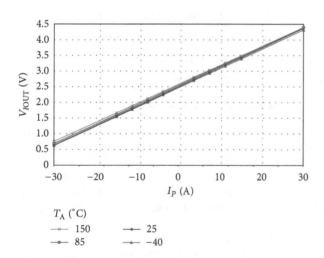

FIGURE 4: ACS712 output voltage versus sensed current for −10 A to 10 A range [6].

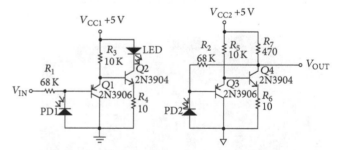

FIGURE 5: HCNR200 basic topology for voltage optoisolation. The voltage signal is prior limited to the appropriate levels by means of a preceding circuit with resistor dividers [7].

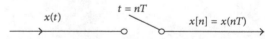

FIGURE 6: Sampling switch of analog signal and converting into discrete time sequence.

obtained thanks to UM232H, a single channel Hi-Speed USB development module by FTDI. This module utilizes the FT232H bridge IC which provides glue less and transparent transfer of data of the 8-bit parallel FIFO to the Host PC.

Below can be seen the main UM232H features [5].

(i) USB 2.0 Hi-Speed (480 Mbits/Second) and Full Speed (12 Mbits/Second) compatible.

(ii) Entire USB protocol handled on the chip—no USB-specific firmware programming required.

(iii) Small USB Type B connector common on many commercial devices.

(iv) USB bus or self-powered options.

(v) Support for USB suspend and resume.

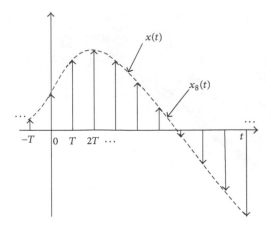
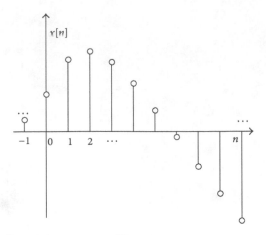

FIGURE 7: Converted analog signal into discrete time sequence diagrams.

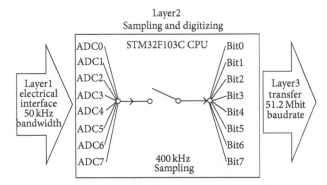

FIGURE 8: Sampling and digitizing schematic of Layer 2.

FIGURE 9: Olimex STM32-P106 with STM32F103C Prototyping Board used for sampling and digitizing.

(vi) UHCI/OHCI/EHCI host controller compatible.

(vii) FTDI's royalty-free VCP and D2XX drivers eliminate the requirement for USB driver development in most cases.

(viii) 1 kByte receive and transmit buffers for high data throughput.

(ix) Transmit and receive LED drive signals.

(x) Adjustable receive buffer timeout.

(xi) Synchronous and asynchronous bit bang mode interface options with RD# and WR# strobes.

(xii) Support for USB suspend and resume.

(xiii) Integrated 3.3 V level converter for USB I/O.

(xiv) USB bulk transfer mode.

(xv) +2.97 V to +5.25 V Single Supply Operation.

(xvi) Low operating and USB suspend current.

(xvii) Low USB bandwidth consumption.

(xviii) −40°C to + 85°C operating temperature range.

(xix) Rapid integration into existing systems.

FT232H can be configured in various modes of operation. For fast data transfer, the most convenient way is to configure the system in FT1248 mode [8]. In this mode the user CPU of layer 2 is the FT 1248 Master and the FT232H chip is the slave [4]. The protocol describes a synchronous transfer mode, where the synchronization signals are control from the Master. The transfer speed on the USB bus during this mode can reach up to 400 Mbit/sec (Figure 10).

The FT232H IC collects the data from Layer 2 in its 8-bit wide FIFO and then pumps the data automatically through the Hi-Speed USB 2.0 interface to the upper driver layer, the driver layer. The mode of operation and other configuration parameters are controlled via the driver and some special manufacturer software provided for free (Figure 11).

2.4. Layer 4: River. It is a fact that many electrical and power engineers working with electrical machines are familiar to embedded programming with microcontrollers for pulse generation, control and measurements. But when it comes to real time big data transfer to host computers, only a few non software engineers are able to complete this task. Usually

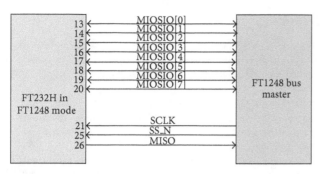

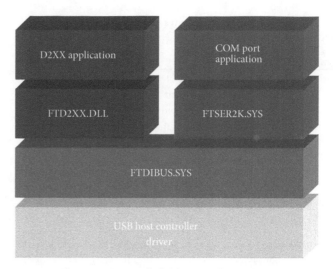

FIGURE 10: FT1248 Mode of operation master-slave connection signals [4].

FIGURE 12: FTDI Driver stack for Microsoft Windows Operating System.

FIGURE 11: UM232H Parallel FIFO to Hi-Speed USB Module with FTDI FT232H chipset onboard.

people are connected to host computers with slower interfaces such as RS-232 and other protocols only for control and monitoring signaling. This happens because in order to send data through a fast interface such as USB, an intermediate piece of software needs to be written in privileged mode that accesses hardware resources directly and collects the data to be used by higher level applications. This software chunk is the driver that accompanies every USB device one can buy on the market. To be able to write USB drivers, professional software suites are necessary and a lot of practice (Figure 12).

On the other hand everyone can develop a high level application using high level programming languages such as C++, Java, Visual Basic, Python, and others. For this reason, the FT232H is provided with its driver ready and royalty-free by the manufacturer, along with a Dynamically Linked Library (DLL) file with which users can easily connect high level applications or a third party package such as MATLAB and LabView for further processing and presentation. This scheme is valid for Microsoft Windows operating systems but similar schemes exist for other popular OS such as Linux and MAC OS.

2.5. Layer 5: Application. The top layer of the proposed stack is the application Layer 5. This can be any high level software application written by the designer in any programming language one prefers, as long as he can access the D2XX DLL library functions provided by the manufacturer. Programmers and engineers can get from the FT232H IC manufacturer royalty free software examples for most popular

high level programming languages that demonstrate how to exchange data with the FT232H using the D2XX library.

In this work, the test implementation of the 5th Layer was done in MATLAB parsing the data directly from the D2XX library file receive buffer. When parsing data for all 8-channels the storage volume needed is 6.4 Mbyte/sec on the hard disk.

3. Conclusions

This paper presents an open architecture with very low cost and simple implementation for fast electrical machines voltage and current measurements in the laboratory. The main advantages of the proposed system are the replacement of expensive and big LEM hall effect transducers for current measurements with a low cost and very efficient hall effect transducer IC, the ACS712, as well as the use of a commercial low price hardware and software kit for the fast USB data transfer to a Host PC. The use of these two alternative solutions in measurement and transfer for signal sampling and the low design cost can be considered as original contribution of this work that could help many engineers to design their own measuring and sampling systems. The researcher or engineer will have to program no software for the transfer other than the usual microprocessor ADC parser and the top application layer data processing. The product is the UM232C Hi-Speed USB 2.0 to 8-bit parallel FIFO bridge module. The above described five layers' hardware and software stack for electrical machine measurements can be easily built in the laboratory and can replace expensive data acquisition cards and software that are widely used today.

In Tables 1 and 2 a short Billing of Material (BOM) analysis shows that the system can be built for about 112 USD in retail prices if ready development kits are used and around 59 USD if a special PCB board is designed to host the FT232H and STM32F103C ICs.

TABLE 1: Proposed measuring system architecture Billing of Material (BOM) for laboratory construction.

Item	Component	Qty	Value ($)
1	ACS712 Current Sensor	4	3.80
2	HCNR200 OptoIsolator	4	2.41
3	UM232H USB toFIFO Bridge	1	32.00
4	STM32-P106 Development Board	1	44.00
5	Other Components and Materials	1	3
6	Breakout Motherboard PCB	1	8
		Total	111.84$

TABLE 2: Proposed measuring system architecture Billing of Material (BOM) for laboratory construction without ready development modules including PCB design.

Item	Component	Qty	Value ($)
1	ACS712 Current Sensor	4	3.80
2	HCNR200 OptoIsolator	4	2.41
3	FT232H USB toFIFO Bridge IC	1	6.66
4	STM32F103C microprocessor	1	5.15
5	Other Components and Materials	1	10
6	Breakout Motherboard PCB	1	12
		Total	58.65$

If the implementation includes the PCB design to host the FT232H chip and the STM32F103C microprocessor on board without need to buy the ready development modules the cost is even lower as show in Table 2.

Although some initial tests for the proposed measurements system architecture were conducted in the laboratory, a lot has yet to be done to verify the system performance [1]. Some testing done included the sinusoidal and step input signals application produced by a frequency generator with variable frequency up to about 50 kHz where the limit of the ACS712 response was met. The jitter, error, and drift were found near the theoretically expected levels while further tests are to be done in order to verify the operation of the measurement platform. The next step will be to connect the system to a laboratory pair of electrical machines mechanically connected in common shaft. The pair consisted of a DC machine functioning as the prime mover and a three-phase synchronous machine that is the generator. The target is to measure all generator's terminal voltages and currents and compare them to measurements taken by another commercial measuring platform that will be used as golden unit.

Conflict of Interests

The authors declare that there is no conflict of interests regarding the publication of this paper.

References

[1] S. Bartknecht, H. Fischer, F. Herrmann et al., "Development of a 1 GS/s high-resolution sampling ADC system," *Nuclear Instruments and Methods in Physics Research A*, vol. 623, no. 1, pp. 507–509, 2010.

[2] S. Bartknecht, H. Fischer, F. Herrmann et al., "Development and performance verification of the GANDALF high-resolution transient recorder system," *IEEE Transactions on Nuclear Science*, vol. 58, no. 4, pp. 1456–1459, 2011.

[3] H. Zhang and L. Wang, "Study on high precision data acquisition system for transformer test," in *Proceedings of the 8th International Conference on Electrical Machines and Systems (ICEMS '05)*, vol. 3, pp. 1731–1735, September 2005.

[4] Future Technology Devices International, "Establishing Synchronous 245 FIFO Communications using a Morph-IC-II," AN165/FT_000387, 2012.

[5] Future Technology Devices International, "UM232H single channel USB hi-speed FT232H development module," Datasheet/FT 000367, 2012.

[6] Allegro Microsystems, *Fully Integrated, Hall Effect-Based Linear Current Sensor with 2.1 kVRMS Voltage Isolation and a Low-Resistance Current Conductor*, ACS712 Datasheet, Allegro Microsystems, Worcester, Mass, USA, 2012.

[7] Avago Technologies, "High-linearity analog optocouplers," HCNR200 Datasheet HCNR200/201, 2008.

[8] Future Technology Devices International, "FT1248 dynamic parallel/serial interface basics," Tech. Rep. AN167/FT_000390, Future Technology Devices International, 2011.

15

High Efficiency Driver for AMOLED with Compensation

Said Saad[1] and Lotfi Hassine[2]

[1]*Group of Electronics and Quantum Physics, Laboratory of Advanced Materials and Quantum Phenomena, Faculty of Sciences of Tunis, Tunis EL Manar University, 2092 Tunis, Tunisia*
[2]*National Institute of Applied Sciences and Technology, University of Carthage, 1080 Tunis, Tunisia*

Correspondence should be addressed to Said Saad; saad.said.bechir@gmail.com

Academic Editor: Liwen Sang

A new proposed compensation driver circuit of flat-panel display (FPD) based on organic light emitting diodes (OLEDs) and on poly-crystalline silicon thin-film transistors (poly-Si TFTs) is presented. This driver circuit is developed for an active-matrix organic light-emitting-diode (AMOLED) display and its efficiency is verified compared with the conventional configuration with 2 TFTs. According to results, this circuit is suitable to achieve acceptable level for power consumption, high contrast, maximum gray levels, and better brightness. And, to show this, a stable driving scheme is developed for circuit with much compensation such as against the data degradation, the threshold voltage dispersions of TFT drive, and suppression of TFT leakage current effect.

1. Introduction

The new generation of display, organic electronic display, based on organic light emitting diodes (OLEDs) has established to eliminate the defects reported for the other technologies (LCD and PLASMA): low contrast and high consumption [1, 2] with speed conditions that may be unacceptable for some displays of 3 dimensions due to addressing type. This new technology meets the needs of users in terms of pure picture quality and functioning level especially for mobile devices; it offers new possibilities previously unattainable as the deposition on large surfaces or on flexible substrates because of low temperature of the OLED treatments [3, 4]; also the vision affects reality in terms of quality. On the other hand, the current driven of OLED device can be provided by a passive matrix or active matrix backplane architecture [5]. In the latter case the colour adjustment is determined by a command based on thin film transistor (TFT) [1, 5]. This solution is preferred, especially when the size of the display increased where we have technical problems [6]. The backplane of the active matrix is like a group of switchers or circuits which controls the current intensity flowing through each OLED pixel and does not let electricity only when this is necessary. These circuits' designs are based on amorphous silicon (a-Si)

[7], polycrystalline silicon (poly-Si) [2], organic TFT (O-TFT) [8], and circuits' designs complementariness. According to the manufacturers, several technologies backplanes in terms of structure and level of fitness, which are used for uniformity and stability sufficient for brightness which differ in their driving speed, power consumption, area occupied, and the accuracy needed to set the current level, have been presented. In particular, these driver circuits can be classified into two programming modes in accordance with the data type: voltage-programming circuit and current-programming circuit [6]. However for both types of circuits at the driving scheme, the variations in threshold voltage of TFTs, due to the change in mobility under the influence of operating time and under abnormal thermal conditions which can attain, respectively, 10% to 50%, the data degradation, and the change in supply voltage, the leakage current [9], and the speed, generate degradation and nonuniformity in brightness over time in the pixel itself and in many cases there is a fluctuation in brightness in the surrounding pixels. These disturbances add up over time and may be the cause of poor vision. To avoid these problems, the manufacturers are using these transistors with adequate compensation methods [5]. Nevertheless, high-quality displays, low power consumption, and improvement of the nonuniform brightness and the efficiency of the driver circuits require several driving transistors

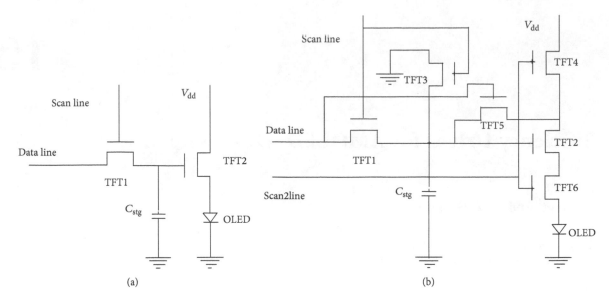

FIGURE 1: AMOLED: (a) conventional pixel circuit and (b) proposed pixel circuit.

per pixel to allow compensation for major technical problems. Specifically using compensation circuits based on poly-Si technology has been a considerable progress in providing stable and uniform brightness with great lifetime. Also it is not cost-effective in comparison with other TFT technologies (a-TFT and O-TFT) and it can provide constant current to OLED and excellent mobility; compensation circuits reached the requirements of OLED driving current under minimum power supply respecting the speed condition and also assists in the direct integration of the driver circuit on the flexible substrates [2, 5–7, 10]. In this paper we privilege the use of poly-Si transistor for the proposed driver circuit. And generally we can say that the choice of one of these transistors is closely related to the manufacture which has its own parameters and its driving schemes and the technology used to make the AMOLED screen.

2. The Proposed Driver Pixel Circuit

As we have said before, we have chosen the use of poly-Si for the proposed driver circuit. And to ensure high speed; it will be preferable to use N-type transistors rather than P-type. Also for applications of high frequency, it is better to use those of N-type instead of P-type. For these reasons we chose using an N-type transistor as a scan transistor. And for application, we use enrichment MOSFET models thanks to the performances that they give especially under abnormal thermal conditions. The proposed design is explained compared with the conventional circuit based on 2-TFTs, Figure 1(a) [6, 8, 11].

This last is established to depart from the approach of passive matrix and to better enhance the performances of pixel with one transistor, where high current, I_{OLED}, to achieve the desired brightness is required with nonuniform levels where the pixel is almost always active. It contains an embedded memory, C_{stg}, and tow transistors. TFT1 is used to select a specific pixel and to transfer the data through the data

line. The data make the loading of the storage capacitance, C_{stg}, during one period of operation when the scan line is in high state. The current is injected to the organic diode which emits light. It is adjusted by the TFT2, the driver transistor, and is expressed by [5]:

$$I_{\text{OLED}} = \frac{K_{\text{TFT2}}}{2} \times \left(V_{\text{GS-TFT2}} - V_{\text{Th-TFT2}} \right)^2, \qquad (1)$$

where K_{TFT2} is the transconductance factor of TFT2, $V_{\text{GS-TFT2}}$ is the voltage applied to the TFT2 gate-source terminal, and $V_{\text{Th-TFT2}}$ is the threshold voltage of TFT2. The simulation for the current delivered to the OLED of this conventional circuit is presented in Figure 2. From the curve, the maximum value of I_{OLED} is 3.798 μA. This last value does not represent truly the data voltage because the voltage level representing data has dropped, and this is due to the threshold voltage $V_{\text{Th-TFT1}}$ of TFT1, and the recovered voltage is $V_{\text{Data}} - V_{\text{Th-TFT1}}$. Also, this conventional configuration presents a variation in the threshold voltage of TFT2 [6] and TFT1. All these problems lead to a nonuniform brightness during the display phase and so have a direct influence on the gray levels. So, we must think about compensation methods to avoid these problems.

Firstly, to well compensate the loss in data voltage, we add another transistor of P-type: a restoration transistor TFT3, Figure 1(b). With this transistor, the charge stored in the capacity is exactly the data voltage, but the most important issue is the size of this restoration transistor. In reality, it acts as capacitance, and their limited size is directly related to the loading time of C_{stg} with TFT1 and their internal capacitance, C_{gb} (gate-bulk capacitance); this is a very important condition for calculating their capacitance. Therefore, this transistor provides a load current in capacitance C_{stg} and decreases the charging time to make it equal to the time of loading capacitance C_{stg} through TFT1, so it must reduce its internal resistance (R_{sh}: drain, source diffusion sheet resistance), resulting from an increase in their ratio $W_{\text{TFT3}}/L_{\text{TFT3}}$.

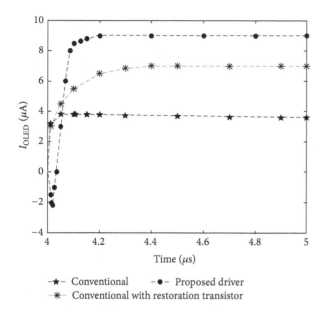

FIGURE 2: OLED current during the display phase in conventional configuration of 2 TFTs, driver circuit with restoration transistor, final model of proposed driver circuit.

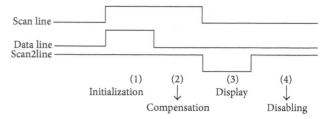

FIGURE 3: Timing scheme.

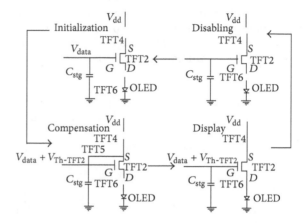

FIGURE 4: Equivalent circuits in each step of operation of the proposed design.

Subsequently we have two conditions that must be respected: we increase W/L to reduce the resistance and we increase $W \times L$ to increase the internal capacitance, C_{gb}, of this transistor. The optimal values that allow us to offset the data degradation are threshold voltage $V_{Th\text{-}TFT3} = -V_{Th\text{-}TFT1}$ and $W = 4 \times L$. In this condition, the maximum current delivered to the OLED becomes $7\,\mu A$; see Figure 2, and therefore we have successfully corrected the data degradation and increase the current I_{OLED}. In addition, despite the compensation of the data degradation, we have always the problem of threshold voltage variation of TFT2 with increased energy consumption due to the presence of anther capacitance which is represented by the transistor TFT3. So we seek a method to resolve these problems. First, to minimize power consumption, the OLED is active only if necessary. For this a new transistor of P-type, TFT4, ordered by a new control line, Scan2line, is added as shown Figure 1(b); this transistor does not leave the driver circuit works only if that is necessary even if the V_{Data} is already stored in the capacitance C_{stg}. Second, to avoid the problem of the threshold voltage variation of TFT2, we add another transistor of P-type, TFT5, Figure 1(b). The simulation gives an increase in the maximum current to $9\,\mu A$; see Figure 2. This increase of $2\,\mu A$ is due in fact to the storage of the threshold voltage $V_{Th\text{-}TFT2}$ of TFT2 in the capacitance C_{stg}. Moreover the presence of leakage current in the OLED during the reset phase requires us to add another transistor to eliminate it, because this current becomes critical for the lifetime of OLED so in screen over time where the screen colours will become darker if we exceed certain hours of operation and has a direct effect on the contrast and on the stored charge in C_{stg} [9]. So it must be reducing leakage current. To do this, we add another transistor, P-type TFT6, to block the emission current through the OLED during the reset phase

and ensure greater stability for the pixel, Figure 1(b). Therefore, the final proposed driver circuit is able to reduce the problem of nonuniformity of brightness, it has a large output current with faster response time, it can block the current circulating in the OLED during the reset period where an increase in contrast ratio is guaranteed, and it reduces energy consumption. Also, the OLED is placed between the drain of TFT6 and the ground to cancel the transient characteristics. So we can say that the proposed circuit pixel is suitable for AMOLED screen, but it requires more precision by the insertion of the new control signal, Scan2line.

3. Driving Scheme

In the proposed design exposed in Figure 1(b) TFT1, TFT4, TFT5, and TFT6 function as switches, TFT3 is the restoration transistor, and TFT2 is the driving transistor; it operates in saturation regime when it is in the passant state. The timing or the driving scheme for this proposed driver is defined as shown in Figure 3. Figure 4 shows the operations steps with the main compensations. From these figures we find the following.

Initialization Phase of Data. It is the activation phase. During it, the signals scan line and data line are in the high state; the TFT1 and TFT3 are passers. The capacitance, C_{stg}, loads up a specific value which is $Q = C \cdot V_{Data}$. But it must pay attention to the threshold voltage of TFT1, $V_{Th\text{-}TFT1}$, because the output of this transistor is also $V_{Data} - V_{Th\text{-}TFT1}$, so the presence of TFT3 is verified, because it enables making the correction for the loss of V_{Data}, so during this phase the capacitance C_{stg} loads to V_{Data} as shown the relationship

$$V_{\text{stg}} = V_{\text{Data}} - V_{\text{Th-TFT1}} + |V_{\text{Th-TFT3}}| . \qquad (2)$$

Compensation Phase of the Threshold Voltage of TFT2. During this phase the signal Scan2line is in the high state, TFT4 and TFT6 are blocked, and data line is in the low state to make the loading of the threshold voltage of TFT2 in the capacitance C_{stg}. At this time the gate of TFT2 is connected with their drain through the transistor TFT5, where we have a diode type connection, and so the voltage across the capacity C_{stg} becomes $V_{\text{Data}} + V_{\text{Th-TFT2}}$. This technique allows cancelling the threshold voltage variation when the TFT2 commands OLED.

Display Phase. After the scan time of pixel and the step of initialization and compensation for the data voltages and for the $V_{\text{Th-TFT2}}$, the signal Scan2line becomes in the low state. During this period, the TFT2 provided to the OLED the current:

$$\begin{aligned} I_{\text{OLED}} &= \frac{K}{2} \times \left(V_{\text{GS-TFT2}} - V_{\text{Th-TFT2}}\right)^2 \\ &= \frac{K}{2} \times \left(V_{\text{Data}} + V_{\text{Th-TFT2}} - V_{\text{Th-TFT2}}\right)^2 = \frac{K}{2} V_{\text{Data}}^2 . \end{aligned} \qquad (3)$$

As indicated in this expression, the drain current of TFT2 is independent of the threshold voltage of TFT2 and is only affected by V_{Data}; therefore uniform brightness of the image can be defined according to the desired gray levels.

Disabling Phase. This is the nonoperation phase of the pixel. It is applied if we want to avoid the operation of the pixel. It is obtained by forcing the signal Scan2line to be in a high state, even in the case where we have a charge stored in the capacitance C_{stg}. This charge remains constant until the next reset phase.

4. Performances and Discussion

In the proposed design of the driver circuit, the OLED current depends only on data. Also there is an increase in the contrast ratio by inserting the transistor TFT6 as the OLED is disconnected from another part of the circuit during the initialization and compensation phase, so it does not emit light during the addressing phase, and therefore a perfect black colour is displayed. In the other hand, the circuit is thermally stressed and its operating temperature increases over time. And to assess how this circuit operates under abnormal thermal conditions, we also perform the simulations at different temperatures. The results are shown in Figure 5 at 27 and 100°C. As a result there is a slight variation in I_{OLED} under thermal conditions due to the changes in mobility in the driving transistor TFT2. In addition, a reduction in energy consumption is accomplished by inserting the transistor TFT4. And generally, the driver circuit behaves as a multitude of capacities that need to load and unload. In general, we have the two energy contributions: static and dynamic. However,

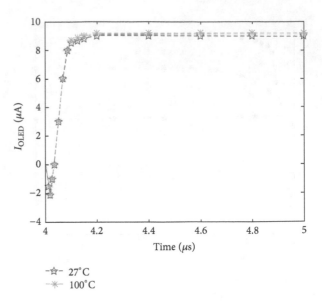

FIGURE 5: Influence of abnormal thermal condition on I_{OLED}.

the static consumption is almost zero, so we need to only consider the dynamic power consumption, given by the following formula [6]:

$$P_{\text{dynamique}} = \frac{1}{2} \times \alpha_{0 \to 1} \times C_L \times V_{\text{dd}}^2 \times f, \qquad (4)$$

where C_L is the equivalent capacitance of loading and unloading in the driver circuit, V_{dd} is the supply voltage, and f is the operating frequency of the driver circuit: position by Scan2line. The $\alpha_{0 \to 1}$ is the probability of having consumption during one clock cycle, this parameter is determined according to the driving scheme used, and its value is between 0 and 1. This expression of power explains the continuing efforts of the circuit designers to reduce the size of transistors, which reduces the value of parasitic capacities, and reduce the supply voltage, in order to increase the working frequency of circuits. For the proposed pixel circuit, we take the values $\alpha = 0.33$ and $C_L = 1$ pF. Hence $P_{\text{dynamique}} = 2.0625 \cdot 10^{-6}$ W. If we calculate this consumption for a screen with the resolution 176×220, we obtain $P = 79.68$ mW. And when a comparison is made between AMOLED screen of 2.2 inch with resolution 176×220 that is presented in [1] and AMOLED screen with the same resolution using the proposed circuit, we note that this consumption has decreased by about 60%. Furthermore, we must note that this consumption is calculated without resistive and capacitive couplings for a total screen, so to make clear consumption we must take into account these capacitances. We must also take into account the overlap capacitances of TFTs (C_{GS}, C_{GD}, and C_{GB}). Also these capacitances have a direct influence on the loading of the data voltage (C_{Dataline}) in the capacitance C_{stg} and on the speed of circuit (C_{Scanline} and R_{Scanline}) for AMOLED screen. But generally these values depend on the manufacturing process of TFTs, and we can neglect them. Moreover, when we talk about gray levels, the parameter which directly affects these levels is the size of the capacitance C_{stg}, because with a very precise adjustment we can oblige the capacitance not to make an error that exceeds the voltage necessary to pass from one gray level to another level.

5. Conclusion

In this work, a new compensation driver circuit based on the technology TFT poly-Si as a support of the organic matrix thanks to its speed and its thermal stability is proposed with its driving scheme, consisting of 6 transistors TFTs and having three input lines (data line and scan line plus one additional control signal: Scan2line) plus one storage capacitance. By results of circuit tests, it is found that the proposed circuit can be successfully operated under optimal timing scheme; it offers a stable output current of high value while keeping a response time relatively fast compatible with the requirements of AMOLED displays under minimal power consumption and abnormal thermal conditions. We have shown for this proposed circuit a driving scheme based on restoration technique for the data degradation and compensation of the threshold voltage variation of TFT2 driver by using the method of loading this voltage in the capacitance C_{stg}. On the other hand, to reduce energy consumption, increase the contrast ratio, and suppress TFT leakage effect, we have introduced two other transistors with additional signal line to prevent the leakage current and to make this pixel operate as needed. This proposed circuit is compared with the conventional circuit of 2 TFTs, and we can simply say that it is very suitable for contrast, consumption, speed, and stability of brightness.

Conflict of Interests

The authors declare that there is no conflict of interests regarding the publication of this paper.

Acknowledgments

This work was supported by Laboratory of Advanced Materials and Quantum Phenomena, Faculty of Sciences of Tunis, Tunis EL Manar University and funded by the Ministry of Higher Education and Scientific Research, Tunisia.

References

[1] J. Y. Lee, J. H. Kwon, and H. K. Chung, "High efficiency and low power consumption in active matrix organic light emitting diodes," *Organic Electronics*, vol. 4, no. 2-3, pp. 143–148, 2003.

[2] K. Park, J.-H. Jeon, Y. Kim et al., "A poly-Si AMOLED display with high uniformity," *Solid-State Electronics*, vol. 52, no. 11, pp. 1691–1693, 2008.

[3] C.-C. Wu, C.-W. Chen, C.-L. Lin, and C.-J. Yang, "Advanced organic light-emitting devices for enhancing display performances," *IEEE/OSA Journal of Display Technology*, vol. 1, no. 2, pp. 248–266, 2005.

[4] G.-F. Wang, X.-M. Tao, and R.-X. Wang, "Fabrication and characterization of OLEDs using PEDOT:PSS and MWCNT nanocomposites," *Composites Science and Technology*, vol. 68, no. 14, pp. 2837–2841, 2008.

[5] B.-T. Chen, Y.-H. Tai, Y.-J. Kuo, C.-C. Tsai, and H.-C. Cheng, "New pixel circuits for driving active matrix organic light emitting diodes," *Solid-State Electronics*, vol. 50, no. 2, pp. 272–275, 2006.

[6] G. Palumbo and M. Pennisi, "AMOLED pixel driver circuits based on poly-Si TFTs: a comparison," *Integration, the VLSI Journal*, vol. 41, no. 3, pp. 439–446, 2008.

[7] M. H. Kang, J. H. Hur, Y. D. Nam, E. H. Lee, S. H. Kim, and J. Jang, "An optical feedback compensation circuit with a-Si:H thin-film transistors for active matrix organic light emitting diodes," *Journal of Non-Crystalline Solids*, vol. 354, no. 19-25, pp. 2523–2528, 2008.

[8] W. F. Aerts, S. Verlaak, and P. Heremans, "Design of an organic pixel addressing circuit for an active-matrix OLED display," *IEEE Transactions on Electron Devices*, vol. 49, no. 12, pp. 2124–2130, 2002.

[9] J.-H. Lee, H.-S. Park, J.-H. Jeon, and M.-K. Han, "Suppression of TFT leakage current effect on active matrix displays by employing a new circular switch," *Solid-State Electronics*, vol. 52, no. 3, pp. 467–472, 2008.

[10] Y.-J. Yun, B.-G. Jun, Y.-K. Kim, J.-W. Lee, and Y.-M. Lee, "Design of system-on-glass for poly-Si TFT OLEDs using mixed-signals simulation," *Displays*, vol. 30, no. 1, pp. 17–22, 2009.

[11] J. O. Lee, H.-H. Yang, W. W. Jang, and J.-B. Yoon, "A new method of driving an AMOLED with MEMS switches," in *Proceedings of the IEEE 21st International Conference on Micro Electro Mechanical Systems (MEMS '08)*, pp. 132–135, Tucson, Ariz, USA, January 2008.

16

Design of CDTA and VDTA Based Frequency Agile Filters

Neeta Pandey,[1] **Aseem Sayal,**[2] **Richa Choudhary,**[2] **and Rajeshwari Pandey**[1]

[1]*Department of Electronics and Communication Engineering, Delhi Technological University, Delhi 110042, India*
[2]*Department of Electrical Engineering, Delhi Technological University, Delhi 110042, India*

Correspondence should be addressed to Neeta Pandey; n66pandey@rediffmail.com

Academic Editor: Weisheng Zhao

This paper presents frequency agile filters based on current difference transconductance amplifier (CDTA) and voltage difference transconductance amplifier (VDTA). The proposed agile filter configurations employ grounded passive components and hence are suitable for integration. Extensive SPICE simulations using 0.25 μm TSMC CMOS technology model parameters are carried out for functional verification. The proposed configurations are compared in terms of performance parameters such as power dissipation, signal to noise ratio (SNR), and maximum output noise voltage.

1. Introduction

The rapid evolution of wireless services has led to demand for one-fits-all "analog" front end solution. These services use different standards and therefore necessitate development of integrated multistandard transceivers as they result in reduction of size, price, complexity, and power consumption. The parameters of integrated transceiver can be modified in order to be able to adapt to the specifications of each standard [1]. Practically, the designs employ either elements handling various standards in parallel or reconfigurable elements. The frequency agile filter (FAF) [1–10] characterized by adjustment range, reconfigurability, and agility may be used in transceivers. The term shadow filters is sometimes used in literature to refer to FAF [11, 12]. The literature survey shows that a limited number of topologies of active FAF are available and are based on op-amp [1] and current mode active block [2, 3] and CMOS [4].

There is a wide range of current mode building blocks available in open literature. Among these blocks current difference transconductance amplifier (CDTA) [11] is most suitable for current mode signal processing owing to its low input and high output impedances, respectively. The VDTA is yet another recently introduced building block which works on a principle similar to that of CDTA except that the input current differencing unit is replaced by the voltage

differencing circuit. Many applications such as filters and oscillators based on CDTA and VDTA are available and have been reported in the literature [13–27] and references cited therein.

The main intention of this paper is to present CDTA and VDTA based frequency agile filter topologies. The proposed filters are suitable for integration as these employ grounded capacitors and a resistor. The paper is organised as follows. The FAF implementation scheme is briefly reviewed in Section 2. The CDTA based Class 0, Class 1, and Class 2 FAF are presented in Section 3. Section 4 deals with the realization of VDTA based Class 0, Class 1, and Class 2 FAF. In Section 5, nonideal analysis of filters is presented. Simulation results are provided in Section 6 to substantiate the proposed FAF topologies. The performance characteristics of filter topologies are described in Section 7. The paper is concluded in Section 8.

2. Implementation Scheme of FAF

The implementation scheme of frequency agile filter (FAF) [3] is briefly reviewed in this section.

2.1. Class 0 FAF. A classical second order filter with band pass (I_{BP}) and low pass (I_{LP}) outputs of Figure 1 is designated as

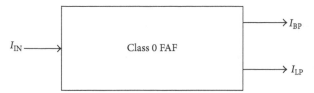

FIGURE 1: Class 0 FAF [3].

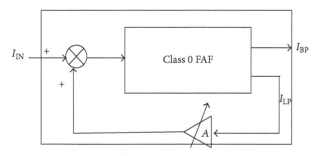

FIGURE 2: Class 1 FAF.

Class 0 FAF [3]. The transfer functions of Class 0 FAF are given by

$$T_{BP}(s) = \frac{I_{BP}}{I_{IN}} = \frac{ks}{1 + \alpha s + \beta s^2},$$

$$T_{LP}(s) = \frac{I_{LP}}{I_{IN}} = \frac{p}{1 + \alpha s + \beta s^2}. \tag{1}$$

The center frequency (f_0) and quality factor (Q) of the filter are represented by (2) and (3), respectively:

$$f_0 = \frac{1}{2\pi\sqrt{\beta}}, \tag{2}$$

$$Q = \frac{\sqrt{\beta}}{\alpha}. \tag{3}$$

2.2. Class 1 FAF. The basic block diagram of Class 1 FAF is shown in Figure 2 wherein the low pass output of the Class 0 FAF is amplified (with variable gain A) and fed back to the input. The characteristic frequency (f_{0A}) and quality factor (Q_A) of Class 1 FAF are given by (4) and (5), respectively:

$$f_{0A} = f_0\sqrt{(1 + Ap)}, \tag{4}$$

$$Q_A = Q\sqrt{(1 + Ap)}. \tag{5}$$

2.3. Class n FAF. The method outlined for Class 1 FAF realization can be extended for Class n FAF implementation as shown in Figure 3. This requires n amplifiers each with gain A ($A_1 = A_2 = \cdots = A_{n-1} = A_n$) to be placed in n feedback paths obtained in the same way as done in Class 1 implementation. The characteristic parameters of Class n FAF are given by

$$f_{0An} = f_0(1 + Ap)^{n/2},$$

$$Q_{An} = Q(1 + Ap)^{n/2}. \tag{6}$$

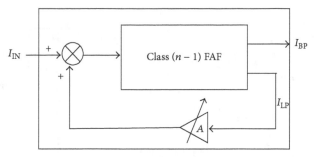

FIGURE 3: Class n FAF.

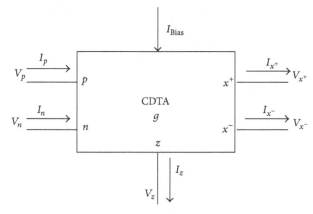

FIGURE 4: Symbol of CDTA.

3. CDTA Based FAF

The CDTA [11–18] consists of a unity-gain current source controlled by the difference of the input currents and a transconductance amplifier providing electronic tunability through its transconductance gain. The CDTA symbol is shown in Figure 4 and its terminal characteristics in matrix form are given by

$$\begin{bmatrix} V_p \\ V_n \\ I_z \\ I_{x+} \\ I_{x-} \end{bmatrix} = \begin{bmatrix} 0 & 0 & 0 & 0 & 0 \\ 0 & 0 & 0 & 0 & 0 \\ 1 & -1 & 0 & 0 & 0 \\ 0 & 0 & g & 0 & 0 \\ 0 & 0 & g & 0 & 0 \end{bmatrix} \begin{bmatrix} I_p \\ I_n \\ V_z \\ V_{x+} \\ V_{x-} \end{bmatrix}, \tag{7}$$

where g is transconductance of the CDTA. The CMOS implementation of CDTA [16] is given in Figure 5. The transistor network comprising transistors Mc1–Mc17 performs [16] current differencing operation on the currents entering at p and n nodes, which is available at Z terminal. The voltage of z terminal drives the source coupled pair (transistors (Mc18–Mc21)) [16] of differential amplifier (Mc18–Mc26) giving a transconductance of g. The value of transconductance (g) is expressed as

$$g = \sqrt{2\mu C_{ox}\left(\frac{W}{L}\right)_{19,21} I_{Bias}}, \tag{8}$$

which can be adjusted by bias current I_{Bias} of CDTA.

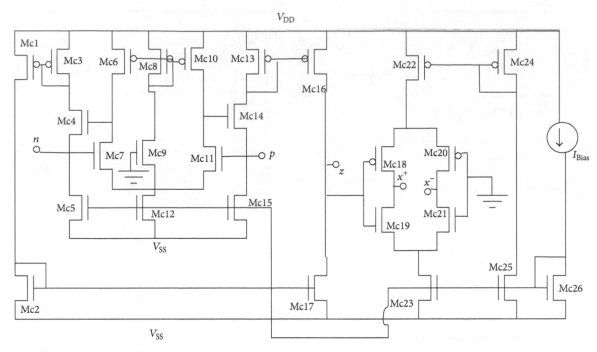

FIGURE 5: CMOS implementation of CDTA [16].

3.1. CDTA Based Class 0 FAF. The CDTA based second order filter employing two CDTA blocks and two grounded capacitors is shown in Figure 6. The second CDTA block uses additional TA block with its current output terminals denoted by x_c^+ and x_c^-. It provides both low pass and band pass responses at high output impedance and can be used as Class 0 FAF. The current flowing through x^+ and x^- is controlled through transconductance g_2 whereas current flowing through terminals x_c^+ and x_c^- is controlled through g_3. The low pass and band pass transfer functions of CDTA based Class 0 FAF are given by (9) and (10), respectively:

$$\frac{I_{LP}}{I_{IN}} = \frac{g_1 g_2}{C_1 C_2 s^2 + sC_1 g_3 + g_1 g_2}, \tag{9}$$

$$\frac{I_{BP}}{I_{IN}} = \frac{sC_2 g_1}{C_1 C_2 s^2 + sC_1 g_3 + g_1 g_2}. \tag{10}$$

The center frequency and quality factor of Class 0 FAF are expressed as

$$f_0 = \frac{1}{2\pi} \sqrt{\frac{g_1 g_2}{C_1 C_2}}, \tag{11}$$

$$Q = \frac{1}{g_3} \sqrt{\frac{g_1 g_2 C_2}{C_1}}. \tag{12}$$

It may be noted that the Q of the Class 0 FAF can be controlled independent of f_0 by varying g_3.

3.2. CDTA Based Class 1 FAF. The CDTA based Class 1 FAF is shown in Figure 7. It employs Class 0 FAF of Figure 6 along with an additional TA block (provides an output current

which is product of its transconductance and voltage difference between noninverting (+) and inverting (−) terminals) and one grounded resistor. The TA block in the feedback path functions as an amplifier with tunable gain A as given in

$$A = g_4 R, \tag{13}$$

where g_4 is the transconductance of TA block and is given by $\sqrt{2\mu C_{ox}(W/L)_{19,21} I_{Bias4}}$.

The low pass and band pass transfer functions of CDTA based Class 1 FAF are given by (14) and (15), respectively:

$$\frac{I_{LP}}{I_{IN}} = \frac{g_1 g_2}{C_1 C_2 s^2 + sC_1 g_3 + g_1 g_2 (1 + Rg_4)}, \tag{14}$$

$$\frac{I_{BP}}{I_{IN}} = \frac{sg_1 C_2}{C_1 C_2 s^2 + sC_1 g_3 + g_1 g_2 (1 + Rg_4)}. \tag{15}$$

The center frequency and quality factor of the CDTA based Class 1 FAF are expressed by (16) and (17), respectively:

$$f_{0A} = \frac{1}{2\pi} \sqrt{\frac{g_1 g_2}{C_1 C_2}} \sqrt{(1 + g_4 R)}, \tag{16}$$

$$Q_A = \frac{1}{g_3} \sqrt{\frac{g_1 g_2 C_2}{C_1}} \sqrt{(1 + g_4 R)}. \tag{17}$$

3.3. CDTA Based Class 2 FAF. The CDTA based Class 2 FAF is shown in Figure 8. It employs two CDTA blocks, two grounded capacitors, three TA blocks, and two grounded resistors. The TA blocks in the feedback path are used as amplifier with tunable gain A. The gain A of TA based

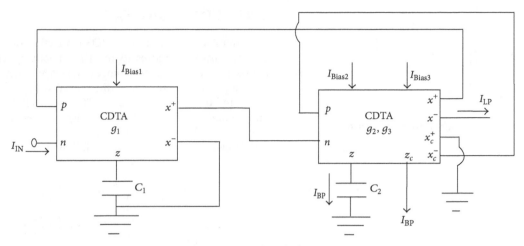

FIGURE 6: CDTA based Class 0 FAF.

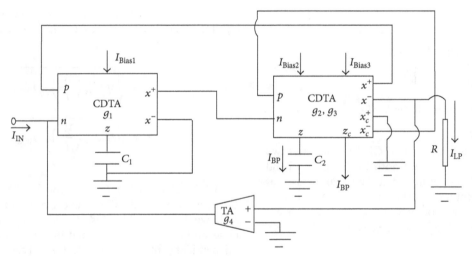

FIGURE 7: CDTA based Class 1 FAF.

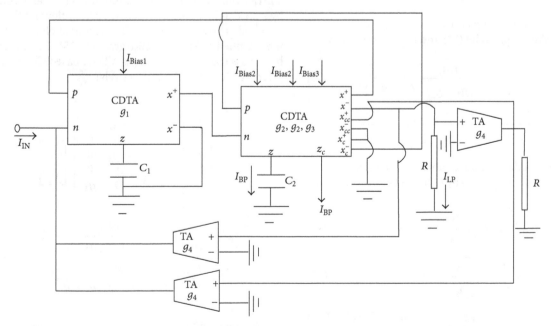

FIGURE 8: CDTA based Class 2 FAF.

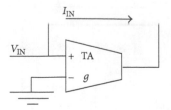

FIGURE 9: TA realization of a grounded resistor.

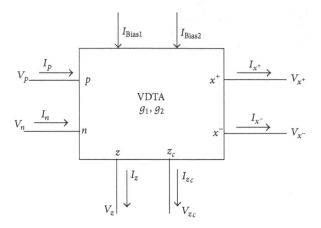

FIGURE 10: Symbol of VDTA.

amplifier is given by (18). The second CDTA block uses two additional TA blocks. It provides both low pass and band pass responses at high output impedance and can be used as Class 0 FAF. The current flowing through x^+, x^-, x_{cc}^+, and x_{cc}^- is controlled through transconductance g_2 whereas current flowing through terminals x_c^+ and x_c^- is controlled through g_3:

$$A = g_4 R. \tag{18}$$

The low pass and band pass transfer functions of CDTA based Class 1 FAF are given by (19) and (20), respectively:

$$\frac{I_{LP}}{I_{IN}} = \frac{g_1 g_2}{C_1 C_2 s^2 + s C_1 g_3 + g_1 g_2 \left(1 + R g_4\right)^2}, \tag{19}$$

$$\frac{I_{BP}}{I_{IN}} = \frac{s g_1 C_2}{C_1 C_2 s^2 + s C_1 g_3 + g_1 g_2 \left(1 + R g_4\right)^2}. \tag{20}$$

The center frequency and quality factor of the CDTA based Class 1 FAF are expressed by (21) and (22), respectively:

$$f_{0A} = \frac{1}{2\pi} \sqrt{\frac{g_1 g_2}{C_1 C_2}} \left(1 + g_4 R\right), \tag{21}$$

$$Q_A = \frac{1}{g_3} \sqrt{\frac{g_1 g_2 C_2}{C_1}} \left(1 + g_4 R\right). \tag{22}$$

The proposed filter uses grounded resistor of value R (= $1/g$) which can easily be implemented using the TA based structure given in Figure 9.

4. The VDTA Based FAF

The circuit symbol and the CMOS realization of VDTA [20, 21] are shown in Figures 10 and 11, respectively. The VDTA consists of two transconductance (TC) stages termed as input and output stages. The input differential voltage ($V_p - V_n$) is converted to current I_z through TC gain (g_1) of input stage and second stage converts the voltage at z terminal (V_z) to current (I_x) through its TC gain (g_2). The port relations of VDTA can thus be defined by the following matrix:

$$\begin{bmatrix} I_z \\ I_{z_c} \\ I_{x^+} \\ I_{x^-} \end{bmatrix} = \begin{bmatrix} g_1 & -g_1 & 0 \\ -g_1 & g_1 & 0 \\ 0 & 0 & g_2 \\ 0 & 0 & -g_2 \end{bmatrix} \begin{bmatrix} V_p \\ V_n \\ V_z \end{bmatrix}. \tag{23}$$

The TC g_1 and TC g_2 are expressed by (24) which can be adjusted by bias currents I_{Bias1} and I_{Bias2}, respectively:

$$g_1 = \sqrt{2\mu C_{ox} \left(\frac{W}{L}\right)_{1,2} I_{Bias1}},$$

$$g_2 = \sqrt{2\mu C_{ox} \left(\frac{W}{L}\right)_{5,6} I_{Bias2}}. \tag{24}$$

4.1. VDTA Based Class 0 FAF. The VDTA based Class 0 FAF employing single VDTA and two grounded capacitors is shown in Figure 12. This circuit configuration is based on second order filter presented in [21]. However, to allow independent control of quality factor and center frequency an additional TA block with transconductance g_2 is included in VDTA. The current flowing through z terminal is controlled by transconductance g_1 whereas current flowing through z_c terminal is controlled by g_2. The terminal characteristics of the modified VDTA block are given by (25). The low pass and band pass transfer functions of VDTA based Class 0 FAF are given by (26) and (27), respectively:

$$\begin{bmatrix} I_z \\ I_{z_f} \\ I_z^f \\ I_{z_c}^f \\ I_{x^+} \\ I_{x^-} \end{bmatrix} = \begin{bmatrix} g_1 & -g_1 & 0 \\ -g_1 & g_1 & 0 \\ g_2 & -g_2 & 0 \\ -g_2 & g_2 & 0 \\ 0 & 0 & g_3 \\ 0 & 0 & -g_3 \end{bmatrix} \begin{bmatrix} V_p \\ V_n \\ V_z \end{bmatrix}, \tag{25}$$

$$\frac{I_{LP}}{I_{IN}} = \frac{g_1 g_3}{C_1 C_2 s^2 + s C_2 g_2 + g_1 g_3}, \tag{26}$$

$$\frac{I_{BP}}{I_{IN}} = \frac{s C_2 g_1}{C_1 C_2 s^2 + s C_2 g_2 + g_1 g_3}. \tag{27}$$

The center frequency and quality factor of Class 0 FAF are expressed by (28). The center frequency can be controlled by

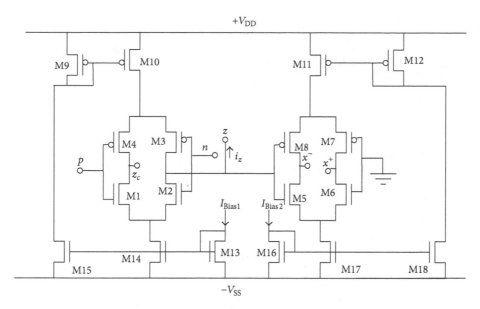

FIGURE 11: CMOS implementation of VDTA [21].

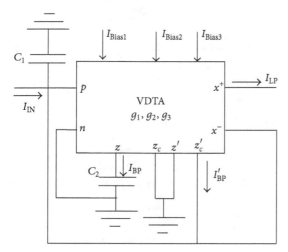

FIGURE 12: The VDTA based Class 0 FAF.

I_{Bias1} and I_{Bias3} whereas quality factor can be independently controlled by I_{Bias2}:

$$f_0 = \frac{1}{2\pi}\sqrt{\frac{g_1 g_3}{C_1 C_2}},$$

$$Q = \frac{1}{g_2}\sqrt{\frac{g_1 g_3 C_1}{C_2}}. \tag{28}$$

4.2. VDTA Based Class 1 FAF. The VDTA based Class 1 FAF is shown in Figure 13. It employs two VDTA blocks and two grounded capacitors. The second VDTA block is used as amplifier with tunable gain A. The gain A of VDTA based amplifier is given by

$$A = \frac{g_4}{g_3} \tag{29}$$

and can be adjusted by varying I_{Bias3} and I_{Bias4}.

The low pass and band pass transfer functions of VDTA based Class 1 FAF are given by (30) and (31), respectively:

$$\frac{I_{\text{LP}}}{I_{\text{IN}}} = \frac{g_1 g_3}{C_1 C_2 s^2 + sC_2 g_2 + g_1 g_3\left(1 + \left(g_4/g_3\right)\right)}, \tag{30}$$

$$\frac{I_{\text{BP}}}{I_{\text{IN}}} = \frac{sg_1 C_2}{C_1 C_2 s^2 + sC_2 g_2 + g_1 g_3\left(1 + \left(g_4/g_3\right)\right)}. \tag{31}$$

The center frequency and quality factor of the CDTA based Class 1 FAF are expressed by (32) and (33), respectively. The center frequency can be independently controlled by varying I_{Bias2} without changing center frequency:

$$f_{0A} = \frac{1}{2\pi}\sqrt{\frac{g_1 g_3}{C_1 C_2}}\left(\sqrt{1 + \frac{g_4}{g_3}}\right), \tag{32}$$

$$Q_A = \frac{1}{g_2}\sqrt{\frac{g_1 g_3 C_1}{C_2}}\left(\sqrt{1 + \frac{g_4}{g_3}}\right). \tag{33}$$

4.3. VDTA Based Class 2 FAF. The VDTA based Class 2 FAF is shown in Figure 14 which employs three VDTAs, two grounded capacitors and one grounded resistor. The second VDTA block is used as amplifier with tunable gain A. The gain A of VDTA based amplifier is given by (34). The proposed filter uses grounded resistor which can easily be implemented using the TA with transconductance equal to g_3 based structure given in Figure 9. To realize second order filter, I_{Bias7} is set to value of I_{Bias4} such that g_7 is equal to g_4 and I_{Bias6} is set to value such that g_6 is equal to sum of g_3 and g_4; that is, $g_6 = g_3 + g_4$. Consider

$$A = \frac{g_4}{g_3}, \tag{34}$$

which can be adjusted by varying I_{Bias3} and I_{Bias2}, thereby making f_{0A} tunable.

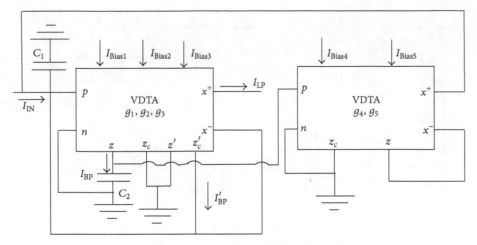

FIGURE 13: VDTA based Class 1 FAF.

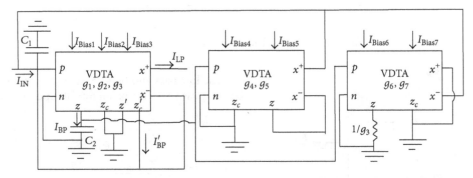

FIGURE 14: VDTA based Class 2 FAF.

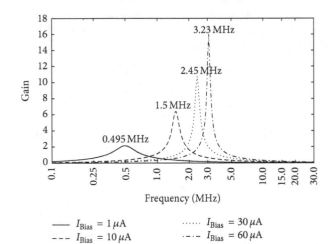

FIGURE 15: Frequency response of CDTA based Class 0 FAF with $I_{Bias1} = I_{Bias2} = I_{Bias}$.

The low pass and band pass transfer functions of CDTA based Class 1 FAF are given by (35) and (36), respectively:

$$\frac{I_{LP}}{I_{IN}} = \frac{g_1 g_3}{C_1 C_2 s^2 + s C_2 g_2 + g_1 g_3 \left(1 + (g_4/g_3)\right)^2}, \quad (35)$$

$$\frac{I_{BP}}{I_{IN}} = \frac{s g_1 C_2}{C_1 C_2 s^2 + s C_2 g_2 + g_1 g_3 \left(1 + (g_4/g_3)\right)^2}. \quad (36)$$

The center frequency and quality factor of the CDTA based Class 1 FAF are expressed by (37) and (38), respectively:

$$f_{0A} = \frac{1}{2\pi} \sqrt{\frac{g_1 g_3}{C_1 C_2}} \left(1 + \frac{g_4}{g_3}\right), \quad (37)$$

$$Q_A = \frac{1}{g_2} \sqrt{\frac{g_1 g_3 C_1}{C_2}} \left(1 + \frac{g_4}{g_3}\right). \quad (38)$$

5. Nonideal Analysis

In this section, nonideal analysis of CDTA and VDTA based Class 0 FAF is presented.

5.1. Nonideal Analysis of Class 0 CDTA Based FAF. In practice, the transfer functions (9) and (10) modify due to nonidealities which are classified as tracking errors and parasites. The tracking errors cause current transfer from p and n ports to z port to differ from unity value and are represented by α_p and α_n. There is deviation in transconductance transfer from z to x^+ and x^- ports which is modeled by $\beta g V_z$. The parasites denoted by resistances R_p and R_n are at p and n terminals; shunt output impedances ($R//C$) are present at terminals z, z_c, x^+ and x^-, and x_c^+ and x_c^-. The effect of the parasites is highly dependent on the topology. A close inspection of the circuit of Figure 6 shows that the parasitic capacitances

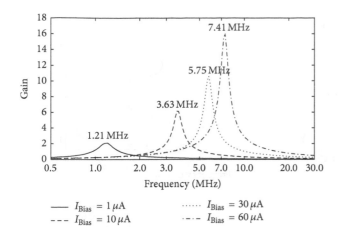

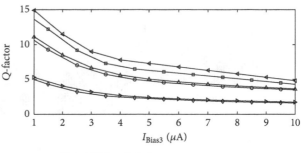

FIGURE 16: Frequency response of CDTA based Class 1 FAF with $I_{Bias1} = I_{Bias2} = I_{Bias}$.

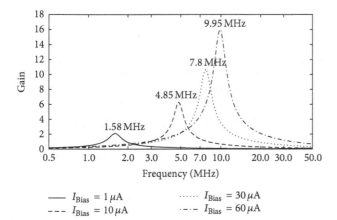

FIGURE 17: Frequency response of CDTA based Class 2 FAF with $I_{Bias1} = I_{Bias2} = I_{Bias}$.

FIGURE 18: Electronic Q-factor control of CDTA based FAF.

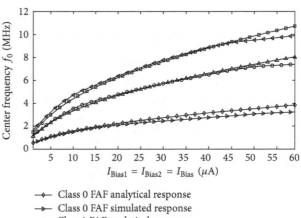

FIGURE 19: Electronic center frequency control of CDTA based FAF.

present at z terminal can be easily accommodated in external capacitances.

Reanalysis of the proposed circuit (Figure 6) yields the following nonideal transfer functions:

$$\frac{I_{LP}}{I_{IN}} = \frac{\alpha_n^2 \beta^2 g_1 g_2 Q_1}{D_{n1}(s)}, \tag{39a}$$

$$\frac{I_{BP}}{I_{IN}} = \frac{\alpha_n^2 \beta \, g_1 \left(s C_{2eq} + G_z \right) Q_1}{D_{n1}(s)}, \tag{39b}$$

where

$$
\begin{aligned}
D_n(s) &= P_1 Q_1 \left(s C_{1eq} + G_z \right) \left(s C_{2eq} + G_z \right) \\
&\quad + \alpha_p \beta g_3 P_1 \left(s C_{1eq} + G_z \right) + \alpha_p \alpha_n \beta^2 g_1 g_2, \\
P_1 &= \left(1 + G_X R_n + s C_X R_n \right); \\
Q_1 &= \left(1 + G_X R_p + s C_X R_p \right); \\
C_{1eq} &= C_1 + C_z; \qquad C_{2eq} = C_2 + C_z; \\
G_z &= \frac{1}{R_Z}; \qquad G_X = \frac{1}{R_X}.
\end{aligned}
\tag{39c}
$$

Choosing operating frequencies below $\min(1/C_X R_p, 1/C_X R_n)$ (as $G_X R_n \ll 1$ and $G_X R_p \ll 1$) the terms P_1 and Q_1 would not affect the transfer function. For frequencies below $\min(G_Z/C_{1eq}, G_Z/C_{2eq})$, (39c) modifies to

$$D_n(s) = s^2 C_{1eq} C_{2eq} + \alpha_p \beta g_3 s C_{1eq} + \alpha_p \alpha_n \beta^2 g_1 g_2, \tag{39d}$$

and transfer functions (39b) and (39c) change to

$$\frac{I_{LP}}{I_{IN}} = \frac{\alpha_n^2 \beta^2 g_1 g_2}{D_n(s)}, \tag{40a}$$

$$\frac{I_{BP}}{I_{IN}} = \frac{\alpha_n^2 \beta g_1 s C_{2eq}}{D_n(s)}. \tag{40b}$$

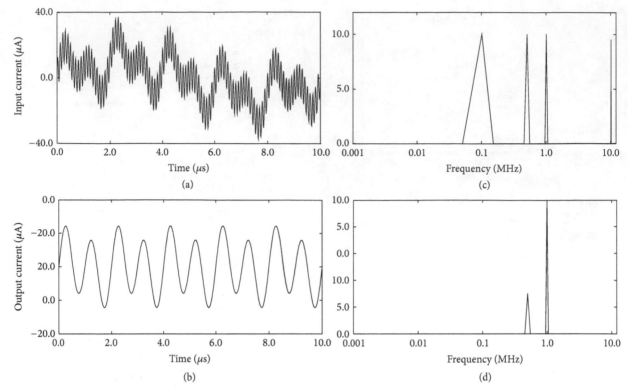

FIGURE 20: ((a) and (c)) Input and its frequency spectrum. ((b) and (d)) Output and its frequency spectrum for Class 0 FAF.

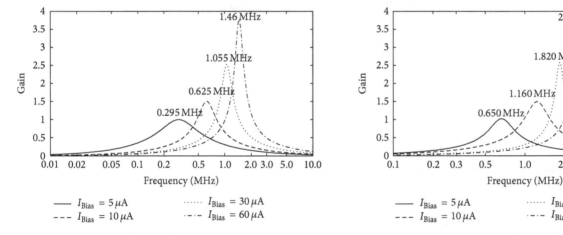

FIGURE 21: Frequency response of VDTA based Class 0 FAF with $I_{\text{Bias1}} = I_{\text{Bias3}} = I_{\text{Bias}}$.

FIGURE 22: Frequency response of VDTA based Class 1 FAF with $I_{\text{Bias1}} = I_{\text{Bias3}} = I_{\text{Bias}}$.

The center frequency, quality factor of Class 0 FAF can be expressed as

$$f_0 = \frac{1}{2\pi} \sqrt{\frac{\alpha_n \alpha_p \beta^2 g_1 g_2}{C_{1eq} C_{2eq}}}, \tag{41a}$$

$$Q = \frac{1}{g_3} \sqrt{\frac{\alpha_n \alpha_p g_1 g_2 C_{2eq}}{C_{1eq}}}. \tag{41b}$$

It is clear that the transfer functions and filter parameters ((40a), (40b) and (41a), (41b)) deviate from the ideal value in presence of nonidealities. The change can, however, be accommodated by adjusting bias currents.

5.2. *Nonideal Analysis of Class 0 VDTA Based FAF.* Considering the nonideal characteristics of the VDTA, the port relations of current and voltage in (25) can be rewritten as

$$\begin{bmatrix} I_z \\ I_{z_f} \\ I_{\tilde{z}}^f \\ I_{z_c}^f \\ I_{x^+} \\ I_{x^-} \end{bmatrix} = \begin{bmatrix} \beta g_1 & -\beta g_1 & 0 \\ -\beta g_1 & \beta g_1 & 0 \\ \beta g_2 & -\beta g_2 & 0 \\ -\beta g_2 & \beta g_2 & 0 \\ 0 & 0 & \beta g_3 \\ 0 & 0 & -\beta g_3 \end{bmatrix} \begin{bmatrix} V_p \\ V_n \\ V_z \end{bmatrix}, \tag{42}$$

where β represents the tracking error. Apart from tracking error, the parasites appear as shunt impedances ($R//C$) at

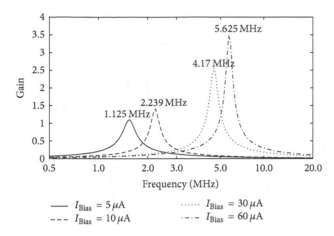

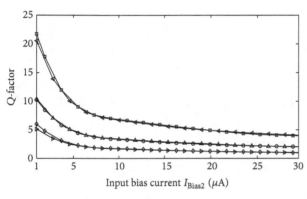

FIGURE 23: Frequency response of VDTA based Class 2 FAF with $I_{Bias1} = I_{Bias3} = I_{Bias}$.

FIGURE 24: Electronic Q-factor control of VDTA based FAF.

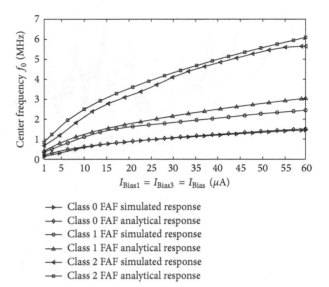

FIGURE 25: Electronic center frequency control of VDTA based FAF.

ports p, n, z, z_c, and x^+ denoted by $(R_p//C_p)$, $(R_n//C_n)$, $(R_z//C_z)$, $(R_z//C_z)$, and $(R_x//C_x)$, respectively. The parasitic capacitances present at p, z, z_c, and x terminal can be easily accommodated in external capacitances.

Reanalysis of the proposed circuit in Figure 12 yields the following nonideal transfer functions of Class 0 VDTA based FAF.

Then

$$\frac{I_{LP}}{I_{IN}} = \frac{g_1 g_3 \beta^2}{D_{n2}(s)}, \tag{43a}$$

$$\frac{I_{BP}}{I_{IN}} = \frac{\beta g_1 \left(sC_{2eq} + G_Z\right)}{D_{n2}(s)}, \tag{43b}$$

where

$$D_{n2}(s) = \left(sC_{2eq} + G_z\right)\left(sC_{1eq} + G_x + G_z + G_p + \beta g_2\right)$$
$$+ \beta^2 g_1 g_3,$$

$$C_{1eq} = C_1 + C_x + C_z + C_p; \qquad C_{2eq} = C_2 + C_z;$$

$$G_z = \frac{1}{R_Z}; \qquad G_X = \frac{1}{R_X};$$

$$G_p = \frac{1}{R_p}. \tag{43c}$$

As $G_x + G_z + G_p \ll \beta g_2$, (43c) modifies to

$$D_{n2}(s) = \left(sC_{2eq} + G_Z\right)\left(sC_{1eq} + \beta g_2\right) + \beta^2 g_1 g_3. \tag{44}$$

Choosing operating frequencies below min $(G_Z/C_{1eq}, G_Z/C_{2eq})$ (44) reduces to

$$D_{n2}(s) = s^2 C_{1eq} C_{2eq} + sC_{2eq}\beta g_2 + \beta^2 g_1 g_3, \tag{45}$$

and the transfer function (43b) simplifies to

$$\frac{I_{LP}}{I_{IN}} = \frac{g_1 g_3 \beta^2}{D_{n2}(s)}, \tag{46a}$$

$$\frac{I_{BP}}{I_{IN}} = \frac{\beta g_1 sC_{2eq}}{D_{n2}(s)}. \tag{46b}$$

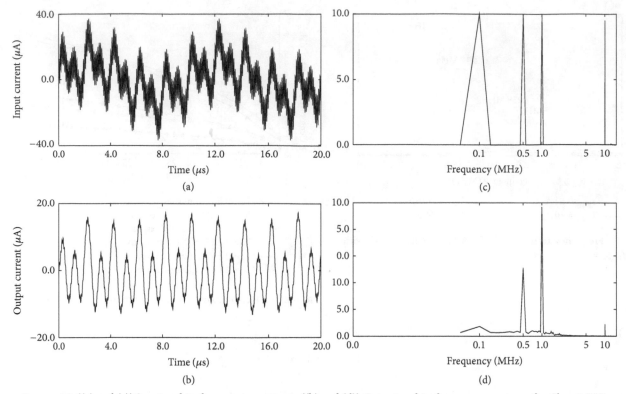

FIGURE 26: ((a) and (c)) Input and its frequency spectrum. ((b) and (d)) Output and its frequency spectrum for Class 0 FAF.

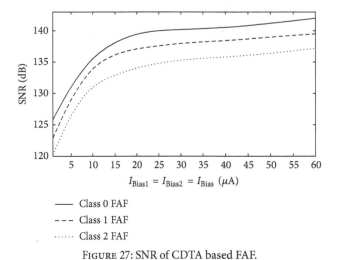

FIGURE 27: SNR of CDTA based FAF.

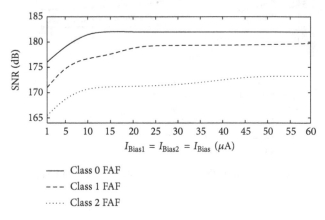

FIGURE 28: SNR of VDTA based FAF.

And the filter parameters are calculated as

$$f_0 = \frac{1}{2\pi} \sqrt{\frac{\beta^2 g_1 g_3}{C_{1eq} C_{2eq}}}, \tag{47a}$$

$$Q = \frac{1}{g_2} \sqrt{\frac{g_1 g_3 C_{2eq}}{C_{1eq}}}. \tag{47b}$$

It is clear that the transfer functions and filter parameters ((46a), (46b) and (47a), (47b)) deviate from the ideal value in presence of nonidealities. The change can, however, be accommodated by adjusting bias currents.

6. Simulation Results

In this section, the functionality of the proposed filters has been verified. The SPICE simulations results for CDTA and VDTA based filters have been presented using TSMC 0.25 μm CMOS process model parameters and supply voltages of $V_{DD} = -V_{SS} = 1.8$ V.

6.1. Simulation of CDTA Based FAF. The CMOS schematic of Figure 5 is used for verifying CDTA based FAF and the aspect ratios of the MOS transistors are given in Table 1. The additional TA blocks in CDTA providing current ports (x_C^+, x_C^-, x_{cc}^+, and x_{cc}^-) use aspect ratios same as that for x^+ and x^-. The capacitors C_1 and C_2 are chosen as 50 pF each.

TABLE 1: Aspect ratios of MOS transistors used in CDTA.

MOSFETs	$W\,(\mu m)/L\,(\mu m)$
Mc26	7.0/0.7
Mc1	9.8/0.7
Mc2, Mc3, Mc13, Mc16, Mc17	10.5/0.7
Mc19, Mc21	16.1/0.7
Mc6, Mc20	28.0/0.7
Mc8, Mc10, Mc18	28.7/0.7
Mc15, Mc12, Mc5	35.0/0.7
Mc4, Mc14	42.0/0.7
Mc22, Mc23, Mc25	56.0/0.7
Mc24	58.8/0.7
Mc7, Mc9, Mc11	70.0/0.7

TABLE 2: Aspect ratios of MOS transistors used in VDTA.

MOSFETs	$W\,(\mu m)/L\,(\mu m)$
M1, M2, M5, M6	16.1/0.7
M3, M4, M7, M8	28/0.7
M9–M12	21/0.7
M13–M16	7/0.7
M14, M15, M17, M18	8.5/0.7

The grounded resistor of Figure 7 is realized using TA block. The bias current is set as $0.85\,\mu A$ to realize a resistor of value $10\,k\Omega$. The frequency responses of CDTA based Class 0, Class 1, and Class 2 FAF topologies are depicted in Figures 15, 16, and 17, respectively. The responses are obtained by varying bias currents I_{Bias1} and I_{Bias2} ($I_{Bias} = I_{Bias1} = I_{Bias2}$) to $1\,\mu A$, $10\,\mu A$, $30\,\mu A$, and $60\,\mu A$ while keeping I_{Bias3} and I_{Bias4}, respectively, at $0.5\,\mu A$ and $10\,\mu A$. It can be clearly noticed that center frequency f_0 increases on increasing the bias current.

The electronic tunability of quality factor and center frequency for proposed Class 0, Class 1, and Class 2 CDTA based FAF is plotted in Figures 18 and 19, respectively. The analytical and simulated responses describing variation of quality factor are shown in Figure 18 for different values of I_{Bias3} while setting I_{Bias1} and I_{Bias2} to $30\,\mu A$. Figure 19 depicts the analytical and simulated responses for center frequency variation for different values of I_{Bias1} and I_{Bias2} ($I_{Bias1} = I_{Bias2}$) while setting I_{Bias3} to $0.5\,\mu A$. To plot responses for Class 1 and Class 2 CDTA based FAF the bias current I_{Bias4} is taken as $10\,\mu A$.

The transient behaviour of proposed CDTA based FAF is also studied by applying input signals of frequencies 100 KHz, 500 KHz, 1 MHz, and 10 MHz, each having an amplitude of $10\,\mu A$. The responses for Class 0 FAF are obtained by setting bias currents I_{Bias1} and I_{Bias2} each to $10\,\mu A$ and I_{Bias3} to $0.5\,\mu A$. Figure 20 shows the input and output waveforms along with their frequency spectrum for CDTA based Class 0 FAF. It may clearly be noted that the CDTA based Class 0 FAF allows only 1 MHz signal to pass and significantly attenuates signals of frequencies 100 KHz, 500 KHz, and 10 MHz. Similar responses for Class 1 and Class 2 FAF were also obtained.

6.2. Simulation of VDTA Based FAF. The CMOS schematic of Figure 12 is used for verifying VDTA based FAF and the aspect ratios of the MOS transistors are given in Table 2. The capacitors C_1 and C_2 are taken as 50 pF each. In the realization of Class 2 FAF, the grounded resistor is implemented by TA block. The frequency responses of VDTA based Class 0, Class 1, and Class 2 FAF topologies are shown in Figures 21, 22, and 23, respectively. The responses are obtained by keeping I_{Bias2} to $5\,\mu A$ and setting bias currents I_{Bias1} and I_{Bias3} ($I_{Bias} = I_{Bias1} = I_{Bias3}$) to $5\,\mu A$, $10\,\mu A$, $30\,\mu A$, and $60\,\mu A$. In

realization of Class 1 FAF, I_{Bias4} is set to obtain $g_4/g_3 = 3$ while keeping I_{Bias5} equal to I_{Bias1}. In realization of Class 2 FAF, I_{Bias6} is selected such that $g_6 = g_3 + g_4$ while I_{Bias7} is equal to I_{Bias4}.

The electronic tunability of quality factor and center frequency for proposed Class 0, Class 1, and Class 2 VDTA based FAF is plotted in Figures 24 and 25, respectively. The analytical and simulated responses describing variation of quality factor are shown in Figure 24 for different values of I_{Bias2} while setting I_{Bias1} and I_{Bias3} to $30\,\mu A$ in Class 0. To plot responses for Class 1, I_{Bias4} is set to get $g_4/g_3 = 3$ and I_{Bias5} is set equal to I_{Bias1}. Class 2 FAF responses are plotted by selecting I_{Bias6} to obtain $g_6 = g_3 + g_4$ while I_{Bias7} is equal to I_{Bias4}. Figure 25 depicts the analytical and simulated responses for center frequency variation for different values of I_{Bias1} and I_{Bias3} ($I_{Bias1} = I_{Bias3}$) and keeping I_{Bias2} to $5\,\mu A$. To plot responses for Class 1, I_{Bias4} is set to a value in such a manner that $g_4/g_3 = 3$ whereas I_{Bias5} is set equal to I_{Bias1}. The Class 2 FAF responses are plotted by setting I_{Bias6} to value such that $g_6 = g_3 + g_4$ whereas I_{Bias7} is equal to I_{Bias4}.

The transient behaviour of proposed agile filter is also studied by applying input signals of frequencies 100 KHz, 500 KHz, 1 MHz, and 10 MHz, each having an amplitude of $10\,\mu A$. The responses for Class 0 FAF are obtained by setting bias currents I_{Bias1}, I_{Bias2}, and I_{Bias3} each to $30\,\mu A$. The input and output waveforms along with their frequency spectrum for VDTA based Class 0 FAF are shown in Figure 26. It may clearly be noted that the VDTA based Class 0 FAF allows only 1 MHz signal to pass, partially attenuates signal of frequency 500 KHz, and significantly attenuates signals of frequencies 100 KHz and 10 MHz. Similar responses for Class 1 and Class 2 FAF are also obtained.

7. Performance Evaluation

The performance of proposed CDTA and VDTA based FAF circuits is studied in terms of power dissipation, output noise voltage, and SNR. The overall performance characteristics are summarized in Table 3. Figures 27 and 28 depict the signal to noise ratio (SNR) for the proposed CDTA and VDTA based filter topologies for Class 0, Class 1, and Class 2, respectively. The VDTA based FAF proved to be optimum concerning the power dissipation and signal to noise ratio. The maximum output noise voltage is better in VDTA based FAF.

8. Conclusion

In this paper CDTA and VDTA based frequency agile filters are presented. The proposed FAF configurations employ

TABLE 3: Performance characteristics of CDTA and VDTA based Class 0, Class 1, and Class 2 FAF.

Performance characteristics	Type of FAF	$I_{Bias} = 1\ \mu A$			$I_{Bias} = 10\ \mu A$			$I_{Bias} = 30\ \mu A$			$I_{Bias} = 60\ \mu A$		
		Class 0	Class 1	Class 2	Class 0	Class 1	Class 2	Class 0	Class 1	Class 2	Class 0	Class 1	Class 2
Power dissipation (mW)	CDTA	0.359	0.997	1.63	3.34	3.99	4.63	9.98	10.7	11.3	19.9	20.6	21.2
	VDTA	0.089	0.177	0.405	0.32	1.19	3.45	0.835	3.34	9.57	1.60	6.28	17.4
SNR (dB)	CDTA	124.9	122.1	119.2	135.5	134.2	131.3	140.2	137.9	135.3	142.1	139.5	137.2
	VDTA	175.82	170.4	165.0	181.7	180.5	170.96	182.0	179.3	171.6	182.0	179.75	173.2
Max. output noise voltage (nV)	CDTA	79.37	74.92	75.49	155.02	82.99	58.97	210.46	70.69	40.85	290.59	60.56	31.92
	VDTA	28.5	28.85	28.1	38.92	37.43	51.3	46.10	44.4	38.1	51.62	52.63	25.6

grounded passive components and are suitable for integration. The filter configurations are designed in such a way that quality factor can be independently controlled without changing the center frequency. The simulation results are included to demonstrate the workability of the circuits. The performance of the proposed FAF is evaluated in terms of power dissipation, SNR, and noise performance. The VDTA based FAF proved to be optimum concerning the power dissipation and signal to noise ratio.

Conflict of Interests

The authors declare that there is no conflict of interests regarding the publication of this paper.

References

[1] P. I. Mak, U. Seng-Pan, and R. P. Martins, *Analog-Baseband Architecture and Circuits for Multistandard and Low Voltage Wireless Transceivers*, Analog Integrated Circuits and Signal Processing, 2007.

[2] Y. Lakys and A. Fabre, "Multistandard transceivers: state of the art and a new versatile implementation for fully active frequency agile filters," *Analog Integrated Circuits and Signal Processing*, vol. 74, no. 1, pp. 63–78, 2013.

[3] Y. Lakys and A. Fabre, "A fully active frequency agile filter for multistandard transceivers," in *Proceedings of the International Conference on Applied Electronics (AE '11)*, pp. 1–7, September 2011.

[4] S. Kaehlert, D. Bormann, T. D. Werth, M.-D. Wei, L. Liao, and S. Heinen, "Design of frequency agile filters in RF frontend circuits," in *Proceedings of the IEEE Radio and Wireless Symposium (RWS '12)*, pp. 13–16, Santa Clara, Calif, USA, January 2012.

[5] A. J. X. Chen, Y. Wu, J. Hodiak, and P. K. L. Yu, "Frequency agile digitally tunable microwave photonic filter," in *Proceedings of the International Topical Meeting on Microwave Photonics (MWP '03)*, pp. 89–92, September 2003.

[6] G. Subramanyam, F. W. van Keuls, and F. A. Miranda, "A K-band-frequency agile microstrip bandpass filter using a thin-film hts/ferroelectric/dielectric multilayer configuration," *IEEE Transactions on Microwave Theory and Techniques*, vol. 48, no. 4, pp. 525–530, 2000.

[7] H. Chandrahalim, S. A. Bhave, R. G. Polcawich, J. Pulskamp, and R. Kaul, "A Pb($Zr_{0.55}Ti_{0.45}$) O_3-transduced fully differential mechanically coupled frequency agile filter," *IEEE Electron Device Letters*, vol. 30, no. 12, pp. 1296–1298, 2009.

[8] M. W. Wyville, R. C. Smiley, and J. S. Wight, "Frequency agile RF filter for interference attenuation," in *Proceedings of the 6th IEEE Radio and Wireless Week (RWW '12)*, pp. 399–402, Santa Clara, Calif, USA, January 2012.

[9] H. H. Sigmarsson, J. Lee, D. Peroulis, and W. J. Chappell, "Reconfigurable-order bandpass filter for frequency agile systems," in *Proceedings of the IEEE MTT-S International Microwave Symposium (MTT '10)*, pp. 1756–1759, Anaheim, Calif, USA, May 2010.

[10] Y. Lakys, B. Godara, and A. Fabre, "Cognitive and encrypted communications, part 2: a new approach to active frequency-agile filters and validation results for an agile bandpass topology in SiGe-BiCMOS," in *Proceedings of the 6th International Conference on Electrical and Electronics Engineering (ELECO '09)*, pp. II16–II29, November 2009.

[11] V. Biolkova and D. Biolek, "Shadow filters for orthogonal modification of characteristic frequency and bandwidth," *Electronics Letters*, vol. 46, no. 12, pp. 830–831, 2010.

[12] Y. Lakys and A. Fabre, "Shadow filters—new family of second-order filters," *Electronics Letters*, vol. 46, no. 4, pp. 276–277, 2010.

[13] D. Biolek, "CDTA-building block for current-mode analog signal processing," in *Proceedings of the European Conference on Circuit Theory and Design (ECCTD '03)*, vol. III, pp. 397–400, Krakow, Poland, 2003.

[14] D. Biolek, V. Biolkova, and Z. Kolka, "Current-mode biquad employing single CDTA," *Indian Journal of Pure and Applied Physics*, vol. 47, no. 7, pp. 535–537, 2009.

[15] M. Kumngern, P. Phatsornsiri, and K. Dejhan, "Four inputs and one output current-mode multifunction filter using CDTAs and all-grounded passive components," in *Proceedings of the 10th International Conference on ICT and Knowledge Engineering, ICT and Knowledge Engineering*, pp. 59–62, Bangkok, Thailand, November 2012.

[16] S. K. Pandey, A. P. Singh, M. Kumar, S. Dubey, and P. Tyagi, "A current mode second order filter using dual output CDTA," *International Journal of Computer Science & Communication Networks*, pp. 210–213, 2012.

[17] F. Kacar and H. Kuntman, "A new cmos current differencing transconductance amplifier (CDTA) and its biquad filter application," in *Proceedings of the IEEE (EUROCON '09)*, pp. 189–196, St Petersburg, Russia, May 2009.

[18] D. Biolek, E. Hancioglu, and A. Ü. Keskin, "High-performance current differencing transconductance amplifier and its application in precision current-mode rectification," *International Journal of Electronics and Communications*, vol. 62, no. 2, pp. 92–96, 2008.

[19] A. Uygur, H. Kuntman, and A. Zeki, "Multi-input multi-output CDTA-based KHN filter," in *Proceedings of the IEEE International Microwave Symposium Digest*, pp. 1756–1759, 2010.

[20] W. Chiu, S. I. Liu, H. W. Tsao, and J. J. Chen, "CMOS differential difference current conveyor and their applications," *IEE Proceedings-Circuits Devices Systems*, vol. 143, no. 2, pp. 91–96, 1996.

[21] Z. Wang, "2-MOSFET transresistor with extremely low distortion for output reaching supply voltages," *Electronics Letters*, vol. 26, no. 13, pp. 951–952, 1990.

[22] D. Biolek, M. Shaktour, V. Biolkova, and Z. Kolka, "Current-input current-output universal biquad employing two bulk-driven VDTAs," in *Proceedings of the 4th International Congress on Ultra Modern Telecommunications and Control Systems (ICUMT '12)*, pp. 484–489, St. Petersburg, Russia, October 2012.

[23] J. Satansupa and W. Tangsrirata, "Single VDTA based current mode electronically tunable multifunction filter," in *Proceedings of the 4th International Science, Social Science, Engineering and Energy Conference (ISEEC '12)*, pp. 1–8, 2012.

[24] P. Phatsornsiri, P. Lamun, M. Kumngern, and U. Torteanchai, "Current-mode third-order quadrature oscillator using VDTAs and grounded capacitors," in *Proceedings of the 4th Joint International Conference on Information and Communication Technology, Electronic and Electrical Engineering (JICTEE '14)*, pp. 1–4, IEEE, Chiang Rai, Thailand, March 2014.

[25] T. Pourak, P. Suwanjan, W. Jaikla, and S. Maneewan, "Simple quadrature sinusoidal oscillator with orthogonal control using single active element," in *Proceedings of the International Conference on Signal Processing and Integrated Networks (SPIN '14)*, pp. 1–4, 2014.

[26] M. Srivastava, D. Prasad, and D. R. Bhaskar, "New Parallel R-L impedance using single VDTA and its high pass filter application," in *Proceedings of the International Conference on Signal Processing and Integrated Networks (SPIN '14)*, pp. 535–537, 2014.

[27] K. Chumwangwapee, W. Jaikla, W. Sunthonkanokpong, W. Jaikhang, S. Maneewan, and B. Sreewirote, "High input impedance mixed-mode biquad filter with orthogonal tune of natural frequency and quality factor," in *Proceedings of the 4th Joint International Conference on Information and Communication Technology, Electronic and Electrical Engineering (JICTEE '14)*, pp. 1–4, Chiang Rai, Thailand, March 2014.

Permissions

All chapters in this book were first published in AELC, by Hindawi Publishing Corporation; hereby published with permission under the Creative Commons Attribution License or equivalent. Every chapter published in this book has been scrutinized by our experts. Their significance has been extensively debated. The topics covered herein carry significant findings which will fuel the growth of the discipline. They may even be implemented as practical applications or may be referred to as a beginning point for another development.

The contributors of this book come from diverse backgrounds, making this book a truly international effort. This book will bring forth new frontiers with its revolutionizing research information and detailed analysis of the nascent developments around the world.

We would like to thank all the contributing authors for lending their expertise to make the book truly unique. They have played a crucial role in the development of this book. Without their invaluable contributions this book wouldn't have been possible. They have made vital efforts to compile up to date information on the varied aspects of this subject to make this book a valuable addition to the collection of many professionals and students.

This book was conceptualized with the vision of imparting up-to-date information and advanced data in this field. To ensure the same, a matchless editorial board was set up. Every individual on the board went through rigorous rounds of assessment to prove their worth. After which they invested a large part of their time researching and compiling the most relevant data for our readers.

The editorial board has been involved in producing this book since its inception. They have spent rigorous hours researching and exploring the diverse topics which have resulted in the successful publishing of this book. They have passed on their knowledge of decades through this book. To expedite this challenging task, the publisher supported the team at every step. A small team of assistant editors was also appointed to further simplify the editing procedure and attain best results for the readers.

Apart from the editorial board, the designing team has also invested a significant amount of their time in understanding the subject and creating the most relevant covers. They scrutinized every image to scout for the most suitable representation of the subject and create an appropriate cover for the book.

The publishing team has been an ardent support to the editorial, designing and production team. Their endless efforts to recruit the best for this project, has resulted in the accomplishment of this book. They are a veteran in the field of academics and their pool of knowledge is as vast as their experience in printing. Their expertise and guidance has proved useful at every step. Their uncompromising quality standards have made this book an exceptional effort. Their encouragement from time to time has been an inspiration for everyone.

The publisher and the editorial board hope that this book will prove to be a valuable piece of knowledge for researchers, students, practitioners and scholars across the globe.

List of Contributors

Isha Goel
Department of Academic and Consultancy Services Division, C-DAC, Mohali 160071, India

Dilip Kumar
SLIET, Longowal, Sangrur 148106, India

Elena A. Plis
Center for High Technology Materials, Department of Electrical and Computer Engineering, University of New Mexico, Albuquerque, NM, USA
Skinfrared, LLC, Lobo Venture Lab 801, University Boulevard, Suite 10, Albuquerque, NM 87106, USA

Rajeshwari Pandey, Neeta Pandey, Romita Mullick and Sarjana Yadav
Department of Electronics and Communication Engineering, Delhi Technological University, Delhi 110042, India

Rashika Anurag
Department of Electronics and Communication Engineering, JSS Academy of Technical Education, C-20/1, Sector 62, Noida, Uttar Pradesh 201301, India

Manoj Sharma
Department of ECE, Mewar University, Rajasthan 312901, India
Department of ECE, BVCOE, Paschim Vihar, New Delhi 110063, India

Arti Noor
SoE, CDAC Noida, Ministry of Communications and IT, Government of India, Noida, Uttar Pradesh 201307, India

Arthur H. M. van Roermund
Mixed-Signal Microelectronics Group, Department of Electrical Engineering, Eindhoven University of Technology, Den Dolech 2, P.O. Box 513, 5600 MB Eindhoven, The Netherlands

D. K. Kamat
SCOE, Pune 411041, India
Department of E & TC, Sinhgad Academy of Engineering, Pune 411048, India

Dhanashri Bagul
Department of E & TC, Sinhgad Academy of Engineering, Pune 411048, India

P.M. Patil
KJ's Educational Institutes, Pune 411048, India

Ziad Alsibai and Salma Bay Abo Dabbous
Department of Microelectronics, Brno University of Technology, 61600 Brno, Czech Republic

Sadeque Reza Khan and M. S. Bhat
Department of Electronics and Communication Engineering, National Institute of Technology Karnataka, Surathkal, Mangalore 575025, India

V. Kokilavani and K. Preethi
Department of PG Studies in Engineering, S. A. Engineering College (Affiliated to Anna University), Poonamallee-Avadi Road, Veeraraghavapuram, Chennai, Tamil Nadu 600 077, India

P. Balasubramanian
Department of Computer Science and Engineering, S. A. Engineering College (Affiliated to Anna University), Poonamallee-Avadi Road, Veeraraghavapuram, Chennai, Tamil Nadu 600 077, India

Gagandeep Singh and Chakshu Goel
ECE Department, Shaheed Bhagat Singh State Technical Campus, Ferozepur, Punjab 152004, India

Debajit Bhattacharya and Niraj K. Jha
Department of Electrical Engineering, Princeton University, Princeton, NJ 08544, USA

Andreas Demosthenous
Department of Electronic and Electrical Engineering, University College London, Torrington Place, LondonWC1E 7JE, UK

Frédéric Broydé and Evelyne Clavelier
Tekcem, 78580 Maule, France

Lucie Broydé
ESIGELEC, Saint-Étienne-du-Rouvray, 76800, France

Constantinos C. Kontogiannis and Athanasios N. Safacas
Department of Electrical and Computer Engineering, Patras University, 26443 Patras, Greece

Said Saad
Group of Electronics and Quantum Physics, Laboratory of Advanced Materials and Quantum Phenomena, Faculty of Sciences of Tunis, Tunis EL Manar University, 2092 Tunis, Tunisia

Lotfi Hassine
National Institute of Applied Sciences and Technology, University of Carthage, 1080 Tunis, Tunisia

Neeta Pandey and Rajeshwari Pandey
Department of Electronics and Communication Engineering, Delhi Technological University, Delhi 110042, India

Aseem Sayal and Richa Choudhary
Department of Electrical Engineering, Delhi Technological University, Delhi 110042, India